Updating Bergson

Updating Bergson

A Philosophy of the Enduring Present

Adam Lovasz

LEXINGTON BOOKS
Lanham • Boulder • New York • London

Published by Lexington Books
An imprint of The Rowman & Littlefield Publishing Group, Inc.
4501 Forbes Boulevard, Suite 200, Lanham, Maryland 20706
www.rowman.com

British Library Cataloguing in Publication Information Available

Library of Congress Cataloging-in-Publication Data

Library of Congress Control Number: 2021933761

ISBN 978-1-7936-4081-9 (cloth)
ISBN 978-1-7936-4082-6 (electronic)

Contents

List of Figures

Acknowledgments

Life is a system of mutual debts we cannot ever hope to fully repay. Acknowledgment is due to all those who have supported my endeavors in the world of philosophy and academia. I would like to specifically express my gratitude to the following persons who have shaped my intellectual path with their kind advice, conversations, encouragement, guidance, and professional assistance: Natalia Artemenko, Thierry Bardini, Ádám Bartha, Gergely Berta, Bianka Boros, Gábor Boros, David Bunting, Raphaëlle Costa de Beauregard, György Czétány, Elena Fell, Benjamin Fraser, Jure Gantar, Arran Garre, Graham Harman, Ottó Hévizi, Simon Kennedy, Géza Kulcsár, Márk Losoncz, Giancarlo Maiorino, Bence Marosán, Nicola Masciandaro, Zsolt Miklósvölgyi, Timothy Morton, Andor Müller, Márió Z. Nemes, Csaba Olay, Imola Részeg, David Roden, Tamás Seregi, Gary J. Shipley, Zsigmond Szabó, Gerda Széplaky, Attila Szűcs, and Tamás Ullmann. With regard to the preparation of this book, I am indebted to the helpful suggestions of Keith Ansell-Pearson and John Ó Maoilearca, two leading scholars of Bergson whom I hold in the highest regard. I also wish to express my gratitude to the staff at Lexington Books who were involved in forming this book. I am infinitely grateful for the friendship of my colleague and coauthor, Márk Horváth. Had we not met, I would not be the thinker I am today. The patience and love of my family has been a source of inspiration and strength in more ways than one, and for this I am grateful. I thank Vivien Bukodi for being an ever-present source of inspiration. Last but not least, I also thank my detractors. Without resistance there can be no striving.

List of Abbreviations

Works by Henri Bergson

CE: *Creative Evolution.* (1944 [1907]) Trans. Arthur Mitchell. New York: Random House.
CM: *The Creative Mind.* (1946 [1934]) Trans. Mabelle L. Andison. New York: The Philosophical Library Inc.
DS: *Duration and Simultaneity: With Reference to Einstein's Theory.* (1965 [1922]) Trans. Leon Jacobson. Indianapolis, New York and Kansas City: The Bobbs-Merrill Company.
MM: *Matter and Memory.* (2005 [1896]) Trans. Nancy Margaret Paul and W. Scott Palmer. New York: Zone Books.
TFW: *Time and Free Will: An Essay on the Immediate Data of Consciousness.* (1950 [1889]) Trans. F. L. Pogson. London: George Allen & Unwin.
TS: *Two Sources of Morality and Religion.* (1935 [1932]) Trans. R. Ashley Audra, Cloudesley Brereton, and W. Horsfall Carter. London: Macmillan.

Introduction

The many varied elements of becoming, emerging at various levels of geophysical time, are all inherently genetic. Differential yet interpenetrating temporalities are seen to accumulate, while the succession of past, present, and future will nonetheless be proven to constitute nothing but an illusion called forth by our practical subjective necessities. Epochal intransitivity is the form of time. Change and eternity, far from contradicting one another, will be found to compose a crystallized whole, an eternity of modification. The banal phrase, "frozen in time," acquires a new meaning. Instead of succession, we have a coexistence of tenses. Through the analysis of time, we may glean information about the nonlinear dynamics of ontological consistency. The mystery of existence can be summarized as the circumstance of *persistence*. Why do things endure? That is the central question animating Bergsonism. Objects exist, of this much we may be sure, even despite their lacking any substance whatsoever apart from the fact of this very perseverance. But what does "endurance" mean? It is this mystery that we must get to the bottom of. Phrases such as "I can," "I shall," "I will" indicate the prior acceptance of a fundamental truth, namely, the persistence and central importance of the duration we happen to be occupying. Any present necessitates a lived time, or so we hold. In our everyday use of language, we speak of things "taking time." For the most part, we do not question the veracity of the sequential view of temporality. We are born, we live, then we die. Things emerge, persist, then vanish. The idea of persistence must be considered basic to any sentence containing the present participle. Our goal is to uncover the basic ontological structure of the "-ing," otherwise far too often occluded by the retrospective unification of different times under the heading of a single succession. Notice in the latter case a difference in tone. The subject has extended itself along a different axis, a differential serving as an analytical postulate substituting for the survival of objects in and of themselves. If we are to gain initiation into the secrets of objects, we must intuit their own becomings. The goal of philosophy is to return to the objects themselves. For Bergson, philosophy is, above all else, an exercise in sympathy. As he states explicitly in no uncertain terms, "We call intuition (. . .) the sympathy by which one is transported into the interior of an object in order to coincide with what there is unique and consequently inexpressible in it" (CM 184). This is the challenge we face when striving to update Bergsonism. How can intuition, a practice with spiritual undertones, penetrate into the material thickness of real, existing, enduring objects? Furthermore, in what manner can a spiritual exercise serve as a philosophical method? Intuition is a rigorous practice that must be nonetheless be distinguished from systematic, reflective knowledge. To quote Gilles Deleuze's famous characterization, the Bergsonian method of intuition constitutes "a superior empiricism, capable of stating problems and of going beyond experience toward concrete conditions" (Deleuze 1997 [1966]: 30). While analysis is, in a sense, a pragmatic solution to the question of an organism's survival, a cognitive method of slicing up the world into more manageable pieces, intuition promises the interpenetration of planes, the promiscuous meeting of durations. A Bergsonian science would be a practice traversing through different semantic currents, achieving a suspension of belief in any and all deterministic laws. Bergson asks us to suspend our judgments. But the Bergsonian reduction is more radical than the Husserlian phenomenological *epokhé*. Far from showing the autonomy of a transcendental consciousness opposed to the world, in Bergson's magnum opus *Matter and Memory*, it will be the self-subtraction of the world which shows the asymmetric dependence of individual consciousness upon the cosmos. In opposition to all correlationist modes of thought, Bergson upholds the mind-independent status of objects and objective, supraindividual materialized durations. Georges Mourélos identifies Bergson's philosophy with a realist metaphysics which attempts to account for all the varied "levels of reality" through an account of the heterogeneity of durations, the latter conceived of in the plural (Mourélos 1964). There is no such thing as "duration" or "change" in general. Rather, each level of reality is endowed with a duration of its own.

Just as the advent of a thermodynamically determined future present in late nineteenth-century physics threatens to evacuate, in advance, the study of all evolutionary processes of meaning, there arrives a philosopher who upholds freedom and indeterminacy. Against those who believe in determinism, Bergson asks us to remain true to the open nature of reality. His universe is an indeterministic cosmos. Evolution as metaprocess is beshadowed by the specter of thermodynamic equilibrium, the postulated return to stasis as envisioned through the organon of late nineteenth-century Occidental physics. It is no exaggeration to say that Bergson's thinking attempts to subvert and even transcend the supposed final truth of thermodynamics. Against all fatalism, Bergson affirms that nothing is truly heading toward annihilation, precisely because no substance can be posited apart from change. Duration is an inescapably hierarchical concept, as it allows for the incorporation of multiple directions and tendencies. The objects obtained through analysis are not unlike snapshots, violently excised from a fundamentally continuous structure. Everyday consciousness functions very much like a camera taking snapshots, reducing cognitive complexity by cutting out extracts from a world far too complicated to comprehend in itself. Against the habitual workings of the intellect, Bergson proposes a new normality, "a health beyond the normal," to quote David Lapoujade (Lapoujade 2018 [2010]: 73). It is against the natural tendency to slice continuity into discrete objects, atoms, or snapshots that Bergson affirms the primacy of continuous *and* heterogeneous Becoming. In order to transcend our attachment to habitual modes of thought, including the scientific categorizations of objects, we are encouraged the "do violence to the mind" and "go counter to the natural bent of the intellect" (CE: 35). The reified image, recoded as an onto-epistemological datum, depends on a general context in three ways: (a) it is a product of satisfaction, that is, the satisfaction of some agent; (b) it constructs a process of sequencing showing an articulatory center; and (c) it is coupled with an oscillatory system, otherwise known as the medium. In particular, the third point lends itself to extrapolation, as a good case can be made that all oscillatory systems, in as much as they process information, must of necessity, be connected with resonant neuronal networks, producing irreducible sets of data. Outside of coding systems, there is naked becoming in itself, the noumenality of change in general. What interests the Bergsonian is how a philosophically amplified consciousness can penetrate into the latter aspect, leaving mediation behind.

Here, we must delineate what we are searching for. First and foremost, the structure of experience, in particular its originality, must be questioned. The problem of accessing the as-yet unforeseen future is beset by a broader problem: the model we are building presently contains no ready-made concept of what phrases such as "originality" or "experience" could mean in the first place. We are afflicted by a poverty that has robbed us of any and all modes of explication. Therefore, the originality of the present is just as daunting for us as the realm of posteriority or priority. Bergsonism is a struggle to transcend what passes under the radar in the guise of a commonsense outlook. Within the present, it would seem that all human and nonhuman agents alike are besieged by constant, unrelenting change, logically entailing the absence of anything truly substantial or existent. To quote José Luis Borges, when faced with impermanence "we intuit that nothing is real" (Borges 1964: 113). If everything changes, nothing may be permitted to persist in itself. Ironically, it is the universal nature of persistence that prevents one from taking all which exists too seriously. Beneath the appearance of persistence, we uncover the absence of all substance. Persistent objects seem to inhere within a spatial container, but persistence, we are given to understand, is characterized by a duplicitous double-sidedness not unlike that of onto-epistemological scientific outlooks in general. The following term, characterized as it is by ironic bracketing, summarizes our problem: "real temporality." If nothing ever did occur, then persistence can only be relative to the collection of objects-in-becoming. Outside of becoming though, there is no unchanging substance to be found. How can one take seriously the existence of time if no object is permitted to inhere within any set space? This is the central problem of Henri Bergson's ontology. "Change," affirms Bergson, "is the very substance of things" (CM: 184). Outside of change, all talk of substance is pointless. If any model of temporality is to do justice to reality, even a bracketed, excluded, derealized, abstracted one, it must produce a representation allowing for the unbinding of relevant productive powers. As opposed to rationalized positivity, we must affirm the productivity of difference. Not everything is natural, but some artificial objects are most definitely more spontaneous than others. The objective of all ontological philosophy is to seize a state as it is, as it persists within its "-ing," to penetrate into the object itself. Under the word "state," we must understand a natural positivity, an impermanent flow whose elements outweigh negative values to such an extent as to make analysis possible. Freedom permeates all processes; the entire cosmos is an indeterminate elaboration lacking necessity. If philosophy is to come to terms with the circumstance of cosmic contingency, it must allow space for the concreteness of time. Sensation too has a place in philosophy. Even conceptual modalities are capable of displaying what may be termed a latent or "underlying" empiricism, once thought arrives at a thoughtstate. When metaphysical writing unbinds itself, the evolution of philosophy arrives at a new level of complexity. Time intervals between signals decrease. Information, the object of communication, the formation that has come to be informed, attains an individuated memory in the form of memory-images differentiated from one another, potentially ad infinitum. The kernel of Bergsonism, is, as we shall see, a resolutely ontological sensitivity to differentiation. Intuition as method implies a "decisive change of orientation" (Mourélos 1964: 68). As Suzanne Guerlac notes perceptively, in the intuitive mode of philosophy inaugurated by Bergson, "time has become concrete. It has become substance" (Guerlac 2006: 3). Duration is for the most part finite and mortal.

Until the first third of the twentieth century, materialism, seeking to exploit the spoils of scientific progress while subtracting the observer from the equation, tended to see everything in the form of mechanisms amenable to practical principles of reverse engineering while being, in themselves, fundamentally unalterable and irreversible. It was supposed in late nineteenth-century

European materialism that there was a reality worth understanding, a mind-independent world full of resources waiting for extraction. The 10,000-year-old intelligence explosion is far from over, and it is doubtful that any human cranium can fathom all its implications (Cochran and Harpending 2009; Eden et al. 2013). During our investigation we cannot avoid revising the concept of "mind." In his epochal and, until fairly recently, relatively rarely cited *Matter and Memory*, Bergson rejects the all too frequent equation of mind with brain, in favor of a panpsychist concept of consciousness. For Bergson, perception, conceived of as a pervasive actuality, is autonomous in relation to the brain and the body alike (MM: 79; Vieillard-Baron 1993: 92). Pure perception overflows the confines of the skull, discharging into the cosmos, like a cloud. With astounding audacity, Bergson affirms that movement in general "already possesses something akin to consciousness, something akin to sensation" (MM: 247). If we seek to truly attain intuition, we cannot assent to a naiive acceptance of various prior beliefs, influenced as they are by climatic fluctuations, social conventions, frustrating orthodoxies, or historical cataclysms. A mode of bracketing is needed that suspends the validity of any knowledge outside pure experience, without reifying this transitory purity into a new idol, metaphysical or otherwise (are not any anti-metaphysical theory a new, duplicitous metaphysics-in-disguise?). Bergsonism can be understood as a mode of unorthodox thought, itself being an affect, the result of "an interruption of habitual action, a delay that is generative of recollection and that can open within habit other ways of seeing and acting," as Alia Al-Saji writes brilliantly (Al-Saji 2009: 386). Objectivity, on an intuitive view, can be incorporated into pure intuitive experience without any unnecessarily psychologistic or culturally biased imperialism resulting from such subsumption. Any assimilation is suspect from the outset. While historically materialism has failed to incorporate spiritual or mental phenomena, with tragic consequences, similarly phenomenology has also failed to transcend its fixation upon subjectivity and representation. Should phenomenology be abandoned as a consequence of its having nothing to say about anything beyond the subject? Phenomena are, after all, appearances. The phenomenological turn cannot but strike us as an unwelcome change, an awful regression as compared with Bergson's idea of images existing and persevering in themselves, independently of consciousness! Far from being connected irremediably to the presence of a disclosing, revealing human subjectivity, the image in Bergsonism is an objective reality that cannot be disclosed by individual forms of sentience. Bergson's image is "a self-existing image" (MM: 10). Images persist, whether a living thing perceives them or not. By definition, phenomenology cannot do without the positing of a subject for whom the appearances manifest themselves. Bergson's universal consciousness, however, is the relaxed perception of matter itself. Tom Sparrow is entirely accurate when he posits the "end" of phenomenology in contemporary philosophy. According to Sparrow's working definition, "phenomenology (. . .) is exclusively committed to investigating only those dimensions of human experience that take shape within the correlation between thought and being" (Sparrow 2014: 2). Such a project is unsustainable. Are not other, ahuman forms of thinking outside of subjectivism imaginable? "A" consciousness by no means denotes anything even remotely similar to human thought. We hope to argue that Bergsonism furnishes us with the conceptual tools needed not just to get back to our own duration, but also to discover all the different types of durations. The entire Bergsonian project is an attempt to see differently, to reach beyond our all too human condition. Bergson says as much when he declares that "philosophy should be an effort to go beyond the human state" (CM: 228). In our present posthuman age, this message is more valid than ever. As Keith Ansell-Pearson observes, commenting upon this Bergsonian imperative, "the restriction of philosophy to the human condition fails to appreciate the extent to which we are not simply creatures of habit and automatism but are also creatures involved in a creative evolution of becoming. The task is not to leave the human behind but rather to broaden the horizon of our experience of life" (Ansell-Pearson 2018: 5). The entire Bergsonian project may be summarized as an attempt to return to immediacy.[1] Intuition strives to reveal the flow of multiple becomings in themselves. It is from outside the change that we must head inward until we reach the immediacy of a moment purified of memory, without of course forgetting our inescapable entrapment within our own time. The ultimate level of temporality is eternity, revealed in a vertiginous movement that demands of us a suspension of credulity and a corresponding dilation or expansion of our perception. An image is infinitely more than what appears as real to human observers. The phenomenological "real," Sparrow holds, what is given is "lived experience"; as a consequence phenomenology is incapable of giving an account of anything outside of human experience. There can be no phenomenology outside of human perception (Sparrow 2014: 12). Because it rejects equating givenness with access to a living consciousness, Bergsonism is irreducible to phenomenology. The perception of duration involves a real coincidence with durations which are not merely our own.

Our task here, apart from a work of commentary, is the construction of an ontological model that, while fundamentally desubjectivized, nonetheless affirms the mind-independent reality of images. According to the realist reading we seek to advance in this book, Bergsonism allows us to account for the autonomy of images, as well as the autonomy of perception from the individual. The task of philosophy—breaking out of the cage of restricted, anthropomorphic thought—necessitates an extension of our perceptive capabilities toward the most distant realms of objectivity. Sympathy, far from denoting some remnant of psychologism, would be a properly ontological term, "an internal communication among tendencies or movements" (Lapoujade 2018 [2010]: 45). What if we take the issue of an observing subject out of the equation altogether? Where would such a radical move leave us? What happens to manifestation and denotation once the subject is decentered? Evolutionary principles will be of assistance, for these are laws that bring interdependence to the forefront. Evolutionary theory can be applied without illegitimately assimilating different world levels. Temporal correlations consistent with evolutionary principles can display relevant signals to the philosopher. Disorder is a geometric aspect, whose primordial mantle plumes emanate from underneath even the most airtight of systems. There is, strictly speaking, no such thing as disorder or nothingness. These are, for the Bergsonian view, false concepts. The void is an illusion created by deluded consciousness, a projection born of disappointment or frustration (CE: 306). Our book is not a commentary upon the work of Bergson, although it certainly may also be read in such terms. Rather, it is an attempt to excavate and recycle, so to speak, the radical potential of Bergsonism, conceived of as a speculative mode of philosophy that can annihilate our own closure. Indeed, for Bergson philosophy as a project is inherently open, an ongoing work of opening ourselves up to other times: "Philosophy can only be an effort to dissolve again into the Whole" (Bergson CE: 210). Over the past decade, there has been renewed interest in speculation within Continental philosophy. Numerous authors even speak of a "speculative turn" (Bryant Harman and Srnicek 2011). For us, Bergsonism as a systematic mode of thinking provides us with a semantics adequate to recapturing the free play of possibilities that composes the fabric of every level of reality. More so, it is an exercise in dissolution, a striving to return to a unity with the immanence of *this* world. The heterogeneity of Bergsonian language is a direct consequence of the real freedom of each and every becoming. Instead of searching for liberty in some Kantian noumenal dimension, Bergson encourages his readers in *Time and Free Will* to return to duration. Freedom is an outflow, originating from its basis in time. Uncertainties activate synergies, feeding into a fundamental care for Being. Ontology is, for us, a special way of caring for the plasmatic membranes of interdependence that envelop *all* beings. The speculative metaphysics we seek to reconstruct, based upon a thoroughgoing reworking of Bergsonism, shall be the articulation of an active, latently universal pure actualization that possesses a wide variety of coefficient connections via the temporal densities it recollects. Equilibrium, within Bergsonism, is the concretion of a complex mixed-mode periodic signaling, a pure quantity. We soon learn, however, that these states are only ever fleeting, arbitrary constructs. Actualization is all that exists. Ontology must come to terms with the decisive importance of movement. If anything, Bergson was the prophet of the flow of thought-images. Ontology is never about grasping something concrete: a thing cannot be pinned down, for all is motion. Rather, what we discover, working through hiddenness, is an infinite variety of becomings, some of which have become fossilized, while others retain some of their liquidity. Possibilities may be held in common: one refraction can survive to influence many different forms and beings, such as viruses that learn to leapfrog from one species to another.

In Bergsonism, being is beset by an almost fatal dualism that threatens to tear this elegant system apart. On the one hand, matter is dominated by the law of entropy. On the other hand, the evolution of life, equated with movement, seeks to free itself of decay and degradation in a negentropic motion of resistance. Being, in so far as it lives, moves. Stasis, on the other hand, is the absolute absence of life, the absence of being. Philosophy, trapped between entropy and negentropy, is not unlike Friedrich Nietzsche's tightrope walker. Once it enters the speculative mode, philosophical ontology shows the chaotic attractors underlying change. This proximity to chaos threatens to destabilize philosophy, shoving it into the realm of destabilized poetic expression. If it ignores the presence of attractors though, ontology risks falling headlong into the pit of a boring scientism, surrendering the articulation of a worldview to the natural sciences. As Bergson reiterates in *Creative Evolution*, we cannot even speak of states or conditions as such, for "the state itself is nothing but change" (CE: 11). There can be no such thing as a "thing." This is not a final consequence, a conclusion to be drawn, but a point of departure for any non-essentialist ontology. For all that, we must not forget an important fact: even the most "realist" of Writing Equations remains just that: a textuality grafted onto becomings which care naught for the act of writing. The drawn line which ostensibly "represents" a real movement never can substitute that actual movement. There is an irreducible gap between the abstract representation of a movement or change and the change itself.[2] In this study, we shall have occasion to creatively blur the boundaries between construction and spontaneous formation, on a textual level as well. The structure of affectivity itself compels us to nothing less than the most thoroughgoing rethinking of reified categories. Every definition of real circuits and movements is only ever provisional. Experience, understanding, sensory nerve stimulation—all these are subsequent constructions that seek to undergird and fix the liquidity of emergence. No endpoint may be identified. Extreme nonequilibrium is the rule, not some kind of exception. If everything is life, to a certain degree, then nothing is ever truly dead. But the corollary also gives food for thought: if death, defined as absolute stillness, is impossible, then the reverse must also hold. Despite our speculative focus, and a reliance upon immediate experience, our formulations must also rest on actual scientific and philosophical data, while remaining true to the concreteness of the speculative ontological framework we have selected. In agreement with Bergson, we too hold that authentically radical philosophy cannot be anything other than "speculative," in the sense that it describes a project of reorienting consciousness toward a coincidence with the reality of objects. The philosophical task rests, in effect, in the development and unbinding of penetrative thinking. Our aim is to reenact what Bergson himself manages so excellently in his article on Félix Ravaisson: to extend the reach of Bergson's own ideas, "adjusting or renewing them," thereby "pushing their development to its farthest possible limits"[3] (CM: 263). It is imperative to always take the object of commentary as far as possible, for the plausibility of a project depends, in large measure, upon the intensity of the author's commitment to their own project. We take this methodological stricture as our own point of departure. Intuition unchained is limitless interpenetration.

Intuitive philosophy operates upon concepts that have undergone extraction from other registers. Similarly to detached portions of living nerves excised by neurosurgeons, these concepts retain various elements of their previous environments. Perhaps they are capable of feeling pain, and so forth. Whatever may be their characteristics, the qualities of the extracted concepts are less relevant than the way we use them. The reworking of Bergsonism we shall attempt will be fundamentally linear, albeit in a reversed form, heading backward in time. This is in line with F. C. T. Moore's understanding of Bergsonism as describing a mode of thinking about time that proceeds backward. If "the distinction between what is present to us, because present in time, and absent from us because past in time, is no longer tenable," as Moore suggests, then we are compelled to commence from the end and proceeds toward the beginning (Moore 1996: 91). As Elena Fell emphasizes, for Bergson "we can talk of an end only retrospectively, tracing back the development that led to a particular result so that there is an end as a result but not as a pre-existing, i.e. already given, model" (Fell 2012: 65). Such will be the character of our own project.

Without detaining the reader further, we must outline the contours of a catalytic and rhythmic philosophy of time. In this book, our task will consist of a careful, concise reading of Bergson's work, with a focus on the theme of time-as-actualization. This will, at times, resemble a commentary. Despite its obvious aesthetic charms, we must not let ourselves be excessively enraptured by Bergson's prose. Our interest is basically instrumental. Bergson provides us with a set of conceptual tools for decrypting the relationship of life and matter, the generally motive nature of being, as well as the status of time-as-actualization. We shall head in reverse-chronological order, seeking out various building blocks from Bergson's main works. Tracing the various effects of Bergsonism through philosophy, literature, and science, using many sources and pathways, we hope to unveil the full contours of a presentist philosophy of time that has undergone a thorough deanthropomorphization. Every lover of wisdom—we mean this in the broadest sense of the word "love"—will hopefully find something of interest in our investigation. Some will undoubtedly be disappointed. Some of Bergson's most important works will not have a place in this present text. For example, Bergson's social philosophy, as contained in the *Two Sources of Morality and Religion* and *Laughter*, relates but obliquely to our present concerns here, and we must therefore defer it to a later systematic investigation.[4] The essays and lectures in the volume translated under the title *Mind-Energy*, for all their interest, also do not relate explicitly to the goal of this book, which is the presentation of an epochal ontology of time informed by Bergson's philosophy. The more intensities a work of speculative philosophy emancipates, the better. Silence may be good for spirituality. For philosophy though, ineffability is the end, the limit of each and every beginning. Our goal is nothing less than a revival and resurrection of Bergsonism, without thereby implying that it had ever been dead to begin with. Today it is impossible to ignore Bergson's critical importance for the history of twentieth-century thought (Garre 2020: 165–166). The death or stasis of a philosophy dedicated to the most wide-ranging affirmation of life would in any case be unbearably ironic. Through the broadening of sense, as conceptualized by Bergsonism, we can construct a presentism that fully accepts the universality of change, without succumbing to the imperialistic imposition of subjective structures upon fluctuations. Of vital importance to this prehuman or posthuman vitalism is the affirmation of the prescientific and the preconceptual, through a radical rejection of the primacy of representation. Instead of the representative, we must affirm the primacy of presentation. Bergsonism is an instrument for accessing, in a speculative manner, the dimension of actualization. This philosophy will serve as our conceptual power source on our journey. We make no pretence to the construction of a system lacking in self-contradictions, for never once did our master and guide himself claim that he was building a coherent, unified system. Ontology is, to a great degree, the seismic detection of individual images while traversing a multiplicity of levels. We hold that absolutely anything can constitute a sensuous flow, provided we adequately refine our understanding of what sensuality ought to look like. Beneath our proposed liberation of sense lies a desire to grasp the omnitemporal givenness of times. If anything, that is the deeper, ethical root of this exercise in speculative ontology.

In the following, we intend to briefly outline the contents of our book, chapter by chapter. Taking Moore's injunction to heart, we proceed backward in time, starting from Bergson's last work, *La Pensée et la Mouvant* (translated and published into English under the relatively unimaginative title *The Creative Mind*). The essays composing this collection are the products of an expansive period of time spanning several decades. Among these, several stand out in particular, such as "Introduction to Metaphysics" (originally published as an essay in 1904) and "The Possible and the Real," as well as the two, quite lengthy Introductions. In these texts, Bergson gives highly important summaries of his entire philosophy. Containing points that are fleshed out in greater detail in the primary works, these writings are nevertheless essential for achieving an accurate understanding of the Bergsonian doctrine. Specifically, in our chapter, we argue that Bergson completely negates the category of "possibility." What our everyday mode of consciousness takes to be the "possible" is, in truth, nothing but a retrospective projection of consciousness, a reworking of a bygone past from the perspective of the present. Instead of thinking in terms of a duality between actuality and possibility, Bergson suggests that we must get back to the duration of the things themselves. With originality and audacity which has found but few philosophical followers in subsequent decades (Gilles Deleuze comes to mind as one example), Bergson shows his readers in "Introduction to Metaphysics" a view of the world which privileges impermanence. Change becomes the substance of things. This metaphysical category undergoes a complete reworking in Bergson's hands, being thoroughly temporalized. It is no exaggeration to say that for Bergson time is everything. But this time has several layers, composing a hierarchy of durations. The elimination of possibility from ontology leaves a multilayered hierarchy of self-actualizing durations in the plural. As opposed to Deleuzian interpretations, we argue that, if anything, Bergsonism privileges the actual, reducing the virtual to a mere parasite of a real duration. The actual, perpetual present will be what gives occasion for memory.

Chapter 2 deals with arguably the most controversial aspect of Bergson's career, one frequently overlooked by many commentators. This is the famous 1922 controversy with Albert Einstein and his followers, as well as the critique of the theory of relativity in *Durée et simultanéité* (*Duration and Simultaneity*). Following the lead of revisionist-minded researchers such as Jimena Canales and Miliĉ Čapek, we argue that the standard view, frequently repeated by many critics of Bergson, to the effect that the philosopher completely misunderstood the scientific import of Einsteinian relativity, is flat out erroneous. While the phrasing of Bergson's book does, on occasion, lend credence to some doubts, on the whole it engages thoroughly with relativity and points to several real inconsistencies in the spatialization of time. That being said, we maintain that Bergson is actually expanding the scope of relativity. Far from seeking to root time inside the subject, what Bergson attempts to do is accentuate the import of relativity. This necessitates the differentiation of multiple times. Differently put, what bothers Bergson in the Einsteinian project is its maintenance of a single, universal, objective time. Temporality is more complex than successivity. Paradoxically, Einstein and his disciples do not go far enough in the affirmation of relativity. Subsequent developments in quantum physics show that time cannot be reduced to space. Continuity is more complicated and perspectival than even Einstein, for all his radicalism, countenanced. In our investigation we must also emphasize the unfortunate consequences of the Bergson-Einstein controversy, which resulted in a great decline in the legitimacy of philosophy that has yet to be recovered even today. Throughout his career, Bergson emphasized the need for dialogue between metaphysics and the sciences. Despite the sad fact that this overture was rarely embraced by representatives of the natural sciences, the idea of metaphysics as an inherently multidisciplinary practice can inspire philosophers even today. Without dialogue between the disciplines, the articulation of a general cosmology, a "theory of everything," is bound to remain a project doomed to either incoherence on the one hand or imprecision on the other.

Chapter 3 represents the first sustained engagement with one of Bergson's three "main" works, namely *L'Évolution créatrice* (*Creative Evolution*). That being said, we highly discourage readers from attempting to skip the preceding chapters. It will be seen from the foregoing chapters that many of the more minor works contain many important insights that allow for a deeper reading of the primary texts. Lacking a firm grounding in the Bergsonian philosophy as a whole, we could not achieve an understanding of the whole. Somewhat unconventionally, we attempt to read Bergson's doctrine integrally, and not as a discontinuous, incoherent collection of different views. Rather than wavering between a position that may loosely be described as "phenomenological" (such as the earliest works) and a more ontological or "cosmological" focus, as exemplified by *Creative Evolution*, we argue that, following the lead of commentators such as Elena Fell, John Mullarkey, and Pete A. Y. Gunter, Bergson is attempting to articulate a multilayered ontology of durations. In *Creative Evolution*, the organic mode of temporality takes center stage. The *élan vital*, loosely translated as "life force," constitutes the entirety of evolution. While distinctly non-teleological—we cannot be sure in which direction this open process of self-elaboration is tending in—the broad tendencies of life's evolution can nonetheless be delineated. The phrase "life force," similarly to a physical tendency, follows the path of least resistance, yet it also opposes matter because it moves itself, displaying an autonomy foreign to the material dimension. Yet, as Messay Kebede has argued, the dichotomy of life versus matter is far from unambiguous. A good case can be made that the inorganic realm, also containing a type of duration (Bergson explicitly says so, albeit material duration is confined to "repetition"), plays a key role in the elaboration of freedom in the universe, acting in the manner of a provocation goading life to strive for new directions. Whatever we may think of its scientific plausibility, *Creative Evolution* does furnish us with a remarkably speculative and realist philosophy of biological time, one that can have a role to play in the construction of new, indeterminist cosmologies. Freedom, defined as contingency, permeates the cosmos in the form of an almost graceful ease of movement. Although interrupted by its finitude, Bergson holds that life, identified with the negentropic ascending movement of complexity, could conceivably triumph over death, enveloping matter, enlivening all things.

Chapter 4 is dedicated to unraveling Bergson's theory of perception, memory, and time in his arguably most systematic work, *Matière et mémoire* (*Matter and Memory*). In this remarkable tour de force, Bergson attempts to solve one of the most intractable of problems, that of mind and its ontological status, as well as the relationship of mind with the physical world. Is there a difference between spirit and materiality? Bergson proceeds by differentiating between "pure perception" on the one hand and "pure memory" on the other. Pure perception is a direct, immediate correspondence with materiality, a condition in which perceiver and perceived cannot be differentiated. Matter is the aggregate of what Bergson calls "images," objects which are capable of being perceived, but need not necessarily enter any particular consciousness. In pure perception nothing is preserved; there is nothing in such a perception apart from the momentary vibration of matter. It corresponds to a present without a past or future. Pure memory would be the conservation of all images and processes, an absolute past. Needless to say, these two extremes are, in Bergson's view at least, merely theoretical constructs. That being said, we make the case for a somewhat unconventional reading of this dualism, connecting duration with the actuality of pure perception and not virtual pure memory. Turning Deleuze's famous reading on its head, we argue that duration is pure actuality, a qualitative moment which is not an instant, situated between past and future. Duration, in other words, is the occasion that gives being to antiquity and posteriority. The Bergsonian dualism between perception and memory is a necessarily asymmetrical one, and in the case of *Matter and Memory*, the actual presence of self-materializing images has the upper hand against memory, the latter being, in itself, powerless and impotent. Along the way, we also deal with such salient themes as the place of embodiment in the Bergsonian ontological framework. Getting back to duration involves the shattering of our own condition, in the context of a reunification with everything.

Our final chapter is a commentary upon Bergson's philosophy of freedom. Reading the philosopher's first systematic work, *Essai sur les données immédiates de la conscience* (translated under the somewhat more accurate title *Time and Free Will*) as an example of moral philosophy, we take the question of free will and moral responsibility to constitute the core problematic of the text. Against both libertarians and determinists, Bergson proposes a novel theory of freedom, one that privileges, once again, the present moment of decision against both past and future. A free act will be an activity stemming from within the individual. Even if we cannot control all aspects which have contributed to us being who we are, a certain degree of moral responsibility can be ascribed to all persons. The return to immediacy, the reunification with the impermanence that is duration, is revealed as a fundamentally moral project, a challenging ascent each free individual must undertake. Freedom, in the final instance, means openness to contingency, as well as a willingness to endure the consequences of our own creations.

In our Conclusion, we seek to envision a mode of duration that is authentically timeless. Drawing on a variety of sources, we conceptualize an "atemporal duration," a persistence inclusive of all pasts and futures, in the form of immediately accessible simultaneous presents.

NOTES

1. For this insight we are indebted to our colleague Gergely Berta's observations, expressed in both private conversation and correspondence. Although we do not share Berta's view that the Bergsonian project is intelligible within the context of any phenomenological philosophy, Bergson shares with Edmund Husserl an emphasis on the immediate. But the pathway the two philosophers take differs fundamentally, for in Bergson's case, philosophical intuition reveals a fundamental inadequacy of subjectivity in relation to other durations. To quote Anthony Feneuil, "intuition" as a philosophical method reveals "a gap between the absolute and one's own individual consciousness," a gap "which is surpassed only in the accomplished mystic," if at all (Feneuil 2012: 49). Differently put, while the phenomenological method reduces externality to subjectivity, Bergsonism shows the inexhaustibility of durations. Violence is done to our perception, hence a certain immediacy with heterogeneous temporalities is achieved, without ever abolishing the real gap between our temporal flow and the different levels of becoming. Bergson makes no hypocritical claim to be giving us unmediated access to any absolute principle. There nevertheless can be imagined a penultimate mode of intuition Bergson chooses to call "mystical" that breaks through to a complete immediacy, but this lies outside any philosophical method.

2. The circumstance of gene editing, the ability of the life sciences to rewrite genetic codes, changes nothing in this regard. Biochemical systems too are always already engineered processes. Editing would merely constitute an artificial acceleration or aggravation of spontaneous formation.

3. In the context of this book we shall not be dealing with this particular text of Bergson's, for it would also demand a thorough analysis of the French Spiritualist school. This goes beyond the scope of the present project. The connections between Bergson and the Spiritualists have been splendidly explored by many authors. Specifically, the influence of Ravaisson upon Bergson is exhaustively explored in Dominique Janicaud's classical monograph. cf. Janicaud (1969: 121–197).

4. For a preliminary insight into our investigations regarding Bergson's social views, we refer readers to our article. Lovasz (2020: 198–244).

Chapter 1

Destabilizing Thought

In the concluding chapter of his work, *Bergson and Philosophy*, a work of commentary that has become a classical piece of philosophy in its own right, John Mullarkey recommends an alternative English title for Bergson's collection of essays, originally entitled *La Penseé et le mouvement* and translated under the rather nondescript name *The Creative Mind*. Mullarkey's alternative title is *Thought and Instability*, as the various writings in question revolve around these two topics: thinking and the inherent ungroundability, instability, and contingency of reality in itself (Mullarkey 2000: 259). Drawing inspiration from Mullarkey's suggestion, we have resolved to entitle this chapter in a manner in line with this view, for our chapter is dedicated to expounding the contents of *La Penseé et le mouvement*, especially insofar as they relate to the topic of time. Despite its apparently varied contents, we hope to prove that *La Penseé et le mouvement* is reducible to a distinctly singular problematic, namely, the metaphilosophical status of speculation and the relation of speculative thought (intuition) to the multiplicity of real temporalities. Epistemology cannot simply be reduced to ontology. While Bergson accepts that there can be no escape from representation, he "is still no idealist as regards ontology" (Mullarkey 2000: 228). This dynamic unity between intuition and reality-as-process shall become abundantly clear through our reading of the most significant portions of *La Penseé et le mouvement*. Specifically, we focus only upon the longest and most fundamental texts composing the book. We must uncover the key to understanding Bergson's concept of understanding and meaning. Material patterns invoke new modes of thought, without presenting us with any general image or form. It must be understood that no theory is exempt from the ever-changing life of methods, the latter being trace forms that mutate themselves into groundless generalities permeating thought. The fundamental activity of the philosopher consists in recognizing and proving the inconsistency of all self-generalizing concepts, without positing a universal matrix underlying the heterogeneous materials of thought. No concept of fact may be permitted to blind us to the heterogeneity of that which thinking refers to. Philosophy, if it is to approach the demiurgic vibration of the real, must resist the temptation to build cathedrals.

Ostensibly, in everyday life, reality gives itself as a series, a multiplicity of arrows, tending in certain directions, accomplishments that can be retraced, reconstructed, fitted into diagrams, and simulated. Tomorrow's achievement would then be prefigured, so to speak, in the effort of today, while yesterday's events were already indications of our present situation. What is Bergson's point of departure? The mystery is that reality and thinking are incommensurable. "Philosophical systems are not cut to the measure of the reality in which we live; they are too wide for reality" (CM: 9). Oriented toward practical activity, our everyday intelligence is geared toward selecting from among a variety of data, transforming the world into convenient slices amenable to practical implementation and the achievement of particular goals. Of this, we are reminded several times in the Bergsonian *oeuvre*. But our present problem is an altogether different one. It pertains to the opposite type of knowledge, the type that in the philosophical tradition has gone by the name "theoretical." The selective ability of the everyday practical orientation still reigns here, but this time, it is an issue of a too wide selectivity. The name "retrograde movement" signifies the construction of a timeline or pattern posterior to the event. Several different elements, events, and actors are refitted into a coherent, progressive narrative. Theoretical knowledge mistakes the accomplished, finished state of reality for the real thing: "By the sole fact of being accomplished, reality casts its shadow (. . .) into the indefinitely distant past: it thus seems to have been pre-existent to its own realization, in the form of a possible" (CM: 23). A possibility is only ever retrospective: it is the by-product of an evolution that has already occurred. "Possibility" is the name we project onto realities which were, at some point in the past, in the making. Never can change be expressed as the unfurling of a possible or the actualization of a preexistent virtuality; evolution is "radically new," and most certainly not "a rearrangement of the pre-existing" (CM: 21). Take Immanuel Kant's famous theory about historical progress. For Kant, all epochs of human history to date are preludes to the construction of a universal, cosmopolitan global civilization, characterized by a high level of culture, the rule of law, and world peace. Every previous stage of history would, in retrospect, have been a preparation of this ideal. The paradox, of course, is that we shall only be able to look back upon history from the very end of human history, that is, the ideal, once realized, will have put an end to the possibility of any further progress. Of course, Kant himself knows all too well that the civilizational progress he describes in the essay "Idea for a Universal

History with a Cosmopolitan Purpose" and elsewhere is not an existent reality, but rather a necessity, a prophylactic intended to guard us against despair. As Louis Dupré notes, Kant "neither shows nor proves the actual presence of design or of historical progress" (Dupré 1998: 815). Kant is fully aware that the idea of a cosmopolitan community is a hopeful projection, a fantasy nevertheless endowed with a supposedly imperative power. We must, from a Kantian perspective, stick to the concept of "progress," because the idea of history thought of as progress promises greater joy than the meager realities actual history presents us with. Over time, the Kantian positing of a mythology of historical progress has enjoyed a highly successful career as the object of real, genuine, at times unshakeable belief. There are countless historical reasons for the breathtaking success of such an improbable concept, but we may emphasize one in particular: it is immensely satisfying to suppose that "we" happen to be on the "correct" path, that the great movements of history flow in a direction favorable to "us" (whomever this "we" happens to be). As if by magic, events are sorted into an order, explicable in terms of the tendency we call "progress," whatever the particular normative contents of this progress happen to be at the moment.

THE FALLACY OF RETROSPECTIVITY

Explaining past events in terms of their contribution to a present is what is known as the *post hoc ergo propter hoc* fallacy (Mullarkey 2000: 249). The belief that the present in any way gives order to the past is an illusion, born of little else apart from wishful thinking. When we make progress into the object of belief, we forget that never has there been anything other than a prophylactic designed to soothe our soul. The concept of progress, like all other existing objects in the world, is not immune to change, nor should it be. Progress is as Protean as anything else. The words of the early twentieth-century American economist Jacob Hollander, while referring to the contents of "liberalism," could be applied to the phrase "progressivism" as well: "Liberalism as a practice conforms to the principle of historical relativity. The spirit remains constant; the expression varies with change in time and in place. An identical institution will express the Liberal ideal in one conjuncture and will deny it in another" (Hollander 1914: 14). Our object here is far from political, yet we find ourselves in the midst of questions relating to social and political theory. The reason for this is that Bergson's object, in the first Introduction written for *La Penseé et le mouvement*, is a general critique of theoretical thinking in general. The lessons have not been learned, for the most cursory glance at everyday communication will display endless examples of muddled, incoherent, and downright deluded thinking. In an article commemorating the anniversary of the momentous D-Day landings of 1944, a professor of history writes that the soldiers engaged in combat fought for today's condition of (relative) social freedom, so as "to secure a pathway for a better future for democracy and humanity. We must always remember the price they paid to create that pathway, and remember that we still travel it today" (Huxen 2019). As far as the stirringly memorable crafting of words goes, this is an excellent example of cosmopolitan progressive patriotism. We share the sentiments of the author, to a greater or lesser degree, a couple of misgivings notwithstanding. But we may share the basic spirit of a text without holding in common its presuppositions and faults. The sentences reproduced above are indicative of a fatal blunder on the part of the author. What Bergson is critiquing under the umbrella term "retrospective movement of truth" is precisely the *post hoc ergo propter hoc* fallacy. According to the static, retrospective view, the present would be but a mere extension, already prefigured, at least in its broad outlines, within the fabric of a past inseparably connected to it. The mistake here consists in not taking the contingency of change seriously enough. There is greater joy in supposing that we are following a preordained pathway opened up by the noble sacrifice of our predecessors. But is this *really* the case? Not for Bergson:

> Radical indeed is the difference between an evolution whose continuous phases penetrate one another by a kind of internal growth, and an unfurling whose distinct parts are placed in juxtaposition to one another. The fan one spreads out might be opened with increasing rapidity, and even instantaneously; it would still display the same embroidery, prefigured on the silk. But a real evolution, if ever it is accelerated or retarded, is entirely modified within; its acceleration or retardation is precisely that internal modification. Its content and its duration are one and the same thing. (CM: 20)

In what we may call the "fan-model" of evolution, every change is already implicated inside of reality. Change is simply the unfurling of a fan, already complete in all of its details. The present is merely the ready-made product of an action begun at some point in the past, which is reaching its fruition at this moment. The type of progressivism which operates with the *post hoc ergo propter hoc* fallacy perversely privileges the present at the expense of past and future alike, while ignoring their respective positions as self-given presents. A certain present, namely our time, is privileged, to the detriment of other times. If we hold that past events must be necessarily connected to our present, that historical meaning only ever pertains to *our* present, the ephemerality and relativity of progress is lost, every event being rewritten into a grand master narrative. Those who commit the retroactive fallacy conspicuously fail to realize the Protean nature of the concept of progress.

Change itself is changing. Real evolution cannot be forecast, for change is not the filling of a preexistent mold. Present properties and constellations hang together loosely with a whole set of relative pasts and futures. As Bergson reminds us, "Taken from another angle, an entirely different reality (not just any reality, it is true) could just as well be linked up to the same circumstances and events" (CM: 23). Take a historical example. Had German National Socialism and Japanese nationalist militarism prevailed in World War II, an entirely different historical narrative would be in place, and society would interpret historical change in a disturbingly different light. In "Crisis on Earth-X," the Arrowverse crossover event depicted in the *Supergirl* TV series, the Nazis have won, ruling despotically over an alternate Earth. For obvious reasons, history books are written differently on that planet, and the same events have different normative meanings. In his religious text, *Exegesis*, sci-fi writer Philip K. Dick speaks of reality as constituting an excess that disturbs our self-reference: "When we talk about reality this is what we're talking about: *a* field—and *now* the significance of 'perturbations' can be appreciated: something *more*, in the field or (more likely *beyond*) the field, disturbing it" (Dick 2011: 335).[1] Similarly, contemporary speculative realist philosopher Graham Harman defines the reality of an object in terms of an inexhaustible, unrelatable excess, albeit one internal to each particular thing: "We could call this [the reality of the object—A.L.] an 'excess' of the relation, except that it lies not outside of the relationship, but nestled deep within" (Harman 2005: 202). More recently, Harman has suggested that the originality of Object-Oriented Ontology—synonymous with speculative realism—lies in its affirmation of "the mutual darkness of objects" (Harman 2017: 12). Inaccessibility is about more than the relationship between knower and known. Relating the concept of excess to historical events, we can recognize not only that past events are inaccessible to us but also each present is inexhaustible in itself. Not everything can be reduced to a single story. There can be no question of ever entirely knowing the full details of an event even after the fact. Ontological liberalism commits us to historical relativity: "It is never the case that all of the details of the history of a thing are inscribed into that thing. The world forgets a great deal, and so too does every object in the world" (Harman 2017: 115). Dick's concept of reality as a "perturbation" arriving from outside a "reality-field," and Harman's idea of reality as the inner excess of the object itself, are both prefigured in Bergson's philosophy. What, after all, is the primary limitation of grand philosophical systems for Bergson? Their self-referentiality prevents them from accessing reality, the latter being an excess, a complexity eluding abstraction. It is for this reason also that Bergson is greatly skeptical regarding the measurement of time. The translation of the flow of time into discrete units instantiates a suppression of duration. It cannot be the case that the time of the clock measures real temporality, because "the line one measures is immobile," while real time "is mobility" (CM: 11). Does not the mere existence of time measurement technologies attest to the empirical validity of the claim: "time is quantity"? Does the existence and apparent success of financial instruments such as options and futures contracts not prove that, in a real sense, "time is money"? Not so, Bergson would claim. The flow of time eerily absconds from any representation. Many things are present in the reality-field of time measurement, but duration is not among them. A properly intuitive method asks us to posit a multiplicity of times and timelines. There are, needless to say, transitory diagrams, representations that do not stake a claim to representation, while nevertheless abstracting from actual movements. Anthropologist Tim Ingold refers to the example of drawing lines:

> Drawing (. . .) subverts the assumptions that underpin the polarity of text and image. Its lines neither solidify into images nor compose themselves into the static verbal forms of the printed text. They do not capture the world in its totality, and render it *back* to the viewer or reader. Rather, they are carried *forward*, in real time, in concert with the movements of the worlding world, in an ever-unfolding relation between observant eyes, gesturing hands and their descriptive trace. (Ingold 2011: 225)

The line, for Ingold, would be a middle way between the diagram that fixes, ossifies, and fossilizes movement, and the vitality of the movement itself. When drawing, we move along with our bodies. Intent upon reproducing yesterday's walk, we can take a pen or open an image-editing software application, and by moving our hands render lines upon a seemingly blank surface. A certain amount of movement will have thereby been created in our world. As opposed to time measurement, drawing is an example of a transitory movement, a second-hand creation of novelty that does not aspire to replace the reality of the object itself. An improvised drawing can remain incomplete, mimicking the incomplete and always open nature of real duration. It could be objected that there is no such thing as pure improvisation. According to Bruce Ellis Benson's working definition, improvisation "is a kind of 'composing' done on the spot" (Benson 2003: 23). Bergson would agree that even apparently spontaneous acts of creation contain a certain element of repetition and solidity. Without its material basis, improvisation could not bear witness to mobility, to the life of movement. This basis is the line itself, movement being the dispersal it engenders. Nothing can represent the chaos arising spontaneously in its purest intensity. Benson identifies in music a tension between performance, identified with repetition (or "re-presentation"), composition, and improvisation. This triple tension is only complicated by the problem of originality versus reproduction. Benson's solution is overcoming the impasse between performance and composition by extending originality to every aspect of a piece of music. There is no hierarchy between "original" creation and performance, as both signify a creative act that adds to the musical tradition: "On my view, both composition and performance are improvisatory in nature" (Benson 2003: 25). Nothing like a repetition can be discerned in the musical tradition, for in truth each individual performance is an individual act of creation, yet these acts are also inseparable from the broader musical genre and tradition they happen to be situated in. A biological example can illustrate further what we are getting at.

If we observe reality in itself, we find nothing like a set of states, segments, or repetitions. The real object is a moving thing, an advent that comes, without any apparent discontinuity being involved. Reality is a set of intersecting states, grains of extreme mobility artificially deconstructed by the dissociative operations of the scientific-analytical mind into distinguished, makeshift categories, or the more appropriate, better differentiated "objects" of scientific study. A study on the "cryptic behavior" of sea urchins breaks down the movements of the predator-prey relation into four different components. As the authors explain, "The predator–sea urchin–kelp interaction is perhaps the best known example of a marine trophic cascade" (Spyksma et. al. 823). A "trophic cascade" is a contingent ecological transformation driven by the presence (or, conversely, absence) of predators. Such processes are mediated by the scarcity or abundance of resources. The authors proceed to dissect the cryptic behavior of sea urchins into a fourfold process of dynamic interaction, wherein the availability of resources, the abundance of prey, the presence of predators, and the behavior of the prey animal all influence each other in complex ways. Unsurprisingly, the study reaches the startling conclusion that sea urchins tend, on the whole, to hide from their predators! At considerable risk of sounding boorish, from a Bergsonian perspective we cannot in the least be surprised by such a result. There is no factual error here, the study in question being of considerable empirical interest for any student of marine biology. If there is no real issue with scientific validity here, then what is the matter? Our point of contention is whether separating living processes into a myriad of components is conducive to a better understanding of reality in the first place. Instead of separating various sub-movements within a larger trophic cascade, it can be said that all of these factors compose a single, indivisible whole. Predator and prey imply one another, their separation being impossible except through an artificial process of analysis. Otherwise, were the relationship of prey and its predator severed, each would be annihilated. The prey species can have no independent existence apart from the predator that feeds upon it. Each constituent of the cascade is dependent upon the other. Life is mutual nourishment.

Perceiving reality in itself, we find a "flux, the continuity of transition; it is change itself that is real. This change is indivisible, it is even substantial" (CM: 16). As Gilles Deleuze puts it, duration is "a case of a 'transition,' of a 'change,' a becoming, but it is a becoming that endures, a change that is substance itself" (Deleuze 1997 [1966]: 37). If anything, Deleuze's characterization is not radical enough, for it associates substance with duration, and duration in turn with memory. But as we shall see, for Bergson "substance is a complexity of change. In other words, processes are processes of other processes, all the way down" (Mullarkey 2000: 204). Any grounding upon memory or any other illegitimate transcendental element will turn out to be spurious. While scientific inquiry must necessarily retrospectively reconfigure reality, separating change into a multitude of discrete units, categories, and entities, in itself movement is indivisible, like a melody. Indivisibility, however, by no means excludes the possibility of differentiating various intensities. Returning to the example of musical improvisation, Benson differentiates eleven different aspects or intensities of improvisation in total. The musical tradition, when viewed in its totality, composes a single, indivisible tradition of sedimented improvisations.[2] Viewed in higher resolution, we find different intensities or shades of improvisation, depending on the radicalness of the artist. Improvisation of the eighth degree, for instance, would bear only the faintest of resemblances to the "original" work: "Using the basic form of the score (such as a typical sixteen-bar blues piece), the performer improvises within those confines. In such a case, there may be no connection to the original melody, or even chords" (Benson 2003: 29). The height of improvisation demands the smallest possible degree of resemblance. Intuition is an ontological method for discovering true differences. Intensities are not quantitative, but qualitative. There is a difference in kind between a habitat overpopulated by sea urchins, with few food sources and many hungry predators, and an ecology with abundant resources and relatively rare predators. One is inhospitable, while the other is a hospitable environment. Deleuze, following Bergson, exhorts us to search for qualitative differences beneath apparent differences in quantity (Deleuze 1997 [1966]: 21). Instead of constituting a characteristic or attribute, "alteration" is "one with the essence or the substance of the thing" (Deleuze 1997 [1966]: 32). Empirical observation necessitates a separation between the various streams we encounter, but such an analytical view will never be capable of descending directly into change. Immediacy evades analysis. It makes a great deal of practical sense to differentiate the growth of a sea urchin spine into separate phases, depending on their degree of solidity and mineral densities. This allows for greater precision. In another study, selected sea urchin specimens were methodically dissected, their spines photographed and cataloged, the contents of their spines meticulously analyzed with the help of the latest software. Yet reality escapes from this slicing: "After removal of the soft tissues, the growing micro-spines are seen (...) although most of them are broken due to their fragility" (Alberic et al. 2019: 4). Our task here is not to criminalize the natural sciences or pick fights with well-funded areas of research, but rather to connect Bergson's concerns with contemporary examples. Analytical reason is as legitimate as theoretical reason and practical intelligence. The problems begin when reality comes to be equated with a hygienic slice that has been desiccated and presented upon the table of a laboratory. Real change has nothing to do with quantitative division; the real object is "an uninterrupted thrust of change" (CM: 16). The growth of the sea urchin is a thrust, a striving of the vital force to gain ascendancy over matter. However, these animals, like all other known life forms, are dependent upon matter. Without a process of biomineralization, during which the animal synthesizes minerals extracted from the environment within its body, the sea urchin would lack any recognizable structure. This process of growth is one indivisible movement, a single creative improvisation.

In denying the ontological validity of retrospective continuity while maintaining the doctrine of the indivisibility of duration, has Bergson run into a self-contradiction? Surely, if becoming is indivisible, then retrospective continuity must pertain. The issue with such a definition is that it rests upon a fundamental confusion between the indivisible nature of *a* becoming and change in general. In Bergsonian cosmology, we cannot really speak of becoming in the singular, because this mode of philosophy is grounded on a rejection of general principles. The words of Alfred North Whitehead, one of the other great process philosophers of the twentieth century, are enlightening on this point: "There is a becoming of continuity, but no continuity of becoming. The actual occasions are the creatures which become, and they constitute a continuously extensive world. In other words, extensiveness becomes, but 'becoming' is not itself extensive" (Whitehead 1978 [1929]: 35). Neither Bergson—or, for that matter, Whitehead—ever claim that becoming must be thought of in the singular. As Ansell-Pearson explains Bergson's idea of change, "we do not have to oppose heterogeneity and continuity. Duration is not a simple indivisible which admits of no division; it is rather that it changes in kind in the very process of getting divided up" (Ansell-Pearson 2002: 72). The challenge is to divide reality according to its own measure. There is no prohibition upon carving up reality. Rather, what is problematic for Bergson is a division that reifies and immobilizes reality, transposing itself between awareness and immediacy. Something about cutting up organisms, halting their process of growth, strikes us as immoral. The sympathetic sentiment within us would prefer that the sea urchins reach their full potential, without being desiccated, their spines snapped, and their bodies transformed into an immobile substratum, a passive data source for human science.

Becoming is epochal, but not extensive (Teixeira 2009: 79). Impressions accommodate themselves to our preferences and desires, but only after a real movement has been made. An authentic division must recognize that in all cases we are only ever in the presence of indivisibles. Deleuze expresses the simultaneity of heterogeneity and indivisibility well: "Duration divides up and does so constantly: That is why it is a *multiplicity*. But it does not divide up without changing in kind, it changes in kind in the process of dividing up: This is why it is a nonnumerical multiplicity, where we can speak of 'indivisibles' at each stage of the division. There is *other* without there being *several*" (Deleuze 1997 [1966]: 42). Indivisibility pertains to every single segment of change. Power, energy, growth, these are all but metaphors for the game of self-production. Being is always already a dynamic multiplicity. Simply put, a view of the world that privileges change is not compelled to abandon the admission of heterogeneity. Objects have a reality for Bergson, one that is not obviated by us taking change seriously. The positing of multiplicity on the ontological level allows the Bergsonian to evade the risk of an all-encompassing elimination of individuality, while also short-circuiting the latent atomism of analytical intelligence. Bergsonism represents a middle way between extremes. Retrospective continuity is illusory, for premonitions only become accessible after the event. Those who reconstruct the past along the lines of a present which has already been achieved trade dynamism for stasis. Had the subsequent events existed ready-made, in the form of latent potentialities, never could we speak of a real duration (CM: 26). There never is any preordained pathway guiding our movements. Trajectories trace themselves, smearing their contours upon the surface of a reality in the making, a reality composed of these very lines. Each trajectory has an endurance of its own, a temporal depth. Events are kinetic forms of departure or arrival at destinations that compartmentalize—in retrospect—into landscapes, territories, patterns, and mental states. Positionality is constituted by the unity of a becoming, which is always more than the sum of its possible involvements. *Existence is more than possibility*. It is in this sense that we must understand Mullarkey's description of the place of actuality in Bergsonism: "Actuality is a creativity neither *ex nihilo* nor *ex potentia*: it is its own ground. The passage of time (or movement) comes from itself at every level, and, as such, is unthinkable in itself" (Mullarkey 2004: 473–474). If we are to regain immediate access to change, we must learn to "forget" the concepts of virtuality and possibility alike, while embracing the continuous game of production, the ceaseless novelty that is creation, evolving within the immanence of the here and now.

If retrospection is an illusion, why then do we stubbornly have recourse to such an instrument? Everyday life is, for Bergson, inherently pragmatically oriented. "The necessity of a retrospection and of a prospection, of a relaxation towards the past and of a tension towards the future imposes itself" upon sentient beings (Lawlor 2003: 125). The absence of any anticipation or preparation whatsoever would make goal-oriented action all but impossible. Recourse to the retrospective imposition of continuity can be an appealing solution to certain complex problems. Improbably, a concept originating from the comic book genre can be of help in better exemplifying our position. We refer here to the idea of *retroactive continuity* or "retconning." According to the following definition, this phrase refers to "the process of revising a fictional serial narrative, altering details that have previously been established in the narrative so that it can be continued in a new direction or so that potential contradictions in previous versions can be reconciled" (Booker 2010: 510). This has very real effects on the narratives of existing stories. An altered narrative has an attenuated duration. "A retcon," writes William Proctor, "*retroactively changes continuity*" (Proctor 2018: 224). The duration of a narrative cannot endure unmodified in the context of hypercomplexified, heterogeneous intertextualities, and transmediations. The reboot—the relaunch of a series—and the retcon tend to work hand in hand, operating with selective deletions, modifications, selections, and strategic omissions. This necessarily has an effect on fandom, conceived of as a repository of collective memory, and also influences the subsequent development of scholarly and critical reception. To cite Proctor's example, "Each variation on the Batman theme is in possession of individual mnemonic circuits, of memory and continuity" (Proctor 2018: 229). The sheer number of iterations, reboots, and retcons makes answering the question of *what is in the process of being remade?* profoundly difficult. Our philosophical effort must be directed at extracting the ontological consequences of this cultural phenomenon. If the "infinite spiral of intertextuality" seems an apt description of immensely extended narratives and imaginary comic book universes of great complexity as Proctor imagines, then surely reality as such may also be described through a similar metaphor. Human beings cannot even keep track of their designs. The very existence of retroactive continuity as a stop-gap narrative device reveals the finitude of human intelligence when faced with complexity, the spontaneous creativity of the mind being superior to its powers of coordination. Creative processes tend to outperform regulatory structures. Try as it might, the retroactive construction of consistency and continuity inevitably falls short of effectively regulating narration, at times disastrously (Proctor mentions a few highly entertaining examples of such failures) (Proctor 2018: 232–233).[3] Previous events are dismissed as mere dreams, characters thought dead miraculously resurrect, parallel worlds are introduced, and so on. Regulation is only ever retrospective, for that which must be rearranged always pre-dates the effort of rational intelligence to bring coherence back into a reality whose runaway growth has outpaced the regulatory faculty. There is a baroque tendency within the complexification of comic book worlds because these products of popular culture too are moments of creation, illuminating the aspects of an *original chance* underlying even the most determined and closed systems.

A HIERARCHY OF DURATIONS

At times, Bergson gives the impression that "pure duration" corresponds to interiority, as in the following remark: "I should be getting back into the flow of the inner life, of which philosophy seemed to me too often to retain only the hardened outer shell" (CM: 28). There can be no question however of psychologizing change, for duration also pertains to the inorganic realm. As Elena Fell puts it, "Bergson presents movement in matter as a case of duration but it lacks the most essential feature of duration, namely growth as an inner accumulation of its own history" (Fell 2012: 156). Anything whatsoever that changes already partakes of duration and, in a sense, of life. We are led to understand that "the living tendency is a tendency to individuation" (Fell 2012: 158). More radically, we can also affirm that differentiation is already an indication of a living tendency, a striving toward becoming-alive. We are surrounded by a world of movement, an accumulating excess of objects. The "innerness" of duration must be understood as corresponding to that aspect of becoming which is not yet finished. On her part, Fell identifies three forms of duration that shall be of fundamental importance for our own investigation as well: *material*, *organic*, and *conscious duration* (Fell 2012: 14). Insofar as an object operates in the manner of a process, it endures. Each process enjoys a certain duration. It is no exaggeration to claim that every object is endowed with an "inner life." Bergson himself, in the Second Introduction to *La Penseé et le mouvement*, speaks of life as a power of selection, defining vitality "as a certain power to act, determined in quantity and quality: it is this virtual action which extracts from matter our real perceptions, information it needs for its own guidance, condensations within an instant of our duration of thousands, millions, trillions of events taking place in *the enormously less drawn-out duration of things*" (emphasis mine, CM: 68–9). Objects too have their own durations, in the form of unextended non-relational inner lives.[4] A tired phenomenology cannot ever do justice to such a view.

"Every relation has a full inner life not exhaustible by any outer perception of it"—this could just as well have been written by Bergson (Harman 2005: 195). The spirit of Bergsonism accords, at least in part, with the spirit of contemporary object-oriented ontology and speculative realism, for both philosophical tendencies are grounded upon taking the excess of real processes seriously. Bergson also shares with the realist philosophical tendencies of the early twenty-first century a keen attentiveness to avoiding the pitfalls of both reductivism and absolutism. Inner life is more than a figural description. Rather, this phrase describes a particular mode of duration. If the First Introduction is dedicated to the task of refuting retrospection and atomism, then the task of the second one is to prepare the way for intuition. We are enjoined to return to a condition of immediacy before the colonization of thinking by ready-made concepts and fixed, static ideas. Intuition is a passive, reverent posture concerning the complexity of being/s that is nevertheless resolutely creative: "Pure change, real duration, is a thing spiritual or impregnated with spirituality. Intuition is what attains the spirit, duration, pure change" (CM: 37). Modifications are made manifest. Each constitution is the novel manifestation of an effort, a striving working outwards from within the reality of every process. Images are primarily given to intelligence in the form of problems; portions of otherwise ambiguous narrative scenes are extracted by minds, carved out into categories. Change in itself is not a universal category however, for it lacks any existence in itself. A change is always the alteration of a specific process or object. Samuel Alexander, a process philosopher and Australian philosophical contemporary of Bergson, summarizes the essence of time in similar terms. For Alexander, "change is not mere difference; but the passage from something to something different. (. . .) Remembering that all existents, no matter what qualities they possess, are in the end complexes of motion, we may describe change as a species of motion which replaces one set of motions by another; it is grounded in motion and may be described as a motion from one motion to another" (Alexander 1966 [1920]: 328–329). If everything is movement, but movement in turn is nothing in itself, duration too cannot have a positive existence apart from its specificity and relativity to a process. Relentlessly, objects change. They are in process. But process in itself is nothing without the objects which are changing, objects are in turn being composed of interpenetrating thrusts of change. What we describe is a moment is "continuous transformation," the "exteriorization" of "an initial interiority" (Mourélos 1964: 93). Bergson describes novelty as "impregnation." Reality is a ceaseless fertilization that every so often creates an entirely new species or an unprecedented, momentous event, a turn. Certainly, the emergence of plant life on Earth approximately 500 million years ago, the invention of nuclear energy, the discovery of metallurgy, or the detection of mysterious repeating radio signals emitted from faraway reaches of the cosmos are all momentous, memorable occasions of qualitative change. We do not seek to disparage the significance of such modifications of reality. But if essence is nothing other than alteration, then the impregnation referred to by Bergson can only but resemble the fertilization of emptiness by emptiness, like the absurdly pointless sexual abuse of a mannequin by the protagonist of Ian McEwan's novella, "Dead As They Come." The absence of any basis to existence, the understanding that there is nothing "inside" apart from more layers of movement, the intuition that nothing possesses any innate value, leads to a panoramic view, a *vertigo*: "I thought I saw my own soul recede from me across a vast black void till it was a pinprick of red light" (McEwan 1977: 82). Not even the depraved and immoral satisfaction of having caused somebody else grave bodily or emotional harm is left to the protagonist. However deeply we would hope to determine at a later point some definite substance or content, impermanence—the protagonist of McEwan's novella has already lived through three failed marriages—is a most adequate description for the groundlessness reality is made of.

As Bergson writes, "An intuition which claims to project itself with one bound into the eternal, limits itself to the intellectual" (CM: 34). Change is not a category, and neither can any real object signify anything apart from an infinite regress of modifications. If intuition is indeed the methodology of Bergsonism as Deleuze contends, we arrive at an understanding of what this constitutes (Deleuze 1997 [1966]: 13). Instead of a strict "philosophical method," intuition must be conceived of as a gyration of the intellect, drilling ever deeper into the inner life of objects: "intuition (. . .) signifies first of all consciousness, but immediate consciousness, a vision which is scarcely distinguishable from the object seen" (CM: 35–6). Recognition of impermanence allows us to remain detached from the alternation of birth and death. We know that the newborn is already "dead as it comes," to paraphrase McEwan's title, but this subtracts nothing from its inherent value as a living thing. The unlikely relationship between man and mannequin, far from being wholly unproductive, results in the creation of new evaluations, affects, outbursts, and violence. It cannot be said to be any less productive than a more "normal," reproductively oriented heteronormative relationship. There is, eerily, no significant difference between fertility and infertility, at least when viewed from the perspective of emptiness. The empty stare of the female mannequin discloses the truth of nonsubstantial inexistence. As a superintellectual flood, intuition teaches obedience to the hidden depths of objects. It is the avenue through which we reach a coincidence with the initial interiority of inanimate, organic, or conscious durations. Intuition, as an exercise in sympathetic speculation, affords momentary access to the coming-to-be generated by a certain actualization: "We find ourselves face to face with the interior of an object, with its internal magma or inner plasm" (Harman 2005: 171). Even the infertile, empty nonliving thing has a charm of its own. McEwan's mannequin too has an inner depth, an excess withdrawing from immediacy. Instead of rejecting presence altogether in the manner of postmodern deconstructionism, Bergson explodes presence in all directions "through a multiplication of presents rather than their dissolution in the past" (Mullarkey 2004: 475). This is a mode of thinking that emerges directly from within movement. Indeed, for Bergson "to think intuitively is to think in duration" (CM: 38). Such representations can solidify into doorways for memory. Every contemplation is at once an ascetic discipline and an enjoyment, a delight guided by an imperative apparently external to "us," whatever this nondeterminate communality may happen to be at the moment of contact.

The intuitive intellect is a soul connected to spirit. In other words, a movement nested in movement, whose credibility is underdetermined. *I am ripening into ripeness*. As we shall discover from our later reading of *Time and Free Will*, freedom consists in its own indeterminacy. An experience of the void need not be a negation. Contemplation of reality is, for Alexander and Bergson alike, an enjoyment, a pleasure. For all the formidable obstacles matter poses for spirit, the undivided motion of intuition guides us back to the real. As Alexander states with majesty, "The mind enjoys itself and contemplates its objects. The act of mind is an enjoyment; the object is contemplated" (Alexander 1966 [1920]: 12). Is there not delight in the pondering of impermanence, the traffic snaking its way through town, the rise and demise of companies, even the extinction of animal species, or the final collapse of the tottering edifice of human civilization? For the meditator, all is fleeting, the entire daunting construct of global culture is but another transitory product of evolutionary selection. This too, all of industrial civilization, will pass. Immobility, the changeless, is but a snapshot; "for intuition," Bergson reiterates, "the essential is change" (CM: 39). Intuition shines its light upon a constantly shifting geomorphology of difference. Needless to say, the intellectual effort of sympathy requires a great expenditure of mental energy on the part of the meditator. This state of mental openness cannot be kept up for long. Neither can movement be entirely reconstructed. Beneath the apparent positivity of becoming, we uncover an unbecoming adversity. As Elizabeth Grosz underlines, in Bergsonism differences of nature are often revealed to be "merely modes of expansion of actualization of internal difference," while differences of degree can also reverse into "differences of nature or kind," each becoming "the slower or faster, compression or dilation of one and the same pulsating unbecoming" (Grosz 2005: 6).[5] Change is the unbecoming between two movements, the absence allowing the advent of two relative presences, present to each other, but, lacking any substantiality, empty in themselves also. "Intuition is arduous and cannot last"—meditation, plumbing down to the very depths, is a moment of impossibility within a world characterized by improbability (CM: 39). What Deleuze and other commentators influenced by him seem to miss, is that intuition cannot be reified into a method. How could it be brought into affiliation with any institution or succession, when intuition pertains to no other temporality apart from the instant?

As we have outlined, for Bergson the retroactive movement of reality is a projection. However he does not claim that the retroactive reconstruction is completely false. As a projection, it has a type of being, but we are cautioned not to confuse this abstraction with real evolution. The absolute novelty of time does raise issues. If it is indeed true that "neither the course," nor the "direction," nor the "end" of creative evolution "were given" when the facts of the present were in the making, if the signs of the future are only revealed as signs a posteriori, then how does the present relate to the past and future? (CM: 25). As Mark Sinclair observes, Bergson sketches "a picture of history without tendencies, movements or forces, and historical change seems to have become a matter of permanent revolution" (Sinclair 2014: 663). For Sinclair, it is difficult to imagine even a local continuity in a world which is completely new. The very newness of the world, conceived of in absolute terms, seems to exclude any continuity or flow of time. How could evolution be a process of maturation, if time is a permanent revolution? Is a "discontinuous evolution" not a contradiction in terms? (Sinclair 2014: 659). We contend that the generalized discontinuity of duration does not in any manner contradict localized continuity. When viewed in its inner aspect, duration presents itself as a flow of heterogeneous, non-localizable states that interpenetrate each other. This is the inner aspect of duration. All of reality is temporal, "duration" being "a non-numerical qualitative multiplicity" (Fell 2012: 86). Instead of a single time, we have identified, following Fell, three different levels of temporality operative within the various works of Bergson (psychological, biological, and material). The supposed primacy of the psychological level, engendered by a misguided phenomenological reading of Bergsonism, is yet another retrospective illusion of the kind we have been describing. The three types of temporality are themselves decomposable into infinite heterogeneities of rhythms and flows. It is not a question of transposing one type of description for another, projecting the same concepts onto any and all objects. Sinclair does not fathom why Bergson "does not transform or restate his account of duration to accommodate his own idea of the retroactivity of the present on the past in history" (Sinclair 2014: 665). At the cost of sounding repetitive, the most probable explanation for this apparent *aporia* is that Bergson has no need for such a restatement because the analyses of durations in the works preceding the publication of *The Creative Mind* relate to different temporal planes than the relatively late analysis of retroactivity. Let us not forget, both examples Bergson gives—the retrospective elongation of democratic ideals into the entire history of mostly undemocratic human civilization, and the remapping of Romanticism onto the Classicism which preceded it—are both historical, social, or cultural constructs fabricated long after the fact. In other words, there need not be a contradiction in describing the evolution of life in itself as displaying the hallmarks of a different temporality from that of cultural evolution. With Suzanne Guerlac, we must say that "Bergson thinks time as force," as the heterogeneous force of productivity (Guerlac 2006: 2). The exact form this productive force takes will always depend on the peculiarities of the level we happen to be investigating. It would be absurd to expect highly accelerated cultural phenomena to evince the same temporal characteristics as slower processes. Cultural evolution is much faster than the evolution of the vital impetus. Memes mutate faster than genes (Dawkins 2006 [1976]: 194).[6] In the essay entitled "The Possible and the Real," Bergson introduces an explicitly ethical dimension to the idea of reality as permanent revolution. Specifically, the contemplation of "the moving originality of things" promises us "a greater joy" than holding on to the illusion of immobility (CM: 124). Taking change seriously means that hope is more than a beautiful but illusory image. Even the most stygian of oppressions is sweetened by a belief in the flow of time. Nothing lasts forever. From an absolute perspective, there is nothing apart from the discontinuity among durations, time being in every case an interval, a moment separating two insubstantial movements. For Sinclair, there is something perverse in finding enjoyment in the fleeting. The constancy of change, "far from joyful," is found by him to be "perfectly disconcerting" (Sinclair 2014: 666). Yet each and every moment, however discontinuous when viewed from an absolutely outer perspective, is a purely continuous intimacy when observed from within. Time bequeaths occasions for contact, including interspecies encounters and interobjective intimacies. Every object is a mirror, a substantive occasion that turns proximate events into media, affording opportunities for new reflections. Countless joys are to be had, even in a fleeting, evanescent world. Desire corresponds to the magnetic occasion, animating our labors, bringing us into coincidence with contemporaries.

At the beginning of his immense tome, Alexander relates the example of "togetherness" between human and horse: "The mind in enjoying itself enjoys its togetherness with the horse" (Alexander 1966 [1920]: 21). We are fully cognizant of the inescapable fact that the horse, this animatedly neighing mass of fur, shall perish at a certain point. From this certainty it does not follow that we must necessarily scorn its company, preferring togetherness with immortal angels or demons to this visible creature. Making a world means accepting its defects. Presentations, as they hit us through bombardments of photons, are multiperspectival bursts. Alexander's togetherness with his horse is the same as his enjoyment, for there is no opposition between contemplation and enjoyment: "I contemplate the horse as together with my enjoyment" means the same thing as "I enjoy myself together with the horse"; furthermore, "if we could suppose the horse to rise to our point of view he would in turn enjoy himself as together with me" (ibid.). Contemplation is a multi-perspectival explosion of enjoyment, a togetherness culminating in a cognition, defined as an affective rekindling of a shared landscape, as well as an accentuation of vision, experienced in the form of an ateleology, a direction without innate meaning, generating complex, self-similar doublings. *Togetherness breeds enjoyment, and joy is community rekindled.* Bergson is adamant that intuition is unavoidably intellectual in its expression, while being supra-intellectual in its fundamental nature (CM: 48). Only an alienated, closed consciousness would see change as problematic. Everyday pragmatic intelligence certainly seeks to reject the intuition of impermanence. It seeks after a place of stable habitation, a hospitable habitat adequate to a stable life. But the predominant characteristic of reality is ceaseless rearticulation. We are mistaken if we seek to make of closure an absolutely relevant ontological category. Whatever kinds of closure happen to pertain, universal becoming and impermanence gnaw upon every structure, until the suitable intensity of becoming or, obversely, relaxed unbecoming, is achieved. There is here a type of annihilation and lack in operation, one that we must reintroduce into Bergsonism. This reintroduction of nothingness shall be achieved in our commentary at a later point. For the moment, however, we must defer this treatment and return to the topic of joy. In the conclusion of "The Possible and the Real," Bergson expresses an optimism regarding change. Acceptance of impermanence, far from crushing the subject, actually makes us stronger:

> Above all we shall have greater strength, for we shall feel we are participating, creators of ourselves, in the great work of creation which is the origin of all things and which goes on before our eyes. By getting hold of itself, our faculty for acting will become intensified. Humbled heretofore in an attitude of obedience, slaves of certain vaguely-felt natural necessities, we shall once more stand erect, masters associated with a greater Master. (CM: 124–5)

Sinclair diagnoses a voluntarism and subjectivism at work in the above passage that he views as inherently problematic and standing in contradiction with the otherwise impersonal aspects of Bergson's concept describing organic duration, the *élan vital* (Sinclair 2014: 666–667). While the metaphors of "mastery" and "strength" do contain masculinist overtones, we must not lose sight of the genuinely naive joyfulness which animates the passage and, more broadly, the entire Bergsonian project. Impermanence humbles even the most powerful of Grand Masters, for nobody, not even Count Dracula, is destined to live forever. Locked up as we are by our individualities, bound by social conventions, separated by differences in stature and rules of bodily and material ownership, antagonized from others by feelings of resentment or jealousy, it is all too simple to arrest the flow of sympathy. But segregation can stake no claim to authenticity. As Alexander writes, "no experience (. . .) ever is isolated or has boundaries which shut it off rigidly from the rest of the world. Rather it is true alike of the enjoyment and of its object that they swim in a surrounding atmosphere or medium" (Alexander 1966 [1920]: 23). Metaphysics does not embrace merely a single pleasure or only one type of time: nothing could be further from the spirit of Bergsonism than the tyranny of a single privileged philosophical object. Rather, it unleashes a whole variety of processes, things, and atmospheres into the philosophical realm. We affirm the variety of all things. The assertion of variety brings with it a profoundly ethical confirmation of the infinite value of individuality and differentiation. As Bergson states explicitly, the metaphysics he seeks to introduce "will embrace realities" (CM: 51). Neither primacy of subjectivity, nor primacy of perception, nor privileging of lived experience: all durations must have their place under the sun. In order to fully utilize our faculty of intuition, we must free ourselves of preexisting prejudices. An intellectual effort is required if intellect is to truly surpass itself. An ever-present temptation, both in the realm of mystical attainment, personal achievement, and politics, is a smugly complacent enumeration of the riches we have already accumulated. Nobody can be enlightened while sticking to the illusion of having actually achieved enlightenment. As Bergson reminds us, "A whole labour of clearing away is necessary in order to open up the way to inner experience" (CM: 53). We have yet to arrive within duration, but this access shall be achieved, in the blink of an eye. After the chaos of a disordered perception arises, the soul ends up with the procurement of its own place, a pure change that is yet to be reunited with new social forms, spiritual dogmas, or structures of cultural categorization and stereotyping. Alia Al-Saji sees in the Bergsonian outlook a radical political possibility: "Critical vision needs to become an attentive effort to hold open the moment of hesitation—in both its powerlessness and generativity—a little longer. This is both a negative effort to delay objectifying seeing and acting and a positive effort to remain affectively open to other historicities and rhythms" (Al-Saji 2009: 387). Seeing "differently" is a method of rejecting reductive and stereotypical objectifications, while the broadening of perception is its happy consequence. It takes effort to perceive differently from our previous habits, both in the relative sense of breaking with our ingrained social habits, and in the broad sense of consciously learning to see difference. By connecting the reversal of perception demanded by Bergsonian intuition and a political-normative concern for respecting cultural differences, Al-Saji shows us the contemporary subversive relevance of "clearing away" the prejudices blocking the pathway to inner experience. A refoundation and a reformation of perception, intuition gifts us with a new, actively contemplative seeing. Intuition is a vision of the new.

If life is about solving problems, then philosophy is about finding them (CM: 58). As an experience of unstoppable alteration, intuition undermines the supposedly immutable. Stating problems correctly "is not simply uncovering, it is inventing," giving being to that which was formerly absent (Deleuze 1997 [1966]: 15). Henceforth, the positing of problems cannot be separated from the self-enjoyment of intellect rising above and beyond its own conditions. For Ansell-Pearson, uplift is the very core of the Bergsonian teaching: "The aim of the enterprise is to expand the humanity within us and allow humanity to surpass itself by reinserting itself in the whole" (Ansell-Pearson 2018: 15). Bergson himself lurches philosophy into a new, posthuman stage when he declares that intuitive philosophy "will have raised us above the human state" (CM: 57). Raising awareness above the present state is more than a mere slogan. Rather, the problematization of the present condition is in itself already a method of transcendence, for such a problem—the problem of the human in its static state—already suggests the necessity of a fundamental fluidification, an unloosening and unleashing of our species into new worlds, or even the invention of ethical possibilities which do not contain a self-contained human race.

What is necessary is that we realize the falsity of the following problem: "How far does intuition go?" (ibid.) We cannot go far enough. There can be no question of delineating a clear-cut borderline between the territory of intuition and other areas of experience. True, the use of words here is, at times, misleading. A couple of sentences prior, Bergson sounds an almost phenomenological tone when he writes, "the truth is that an existence can be given only in an experience. *This experience will be called vision or contact*" (emphasis ours, ibid.). It would be far too simplistic to equate such an affirmation with a phenomenologically grounded emphasis on the primacy of a subjective experience as opposed to objectivity. We must avoid any unnecessary dualism at this point. However strange it may seem, Bergson is equating experience with contact. The very state of commonality, of togetherness, is a mutual enactment which gives birth to a shared ground. Both individuality and commonality are preserved within the enacted encounter. Togetherness inscribes itself into philosophy, like a river bringing disorder, while laying sedimentations—new orders—in the wake of the flooding it has brought to the landscape. Raising unity to a higher power, we are informed by the togetherness we participate in. Participation is already a mode, a preparation for blending. Duality is nothing but a retrospectively construed premonition of the encounter. As Alexander puts it, "We have on the side of mind, flashes of light on a dim background of consciousness; and on the object side, more vivid or interesting particulars rising like peaks out of a continuous range of mountainous country" (Alexander 1966 [1920]: 23). The height of the mountain range corresponds to the greatness of the problems confronting us. We are always capable of positing an entire menagerie of oppositions, dualities, and antagonisms. Individuality can be put side by side with commonality, but both have their rightful places, their respective positionalities, within the sphere of togetherness. A greater intensity of time hangs together with a heightened, expanded commonality. In the moment of encounter, the bond between us and our others is unbreakable. Like pebbles knocking upon a steel plate, accidents occur, but in the moment of togetherness these are but externalities, situated upon the fringes of our perception. Life, it is true, must of necessity generalize, construing catalogs of existents. In a way, it would not be erroneous to say that life is pattern-recognition. To live means to generalize (CM: 62). In togetherness, however, we forget this elementary aspect of life, melting the solidified crust that has hitherto blocked perception and hindered the interflowing of durations. A holistic view allows us to recognize that "everything resembles everything" (CM: 63). We may supplement Bergson's confounding sentence in the following way: everything resembles everything when located *in togetherness*. Returning to Alexander's example, the state of commonality between human and horse is a mutual characterization that creates a background, indeed, a community, an atmosphere of enjoyment. Which self is enjoying whose affects? This question is a false one, for just as intuition cannot be bounded, so the self—in the state of togetherness—cannot, at this present moment, be separated from that whose presence it happens to be enjoying. There is a togetherness-in-horseness on the human side, while on the side of the horse, we would discover a being-in-humanness. Alexander himself maintains that "when we take the deliverance of experience without prepossessions, we realise that our togetherness with our object and the togetherness of two objects are so far forth as togetherness is concerned identical" (Alexander 1966 [1920]: 29). The mutual excitation of elements is what Bergson defines as "contact" or "seeing." As far as togetherness is concerned, all participants in this joyful occasion of mutuality are equal.

Throughout the cosmos there is nothing apart from differences of tension, pressure differentials, differences in temporality, and gradations of density. Each object corresponds to a certain tune or frequency. Illuminatingly, Bergson compares the various types of objects and beings to radio stations, each broadcasting a different program (CM: 69–70). Radically different types of existence can coexist within the same place. Nothing excludes such a possibility. Our own duration, in the moment of contact, opens up to a variety of other durations. Converted into a generative body of tonalities, reflection breaks up into a hazy, fuzzy embeddedness in the present: "my own duration" reveals "other durations that beat to other rhythms, that differ in kind from mine. Duration is always the location and the environment of differences in kind" (Deleuze 1997 [1966]: 32). Far from being composed of a nondescript chaos of mutually ignorant congregations, a concrete reality is here, in the encounter, an ever-present harmonic multitonality of different rhythms, given as an infinite variety of communities. In the foregoing sections of the Second Introduction to *La Penseé et le mouvement*, Bergson diverts our attention to methodological concerns. If intuition is to descend from the heights and become solidified as a method, then it must have some relation with existing philosophies. This relationship is unmistakably one of antagonism. There can be no peaceful coexistence between intuitive and systematic philosophy: "Let us have done with great systems embracing all the possible, and sometimes even the impossible!" (CM: 77). Has not Bergson's desire been borne out, in retrospect, by later developments in the history of philosophy? Few and far between are the systematic experiments today seeking to give a unified, all-encompassing account of reality. Whatever we think about the systematic or nonsystematic nature of Bergsonian philosophy, whether it constitutes a system or not, what matters for us is the attentiveness to reality and real objects we find throughout Bergson's work. An inert, inorganic object too can have a duration, albeit one that is "vastly different" from ours, both in a quantitative and qualitative sense (CM: 82). Every duration is equal. The sentiment of Levi R. Bryant, one of today's speculative realists, corresponds to Bergson's pluralistic ontological intention: "There is an equality of objects, a democracy of objects, in the precise sense that all substances are equally substances" (Bryant 2011: 73). This radical avowal of ontological equality can be translated in the following manner: *there is an equality of times, a democracy of durations, in the precise sense that all durations are equally durations.* In togetherness, there is an absence of inequity. Contact is an activity, the meeting that supports a shift within the world, steering subsystems toward new becomings, or reiterating previous creative repetitions. Every freshened creation is the result of the clearing of a blockage, even repetition has a role to play in the renewal of existence. "Indivisible and indestructible continuity" obviously cannot apply to the entirety of objects in the world, for then change truly would be impossible (CM: 83). There would be no elaboration, no alteration, no enhancement, in a completely unitary world. Indivisibility and indestructible continuity are disclosed within togetherness, defined as a "belonging together to a world" (Alexander 1966 [1920]: 27). Enjoyment flows, in the case of sentient beings, from the mutual occupation of a rightful place. Outside of the moment of togetherness, however, we discern an infinite cascade of durations. Reality, viewed in itself, outside of the privileged homely realm of togetherness, is a chaos, the incoherent elements of which only ever vicariously connect, an aggregation of "movements of movements," and nothing else (CM: 84). Ordinarily, we are conditioned to feel at ease in an environment that is habitually ours, but novelty dissolves the givenness of each now, in favor of an uncertain future, a complexity of coefficients that can be measured and mapped only ever so indefinitely. Far from drilling social norms into students, the goal of education in Bergson's view must consist in a softening and broadening out of perception, so as to aid the adaptive discovery of reality. Pedagogy is as much about stimulating students to make contact as it is about accumulating knowledge (CM: 101). By recomposing education, the mind can also be encouraged, incited to strike out toward the outside, blazing toward an impassioned embrace of alterity, destabilizing all established images in the process.

There cannot be a spatial juxtaposition without some degree of interpenetration. As a mode of cognitive artifice, retroactivity allows for the building up of metaphysical oppositions, such as that of a subject opposed to an object. But the truth of duration's multiplicity will exclude the confusion of this differentiation for the actualized contours of the real. Mullarkey, in distinction from those commentators who interpret Bergson in light of the virtual, prefers to read Bergson as an "actualist." According to the actualist position, "There are no (hidden) forces, no potencies, potentials, ground or substrate to the real, no possibles waiting actualization"; in other words, "there is no virtual" (Mullarkey 2004: 471). If we give ontological priority to actualization at the expense of virtualization, then how can we take change seriously? Alteration must not be taken as expressing the realization of a preexistent potentiality or seed. For Mullarkey, the absence of any potential underlying or preceding change is what allows alteration to occur. "In Bergsonism," writes Mullarkey convincingly, "the possible does not transcend the real but comes after it and is immanent to it" (Mullarkey 2004: 473). Possibility has a second-rate existence, being a derivative of actualization. By demoting potentiality to the status of a derivative, Bergson is able to save genuine novelty from the threat of immobilization. There are few texts which better express Bergson's antipathy to the old notion of potential better than "The Possible and the Real." The goal of this text is to prove the nonreality of what the philosophical tradition has hitherto called "possibility." The whole cannot be a juxtaposition and rearrangement of things, for we are constantly experiencing "the continuous creation of unforeseeable novelty" (CM: 107). The outside is an encouragement, an enticement, a temptation. Things are already imbued with powers. When Gaston Bachelard equates emotional states with interiority, he is reimposing a Cartesian dualism upon experience. In opposition to Bergson, Bachelard writes that "it is not matter that is the obstacle. Things are but opportunities for us to be tempted; temptation is in us" (Bachelard 2000 [1950]: 41). Yes and no! In a temporary sense, the radiations of objects become, verily, inner aspects, burned into our flesh, after the torment of desire has done its damage. On the other hand though, in contact innerness also opens up to the outer, resulting in a forgetfulness of self. Distinctions can only ever be unstable images, reconstructing the immense complexity of reality. Following Bergson, let us renounce the usage of the singular when referring to "reality." We must remember that both discontinuities and continuities are multiple in nature. As David Kreps notes, "Differences in degree exist in space, differences in kind exist in duration. Discontinuity resides in the former, continuity in the latter. *Both, however, are multiplicities*" (emphasis ours, Kreps 2015: 98). Each image is unstable, because the plunge into direct prehension is also unavoidably a dilation or contraction of our own duration, depending on which level our consciousness targets. Under the term *True-First Perception* (TFP), we understand each fresh enactment of consciousness. Creation is at once the springing forth of something unprecedented and the enhancement or buildup of reality. There are constraints to the creative force, as is evident in the example of musical innovation. Even the most radical of innovators must follow previous patterns to some degree: "Composers may be able to conceive new rhythms and chord progressions, but these are usually improvisations upon current rhythms and chord progressions" (Benson 2003: 43). Every repetition begins again through improvisation, yet no repetition is completely reiterative. A duration is enhanced by every supplementary moment added onto it. The emergence of a recognizably new rhythm is a TFP, a sonic self-realization. It is not without some irony that we characterize such a creative emergence as "True-First," for this truth and this eminence are, viewed from the point-of-view of continuity, inseparable from the broader tendency, pattern, or tradition they happen to be embedded in. There is an irony in calling any event a TFP. From the non-relational perspective, the shots we take with our camera do indeed crystallize a moment of artistic innovation. It is between the continuous and the discontinuous that creation is undergone. Several emergences are being born within several continuities on the *plane of consistency*, while on the *plane of actualization* the continuities, unitary in their inner aspect, operate as discrete, heterogeneous tendencies. It is always the melody which orients the subsequent succession. The context of a sensation gives a delineation of its actualizations, but only after the event, "the whole" giving "me an impression at once novel and unique" (CM: 107). Each improvisation is a novel reiteration, a reimagining through which spontaneity makes another advent. For this reason, our book is full of repetitions.

Interestingly, both Bergson and Benson cite the example of Shakespeare. Prior to the advent of the literary personage called William Shakespeare—questions of authorship notwithstanding—it would be absurd to imagine that we could forecast the writing of *Hamlet*. Even the most thorough methods of data and information collection cannot give us more than a vague premonition of what is yet to be made. Shakespeare is manufacturing what *will have been*: "The artist in executing his work is creating the possible as well as the real" (CM: 121). Possibility is not the antithesis of actuality. Rather, it is derivative of a reality that has already come to fruition, a false mask attached to change. The pattern shown is a posteriori, having been transformed into the realization of a potential which had lain hidden until the point in time when Shakespeare finished writing *Hamlet*. As Benson comments, we can speak of the arrival of the "Shakespearian" in literature: "Shakespeare 'discovered' a possible way of shaping the English language.

But he clearly also created something—by 'composing' the line out of preexisting words and by placing it within a particular context" (Benson 2003: 44). Each composition is also a recomposition, and each rearrangement displays a novelty of its own. The error would be to slice up the moment of creation into a spatialized juxtaposition of parts or sets of parts. A vision of a new rhythm is a simple intuition of the moment, the place, so to speak, where concord was broken, where equipoise gave way to change. Each moment is a break located along the crest of the wave. Bachelard's characterization of time is far from alien to the spirit of Bergsonism: "The thread of time has knots all along it" (Bachelard 2000 [1950]: 81). There are knots, micro-discontinuities, and there is a continuity which serves as an environment for these breaks. Concord's natural breakage allows Shakespeare to resituate the line, *To be or not to be: that is the question* sliced out of the English language, into another environment, morphing this thread of words, estranging it from conventional semantics. Already, the inorganic rhythm is contained within the organic structure of the living being: "The repetitions of the inorganic world constitute rhythm in the life of conscious beings and measure their duration. Thus the living being essentially has duration; it has duration precisely because it is continuously elaborating what is new" (CM: 109). Let us note that here too, as elsewhere in Bergson's work, the inorganic blends with the animacy of life. Life is the elaboration of something new. It would be no exaggeration to say that Bergsonism is an esteem of movement. We are only alive to the extent that we create and move. The intensity of elaboration is the measure of life. Praise of movement contains, if only implicitly, the addition of unbecoming to becoming. If it is to be able to continue injecting indeterminacy into existence, life-as-movement must be permitted to periodically undo itself. If one desists from routinely burning the forest, the future shall only bring greater conflagrations. Life "needs to unbecome" if it is to persist in its condition of self-maintenance (Grosz 2005: 10). The image of a knot of time or a momentary peak within the wave's crest captures the negation immanent within change. Lacking constant elaboration, the flow of time would grind to a halt. There is no static "possibility" lying dormant, waiting to be unleashed. A dynamic "whole" we observe is a decentered mobility. What the primacy of the actual allows is a purified description of change. If each object is nothing more than a "movement of movements," there remains the question of how many characteristics of a thing may change without it becoming something else. A. R. Lacey holds that Bergson's position on change can imply "there is no substrate" of change, or "there is a substrate but it is always changing" (Lacey 1989: 110). We must suppose, Lacey reasons, that at least some characteristics or attributes of the changing object are stable, otherwise we would be unable to differentiate it from its environment. Everything would be transformed into a nightmarish primordial soup, or a constantly changing formless mass, a goo the components of which could never be picked apart. In Lacey's objection we discover an ancient confusion of the unity of an object with its being. Pragmatic rationality cannot integrate a Protean concept of objectivity. However paradoxical it may sound, the primacy of actualization necessitates a separation between unity and being. Things are irreducible to the unity they happen to evince. In this, we follow F. C. S. Schiller's position, who writes that real objects possess an "infinity of attributes" (Schiller 1910: 201). Every object also possesses a difference between its reality and its being.

THE INTUITION OF INDETERMINACY

The Bergsonian position temporalizes indetermination. As Bergson asks rhetorically, "would not time be (. . .) indetermination itself?" (CM: 110). Yes and no. The hesitancy separating the present from a future action, on the level of a biological organism, would certainly constitute a type of time. But indeterminacy also pertains to the more fundamental level of a thing's being. The non-absent object, when presented as an instant of a TFP, depends upon its correlation with the nature of the immediate data it is given together with. If not given as perception, its having is "judged" by the immediacy of its coincidence with nature itself. Either way, indeterminacy is immanent to the becoming of an object, because of the infinity of attributes. A certain emptiness must be recognized here. One of the more contentious points in the Bergsonian doctrine, a position that virtually compels a refutation, is Bergson's repeated attempts to evade the issue of nothingness: "Reality, as immediately perceived, is fullness constantly swelling out, to which emptiness is unknown" (CM: 113). To Bergson's credit, this assertion is qualified here by the insertion of the phrase "as immediately perceived." In Gus Van Sant's film, *Gerry*, we see the story of two hikers play out, written as it were, upon the parched windswept New Mexico landscape. The wind, the heat, and the despair of two young men blend into a singular cinematic experience. Immediately perceived, all is one. The death which is in the process of arriving is one with the effervescence of life. An excess of life is already melting away into fatality. Vitality and decadence, the uncaring blue of the sky and the bounty promised (but never delivered) by the incipient rainclouds, the road, all these are one. We, as temporally and spatially distant viewers, are united, tied as it were, through our contemplative power, to the awfully sublime vision of the desert void, sucking in two men, changing both of them forever. The immediate perspective takes us out of ourselves, in a momentary intuition, revealing heterogeneity and continuity. But in the very moment of intuition, a deeper vision or feeling uncovers the emptiness underlying each duration. Objecting to Bergson's one-sided affirmation of positive fullness, Bachelard notes that "there seems to us to be a perfect correlation between emptiness and fullness. One is not clear without the other" (Bachelard 2000 [1950]: 30). If the concept of emptiness does, indeed, depend upon the prior presence of plenitude, as Bergson suggests, then we can also claim that fullness too is dependent on a prior vacuity. Underneath the fullness envisioned by immediate perception, we can outline a completed emptiness, defined as the absence of any substance whatsoever. An absolutely positive world would lack change. Bergson maintains that "in reality there is no vacuum. We perceive and can conceive only occupied space. One thing disappears only because another replaces it. Suppression thus means substitution," a position that is difficult indeed to uphold while also adhering to the absolute nature of change (CM: 114). Significantly, Harman summarizes the entire project of speculative realism in terms of actualism: "What guerrilla metaphysics seeks is the *vacuous actuality* of things" (emphasis ours, Harman 2005: 82). It makes a good deal of sense for us to posit a vacuous actuality between things, defined as processes and durations, thereby avoiding the unnecessary, even damaging privileging of plenitude. Were everything genuinely full, novelty would be stifled. Impermanence could have no place in the world, not even in a relative sense if change were a case of mere substitution. There must be real break or interval between two crests if a second wave of change is to arrive on the scene.

Descending from the outer life into interiority, the contemplative gaze transcends the immediate, hazy perception of continuity. From a metaphilosophical perspective, Mullarkey emphasizes that a totalizing philosophy, one that seeks to exclude any validity to something, would "lack explanatory power, for claiming that everything is (directly or indirectly) *x, y,* or *z* (. . .) they lose the ability to account for their opposite, even as a derivative illusion or error" (Mullarkey 2000: 257). If nothingness were an impossibility, then why would so many brands of philosophy accept its validity? Loss, absence, loneliness, all these are *there*, in the form of corrosive, destructive affects. Emptiness, for Bergson, has a mental existence alone, sharing the same status as retroactivity: the "so-called representation of absolute emptiness is, in reality, that of universal fullness in a mind which leaps indefinitely from part to part, with the fixed resolution never to consider anything but the emptiness of its dissatisfaction instead of the fullness of things" (CM: 115–6). According to this view, negativity is nothing more than the projection of a depressed, discouraged mind which has lost the will to creation. We posit vacuity when stricken with *ennui*. What Bergson does not address is that if reality in itself is an absolutely positive excess, a measureless fullness, how then could the evil thought of nothingness have insinuated itself into the minds of sentient beings? Animated intensities persist, whatever we happen to think about plenitude. Each making must, of necessity, also constitute an unmaking. There are modes of becoming that undo the organism. What is most paradoxical is that such tendencies are frequently felt to be *needs*. As Bachelard writes provocatively, the "shadow of Death" is always "permeating life"; subjects are driven by a death instinct as much as the instinct tending toward life, by "necrophilia" as well as the desire to produce offspring, "the need to lose" along with the need to accumulate, "the need to gamble" as well as the need to earn (Bachelard 2000 [1950]: 41). Each duration must be considered to consist of both positivity and negativity. This truth is expressed by the Jewish mystic, Jacob Frank, in the following line: "The herb of life is concealed in the herb of death" (Frank 2004: 34). There is a relative negation underlying each act of creation, and a positive actualization rising up from even the most heinous of destructions. Beneath, or rather, inside of these two fundamental modes of alteration, revolving, there is the basal underlying emptiness of every object, defined as the lack of any substance apart from endurance. Bergson's absolute is elaboration. In our view, the flow of associativity, the linkage of objects, cannot be achieved without positing absence, and Bergson is mistaken in completely rejecting negativity. The complete destruction of possibility and potentiality accomplished by Bergsonism opens the door to a vacuous causality that is amenable to emptiness, having achieved an evacuation of substance from relations in general. On the relative level, positivity is concealed by negativity. On the absolute level, all is motion. "The possible," Bergson teaches, "is less than the real," as the evolutions of reality are what reveal, in hindsight, what was formerly possible (CM: 117). Possibility has been degraded to the status of a simulacrum, an at best attenuated associativity. The effectiveness of time, on the other hand, privileges the present. As animate intensity, the present is here, whereas the future is not (yet) and the past has already ceased to inhere. Hermann Lotze, another philosophical contemporary of Bergson, writes, "past and future are not, and the representation of them both as dimensions of Time is in fact but an artificial projection" (Lotze 1884 [1879]: 335). From the perspective of insubstantiality the present object too is not, at least not in the sense of constituting an unchanging substrate of change. A thing is the changes it undergoes. Compared to past and future, one can say that *this present*, right now, *is more*. For actualism, the sole temporality which pertains is the animated intensity present at the intersection of past and future. This duration has been contracted to a point in time; it is the epitome, the extreme limit of duration. Indistinct yet efficacious, the fire of the completely contracted present is the sole heat capable of inflaming us with the power of love. In the moment of creation, anything is possible. On Christmas Day, a burning baby appeared to the sixteenth-century Jesuit priest, Saint Robert Southwell, himself waiting to be drawn and quartered.

This improbable vision resulting in the poem, "The Burning Babe." Within the space of creation, no possibility, no attribute, no aspect, can be excluded:

> The metal in this fornace wrought are men's defiled souls:
> For which, as now on fire I am, to worke them to their good,
> So will I melt into a bath to washe them in my bloode. (Southwell 1872: 109–110)

A burning individual's self-sacrifice, the introduction of nothingness into the life of an innocent child, can atone for the defilement of souls, provided that this advent takes place in a suitable context, in a society populated with martyrs, persons in the fullest sense of the word, willing to die for nothing in particular. This generosity is more than a possibility. It is the moment of love, the elucidation of productivity even within death. Inside anguish, the tormenting vision of a baby in flames, there is plenitude. And within each excessive vitality, there is a defilement being prepared, a sinner being cooked by their own transgressions, a demon being emancipated. The possible is only a retrospectively manufactured potentiality, an artificial contrivance born of wishful thinking. The possible is created by a restless mind, unable or unwilling to cope with unpredictability. 1595—the year "The Burning Babe" was written, and also the year of Southwell's execution—was an oppressive time in history, conducive to apparitions, miracles, and religious visions. A flying infant, aflame with love, entering the mind of an imprisoned Jesuit, surely this is an impossibility that makes a mockery of causality! We must remind ourselves that there is no such thing as a possibility in itself, for "the possible is only the real with the addition of an act of mind which throws its image back into the past, once it has been enacted" (CM: 118). If possibility is nothing, this truly means that even improbable events can happen! The evacuation of possibility from being is what allows events to occur. The emergence of a new thing, these are all made possible not by possibility, but by the non-presence of possibility in general. On the one hand, it is the improbability of an object that manifests it. *Actualization is its own precondition.* On the other hand, however, this also means that "one can virtualize without anything existing other than what we call and see as 'the virtual'" (Mullarkey 2004: 474). The virtual and the possible have, in Bergsonian philosophy, no other claim to reality apart from being subjective projections of deluded minds unaware of the real status of time. Any positing of presence has become impossible; reality in itself is composed of limitless actualization. "Difference," Mullarkey emphasizes, is located "within the actual" (Mullarkey 2004: 477). In a similar vein, Mourélos highlights the inseparability of duration from the instant: "Duration and instant are only opposed when seen from the outside, projected onto the plane of linear time" (Mourélos 1964: 233). However intently one virtualizes the possible, sketching it onto the past or extending it into a non-present future, this will always differ from the self-actualization of time in itself. "The possible," Bergson reiterates, is "the mirage of the present in the past" (CM: 119). Only the present contains efficacy, power, aptitude. The virtual is a parasite of the actual.

Each encounter is preceded by a phase of mutual emulation. Those intentionalities that shall be united during the moment of togetherness must be coordinated in some manner. There is a certain danger latent here, namely, that of positing a future "collective intentionality" as an already present possibility, as something which can be separated from the movement of unification. A study on the concept of "togetherness" describes a supposed precondition of the coordination of intentionalities in the following way: "A joint commitment to act as a body is a commitment made by a collection of individuals to perform some present or future action as would a single individual" (Tollefsen et al. 2014: 237). According to the view advanced by the authors, each collective action, to qualify as "action," must be preceded by a prior agreement among all participants. Each collective action has, in Margaret Gilbert's view, a specific "flavour," one that differentiates it from individual or spontaneous but uncoordinated acts (Gilbert 1992: 154). Can togetherness truly be sliced into so many different components of assent? If we accept this hypothesis, some prior principle of unification or preexisting agreement could be found on the part of the various parties involved in such a melting together of intentionalities. Tollefsen, Kreuz, and Dale's stated goal is to prove empirically, through quantitative methods and geometrization, that the everyday intuitions of average person's display the hallmarks of "togetherness." Do people tend to think spontaneously about coordinated activities as being "common" or communal in nature? Their results show that "jointness emerges as the dominant feature by which subjects categorized actions. Subjects seem to be tracking the psychological connectivity of actors rather than simply physical proximity" (Tollefsen et al. 2014: 248–249). The problem with such an analysis is that it proves nothing about the priority of any consensus or agreement prior to the encounter in question. We are not provided with any information regarding the crucial question of whether participants adjust their behavior spontaneously or in response to exterior regulations, rules, or customs. What we dispute here, following the spirit of the Bergsonian critique of analytical reason, is whether such categories as "commitment" or "collective intentionality" are necessary to begin with. After all, it seems immensely difficult to find any intentionality which is wholly individual to begin with! Each individual is also a collectivity. In their introduction to *A Thousand Plateaus*, the second volume of *Capitalism and Schizophrenia*, Gilles Deleuze and Félix Guattari explain that "the two of us wrote *Anti-Oedipus* together. Since each of us was several, there was already quite a crowd" (Deleuze and Guattari 1987 [1980]: 3). The subject itself is a duration, and each duration is a heterogeneous assemblage of different affects. Nothing exists in isolation. Behind relations, there is nothing substantial, no possibility, and certainly no virtuality. Every component of a situation must be conceived of as inherently multiple in nature, for qualitative multiplicity is the fundamental characteristic of each duration, and intentionality, because of its nature as directionality, is at bottom a temporal category. Reproduced, the moment of togetherness certainly can be divided up into an actuality of unity and a potentiality of opportunity. Nevertheless, in practice, the encounter instantiates an abolition of all separateness of parts. "Possible," "virtual," and "preexistence," for Bergson these all belong to the realm of "pure illusion" (CM: 120). To dance the tango, to have sex, to participate in a market, to change the world, to save somebody, all of these are elaborations, elucidated completely in their respective movements of realization. The encounter cannot be grasped through an act of recombination. Far from illuminating newer aspects of the real, the virtual represents an addition, a supplement to reality. Between the ordinary awareness of direction and the limitless ocean of intentionalities, we need not posit a difference in kind. This would be an unacceptable inflation of philosophical categories. "Backwards over the course of time a constant remodelling of the past by the present, of the cause by the effect, is being carried out": this assertion must not be taken to imply that anything is alive, right now, apart from the present itself (CM: 122). Only the present is endowed with a mobile, throbbing duration. The present we speak of here has no single owner. Any privileging of our own duration is an illegitimate and immoral move. Concepts of ownership or identity do not apply to *this* present. The very act of intuition is the process of emerging from one's shell, the capability that allows us to avoid the danger of self-referentiality. "Intuition," writes Deleuze, is "the movement by which we emerge from our own duration, by which we make use of our own duration to affirm and immediately to recognize the existence of other durations, above or below us" (Deleuze 1997 [1966]: 33). Temporalities, far from being mutually exclusive, are nested inside of each other. Gilbert holds that the "plural subject," the subject of togetherness, would be somehow created by the action of unification, constituting a "synthesis *sui generis*" (Gilbert 1992: 268). Instead of being a construct created in retrospect by the encounter, on a Bergsonian basis we must hold that the subject is already a qualitative multiplicity. There is no space in the Bergsonian doctrine for a subject that would inhere prior to its actions. Intentionality is already *given in togetherness*.

Instead of being coproduced by subjects who voluntarily agree to follow a contract or a set of rules lain down in advance, realities are to be thought of as accumulations of spontaneity, energies diverging into multitudes of explosions. The virtual and the possible have no role to play in the self-organization of realities: "It is the real which makes itself possible, and not the possible which becomes real" (CM: 123). To suppose that a collective intention must precede collective action would be to drive a wedge between the preparation of an act and its implementation. In truth, the preparation preceding the act is already an aspect of the movement. We discover in togetherness a simple sensation, a unity insinuating itself into perception as a passing momentum. Change as a condensation of time includes the passages between durations as internal components. Intentionalities come in a variety of scapes. That being said, we must be careful not to erect dividing lines on spurious grounds. One author has thought it appropriate to differentiate "objectscapes" and "subjectscapes." While the former would denote movements, places, and the distances between them, the latter would denote "the elements of subjectivity," (presumably also the localization of said subjectivity), "recognized by us as the capacity to act and to *become*" (Ivakhiv 2013: 11). In and of itself, there is no error in such a differentiation, provided that it does not morph into an all-encompassing ontological difference. The relation between an object and its relations predates any object-subject distinction (Harman 2005: 190). Therefore magnifying this difference into an all-encompassing ontological rift would be a mistake. Any situation can be sliced into a set of intersecting scapes. For Bergson, the undivided mobility of change precedes the separation of objects into organic and inorganic, subjective and objective realms. To be fair, an objectscape is, for the author in question, a cinematic construct: "An establishing shot creates an objectscape" (Ivakhiv 2013: 85). What cinematic vision does is select a group of images from the complexity of undivided, mobile reality. These selections are new worlds, coexisting together with their extra-systematic non-filmic environments. "Cinematic objects," writes Ivakhiv, "carry worlds with them, and set against the objectscape of a world that is perceivable in a film, they set off affective ripples that move in one direction or another depending on the expectations and capacities of viewers" (Ivakhiv 2013: 125). The medium receives directions, imperatives, and intrinsic movements from outside the frame. Capturing these, cinematic vision reduces variation, to the point of appearing to freeze time. This appearance, like the contrast that emerges between subjectscape and objectscape, is a product of audiovisual technology, an image of time, and not temporality in its intrinsic becoming. Our point is that togetherness must not be confused with a static image. The movement of becoming knows no anticipation. All realities are concrete relations and real situations. The actual is immediacy, "time is immediately given" (CM: 124). From our perspective, it is true, possibility appears to inhere. Action happens, we encounter a lizard running along a rocky surface in Rattlesnake Valley, Montana. Our meeting is far from neutral or equal in terms of strength. Both of us have a substance, defined as a duration. Both lizard and human inhere for a certain period, a timespan greatly smaller than that of the rock we shared a moment ago. Togetherness is as much about adversity, conflict, and inequality as harmony or coordination. As psychologist Peter F. Schmid notes beautifully, "encounter is action and passivity, taking and letting go. Encounter happens in the authentic 'game of love' without rules" (Schmid 1998: 83). Encounter is an opening up of our own duration, allowing our time to be infected by other temporalities. Scapes furnish imperatives, as do the beings we meet and the products we are confronted with. The choice lies not before the movement, having no reality outside in its own execution. For Bergson there is no difference between the flow of time and its actualization. There is no interval separating creation from itself, as if the impetus had to first overcome itself before its outflow. We have no idea what species, societies, or becomings the future holds because it is the present which holds all things. Bergson would not be opposed in principle to Harman's assertion that "real objects are more than we think they are; intentional objects are less" (Harman 2005: 200). The virtual is infinitely less than the indivisible movement of a real duration, while the actual is always in excess of access, with the important exception of intuition which, for a fleeting moment, bridges the interval between at least two durations, if not more. We see the lizard and briefly envision something of its own time of life.

The paradigm of togetherness restates the self "as non-substantial," as "a gathering of aggregates, an organismic assemblage of (environmental) composites to which substance is functionally attached" (Bazzano 2014: 209). Change is actualization, without any foundation, a power constantly reworking images, transposing them into new planes. In splicing together slices extracted from the flow, we are recreating becoming in ourselves. Bergson warns against what he calls the "complication of the letter," for discourse tends to occlude the real complexity of change (CM: 126). If it is to truly draw us closer to life, intuition must present itself as a simple act. Beneath the fragmentary, momentary intensity, there is a manifestation, an other vision, uncrowded by superfluous words, overwrought sentences, and overpopulated piazzas. As Schmid would have it, "encounter" is a surplus, containing "a continuous and fertile tension of unity without fusion" (Schmid 1998: 84). Let us imagine a landscape, what biology would call an ecological niche. What else is there, if not a multitude of interspecies encounters, contacts, and challenges. The evolution of one species never takes place in isolation. There exists a convergence between the morphologies of different animals, depending on their functions within a certain niche. Durations interact with one another in ways natural science is only beginning to comprehend and measure. *We hear a rattlesnake*. Hallucinating, we think we have heard it hiss, "*reality is kkkkaaaossss!*," and in a Montanan dialect to boot. Inexplicably, in retrospect, we discover a phrase, "KAoS," which denotes a policy and domain services framework created by the Florida Institute for Human and Machine Cognition, a programming language designed to guide the enforcement of policies relating to human-machine interactions.[7] We speak at this point not of pure chaos, but rather, the mutual coordination and adaptation of becomings, structured by a certain contexture. We borrow the word *KAoS* to describe the coevolution of organisms situated within a shared ecosystem. A recent major study of avian morphology has revealed a convergence between different bird species occupying the same functions in similar ecological niches. As the study highlights, "each trophic niche has the same morphological signature worldwide, highlighting the repeatability of convergence events across multiple evolutionary arenas" (Pigot et al. 2020: 236). This fact of evolution is of momentous importance for our investigation, as it gives a certain empirical grounding to our claim that intentionality is always already collectively given. In the case of avian morphology we see geographically distant bird species evolving similar forms of embodiment. Biological evolution is a dynamic interaction between the duration of the organism and the duration of its ecological niche. These durations in turn are further enhanced, or complicated, by the presence of predator and competitor species. A KAoS, as opposed to pure disorder, is an unforeseeable novelty whose elaboration nevertheless adjusts to a certain ecology. On the whole, the evolution of bird morphologies, because of its status as a time-bound process, is chaotic, but on a local level it is also KAotic, composing a locally ordered, self-guided, domain-limited chaos. The avian morphospace does not take place in a vacuum; it is connected with trophic niches that shape the outlines of its unfurling motion. A scape is a domain framework for change. Before the avian morphospace, there is not any possibility of avian evolution, but rather an ecological framework, a prior duration that proves hospitable for a certain range of bird species. Viewed from an evolutionary perspective this process displays an outstanding simplicity. As a whole, the morphological changes of several bird species, or at least those belonging to the same function, say, that of prey, can be viewed as a continuum. A morphospace, incorporating the development of several species, is also a duration in itself. As David Kreps writes, "a self-organised structure," as opposed to an artificial construct, "*endures*; one might say its complexity incorporates duration" (Kreps 2015: 197). On this view, a duration would be any self-organizing process of actualization whatsoever.

CRYSTALLIZING TIME

At the outset of his 1911 lecture, "Philosophical Intuition," Bergson outlines a method of philosophical interpretation that operates as a type of contraction.

Instead of gradualism, or, worse, recombining and remixing our predecessor's concepts, we must proceed through contraction, compressing all of a philosophy "into a single point" (CM: 128).[8] What would the great thought of Bergsonism be? We advance the following: *all realities are impermanent, and persistence is unchanging*. Duration is impermanence, hence nothing can be posited outside of actuality. *No thing inheres outside of endurance*. This point is infinitely simple, yet Bergson's entire life was dedicated to elaborating upon the implications of this ontological theorem. It takes a prodigious expenditure of words to expand upon such a singular truth. Beneath the concepts, there is always a simple intuition: *being is change*. Bergson expresses his conviction that underneath the most complex doctrines and theologies, we may uncover the intuition of simplicity. The "zigzaggings of a doctrine" are unavoidable in any philosophical work. Truth cannot be simply surgically excised from complexity. Returning to simple, introspective access to the real is implausible without first drawing lines with words. A doctrine, a teaching, a gospel is a movement "which loses itself, finds itself again, and endlessly corrects itself" (CM: 130). Each recomposition cannot give back the simple feeling of submission, passivity, self-surrender. The moment of togetherness is an encounter, a participation that cannot be revivified with texts alone. But philosophy and science can assist in retracing our steps, retracing a path back to simplicity. As Jankélévitch writes regarding intuition, "To understand is this not only to spiritualize the perceived, it is also to humble oneself before the given and to experience its resistance" (Jankélévitch 2015 [1959]: 94). Philosophical realism is the intuition of resistance, the defiance of things felt, interiorized, and then systematized by the thinker. A founder of the contemporary new realist movement, Maurizio Ferraris, in accordance with Bergson and Jankélévitch, describes reality in terms of a dualism between "unamendability" and "resistance": "Reality possesses a structural (and structured) link that not only resists conceptual schemes and perceptive apparatuses (and unamendability consists of this resistance) but precedes them" (Ferraris 2014 [2012]: 39). More often than not, our concepts and ideas bounce back from the resistance of objects. A realist reading of Bergsonism, such as that which we seek to present, must not discount the resistance of real duration to access. There need be no antagonism between a process ontology and a realism, for if there were no change, how could we even speak of the elaboration of things? Disappearance and unbecoming pave the way for novelty, but the actualization unbecoming prepares is always in the making. Extinction, death, and undoing are never opposed diametrically to change by the Bergsonian outlook. No closure is possible, not even dissolution can enclose alteration. Each repetition strikes against the ambiguity of a prior disorder, causing a reverberation. It is in this limited sense that matter, for example, can be imbued with a type of "memory" (Guerlac 2006: 168). Appearances aside, repetition too is a legitimate form of creation. Reiteration has its place in the broader intuitive method. Bergson is quick to defend those who repeat, for he himself returns to the rehabilitation of change almost constantly. "In the very places where the philosopher seems to be repeating things," we find the enactment of originality (CM: 131). Some discrete spatial displacement is unavoidable in the history of thought. Systems better adapted to contemporary concerns tend to replace those philosophies less concordant with the prevailing *Zeitgeist*. If a society is in a volcanic, liquified state, getting ready to explode into war, revolution, or both, then the thought of impermanence shall have its day. On the other hand, if a society remains in a rigidly static or conservative state, the thought of hierarchy will reign supreme, lording it over other ways of thinking and being. Bergson is adamant that each great thinker is a singularity, the embodiment of a thought that can be compressed into a single unit: "A philosopher worthy of the name has never said more than a single thing" (CM: 132). A play upon the same word or line can make the contents of a language flow more freely. An adaptation, as the history of religion proves, can intensify into a full-fledged schism. There is no clear-cut line dividing interpretation and innovation. Philosophy is at once commentary and a movement beyond. Ansell-Pearson sees a shift in Bergson's work, a movement away from psychology toward ontology (Ansell-Pearson 2002: 171). Dividing a whole into phases is a methodically valid approach, but its validity relates to the aspect of historical reconstruction alone. We are not compelled to follow Bergson's methodology when commenting upon his philosophy. Like an organism, the aspects, concepts, and details of a philosophy, at least for Bergson, cannot be arbitrarily selected. We either accept or reject the whole thing (CM: 139). Our own commentary treats Bergson's philosophy as a coherent whole, without making any pretense of our own selectivity. If every great philosopher has only one sole thought, then the singular nature of this thought compensates for any irregularities the system in question happens to contain. The present is always brighter than the past. The latter, as virtuality, draws all of its vitality from a living present.

Our work seeks to advance an image of time as immediate simultaneity. It is instructive to move initially toward a speculative extremity, before retracing our steps and moving backward to the experience of time in its givenness. Time would appear, on first impressions, as a flow or a succession of states. Yet intuition shows us something different altogether. At the point of termination of intuition, we recognize a "centre of force," from which "springs the impulse which gives the impetus, that is to say the intuition itself" (CM: 142). A process philosophy does not exclude the reality of simultaneity or immediacy. Process by definition cannot invalidate the moment, for the instant is also a process. Time is most effective if and when it is operative, and this operativity applies solely to the actual present. Actualism is inseparable from presentism, the teaching that only the present adheres. The future is not (yet), while the past exists in the mode of virtuality, as inoperative passivity. But how then would the past and future impact the present, and vice versa? In what manner can Bergson maintain what appears to be the instantaneous nature of intuition along with a view of the world that maintains the primacy, indeed, exclusivity of change? Simply put, if change is all there is, how can we posit duration as a *moment*? The image of time is a heterogeneous unity, an indivisible simple intuition into which time has been contracted. Prior to temporal concepts, there is image as contraction (CM: 141). We may gain an understanding of what precisely is at stake here by citing Deleuze's idea of *instantaneous becoming*, as summarized in Corry Shores's brilliant exegesis. Shores begins with an apparent self-contradiction in Deleuze's philosophy, not unlike our own Bergsonian paradox. On the one hand, all reality is change and becoming. This would imply that temporal phenomena are continuous. But in the second volume of his work on cinema, *The Time-Image* and elsewhere, Deleuze introduces multiple examples of "motionless temporality," illustrative of "the disjunctive synthesis of virtually simultaneous movements that are given instantaneously" (Shores 2014: 200). Against readings of Deleuze which seek to marginalize the status of the instant, Shores argues that instantaneous becoming forms, in actual fact, the fundamental basis of Deleuze's entire philosophy of time. We can unite process with the idea of a moment that flows. In accordance with our own reading of Bergson, Shores reads Deleuze as a presentist for whom past and present are, in the intensive moment of the present, given in the mode of immediacy. Far from constituting a flow, time is a heterogeneous simultaneity: "Duration (. . .) is not merely something occurring over the course of a series of moments but can rather be what happens between moments brought infinitely close together such that no time extends between them. Duration in this sense is a matter of the *extensive simultaneity* of a 'before' and an 'after,' whose differential relation implies an *intensive succession*" (Shores 2014: 203–204). The transparent image of time reflects a past and a future alike, coinciding in the context of a living present. Traced upon the initial element is a succession which, upon passing, dematerializes, decomposing into pacified memory. Each present moment is a contraction of past and present, an amassing of virtual resources in the service of empowered actualization. There is no contradiction in stating that a process can be infinitely contracted into what effectively amounts to a moment. We hold that duration is infinitely contracted processuality. Because even the purest encounter is shot through with the past, there is no clean moment, through which nothing of the previous moment has not already passed. Each duration is a cone, taking up intensities, authenticities, actions, disordering these until they are virtualized, deposited into the past as memory. Deleuze, in positing the instantaneous simultaneity of past and present *within the synthetic power of the present*, proves to be a faithful student of Bergson, the erroneous privileging of the virtual realm in his earlier book on Bergson notwithstanding.[9] The Deleuzian time-image represents an aspect of time that corresponds to the actualist vision of time as immediacy: "What we call temporal structure, or direct time-image, clearly goes beyond the purely empirical succession of time—past-present-future. It is, for example, a coexistence of distinct durations, or of levels of duration; a single event can belong to several levels: the sheets of past coexist in a non-chronological order" (Deleuze 1997 [1989]: xii). As an image of time, duration corresponds to immediate simultaneity. Maximal intensity is only reached in the moment of becoming. Ceaselessly changing circumstances are being lurched into the future by the violence of actual reality. An abstract representation of temporality as succession would give us nothing beyond a stream of second-hand simulacra. In itself, a linear patterning traced through a present conceived of as a mere hole shows nothing of duration. Writing of Joseph L. Mankiewicz's films, Deleuze discovers a logic of "forking" that serves as an illustration of duration's immediacy: "The straight line" is, in the mode of actualization, a "force of time," a "labyrinth of time" that "forks and keeps on forking, passing through incompossible presents, returning to not-necessarily true pasts" (Deleuze 1997 [1989]: 131). The most fundamental of durations is that which gives itself in the unmediated intuition of time. A moment is a temporality that possesses its attributes in actuality, nested within the most narrow of contractions. To wit, the phrase "synthesis" must be understood here not as the unification of discrete elements, but a description of the incompossibility and indivisibility of present and future within the simultaneity afforded by the moment. Each forking functions as a flashback "that unites past, present and future into a single instant" (Shores 2014: 211). By getting itself twisted into labyrinths, time shows its implication with relation and substance conceived of as change. Intuitive experience suggests the impossibility of conceiving the same in the form of a difference that flies off into imperceptibility.

The past, as organic or geological memory, constitutes a preservation, but the virtual is always of lesser power than the immediate actual. (Our "actuality" need not correlate with a necessarily human consciousness or presence, being a word for that which is presently, *in* relation.) In Shores's view, each forking is eventuated as a flashback, a sense of déjà vu which relates as much to fictional, nonexistent events as real pasts. We would, again, be bolder on this score. On our part, we may equate the virtual with the fictional. The forking of time, the break in memory, points toward the unreal nature of the moment which has passed into oblivion. As fiction, as retrospect, spent time has a reality, one that can even be rehashed to a certain extent, but never again shall it have anything other than an invisible, ineffable life. In becoming, "the present becomes something it is not, the past" (Shores 2014: 218).

Each real becoming is an actualized crystallization of the past and future. As distinct from Shores, we view simultaneity not merely as an aspect of experience alone. This move would constitute a phenomenologization of time. Ever impatient to reduce manifestation to a subjective givenness, phenomenology is always tempted to expand until it has incorporated all relations into the subjectively given. The concept of actuality we are seeking coincides with the entire relational realm, irrespectively of the presence or absence of sentience in the universe. Pure duration is "a reciprocal penetration, refractory to law and measurement" (CM: 146–7). If juxtaposition is time that has been domesticated, then the durational instant is an infinitely concentrated temporal singularity, one that nonetheless endures. Deleuze holds that duration can be solidified into a moment without thereby jeopardizing change. He achieves this philosophical feat through the introduction of the image of the *time-crystal*. "In the crystal," we observe a differentiation "that the crystal itself constantly causes to turn on itself," preventing "completion, because it is a perpetual self-distinguishing, a distinction in the process of being produced; which always assumes the distinct terms in itself, in order constantly to relaunch them" (Deleuze 1997 [1989]: 81–82). Inside of the moment, a complex simplicity is revealed in which we find, beneath the apparent distinction between "the actual and the virtual," "a bit of time in the pure state," in the form of permanent self-reconstitution (Deleuze 1997 [1989]: 82). Inner duration is a situation, a contexture in which the present is almost compressed into the past, but not entirely. The crystal locks in this tension, solidifying the flow into a mesmerizing geological composition. Recent physics has arrived at a recognition of the existence of "space-time crystals," far from equilibrium particles or groups of particles which repeat in time as well as space. These are the closest entities to an ideal object that changes from moment to moment. One could speculate that space-time crystals approximate time itself, in its purest state, which is the moment. The actual, within the time crystal, has been situated closest to the virtual, almost without any interval separating them. Paraphrasing Bergson, Deleuze states that "at each moment time splits itself into present and past, present that passes and past which is preserved" (ibid.). The past, however, has already passed into the virtual, while the evanescent present holds all the efficiency of time. What the time-image, conceived of as a time-crystal, allows us to do is to decompose the idea of time as a meta-continuity. Subverting linearity, each time-image shatters "the empirical continuation of time, the chronological succession, the separation of the before and the after" (Deleuze 1997 [1989]: 155). If we are to intuit the instant of time, we must learn to recognize falsity alongside truth. The impotentiality of the virtual does not detract from its being; as an illusory object, it has a certain substance, but its powers, as compared with vitally vibrant actuality, are diminished. It is within the false memory, the experience of déjà vu, that Deleuze locates the evidence of nonlinearity: the "power of the false" leads us to pure duration, "because it poses the simultaneity of incompossible presents, or the coexistence of not-necessarily true pasts" (Deleuze 1997 [1989]: 131). The concept of the time crystal inscribes the presence of motion at even the most basic level of existence. To quote Alfred Shapere and Frank Wilczek's classic 2012 formulation, "time-independent, conservative classical systems might exhibit motion in their lowest-energy states" (Shapere and Wilczek 2012: 1). Motion is a constant, and this constancy could be taken to repeat endlessly, hence the metaphor of "crystal." If a crystal is a spatial repetition, then a time crystal is a temporal repetition. As Bergson would have it, "what we have is a present which endures" in the form of a "perpetual present" (Bergson CM: 180). Duration is an indivisible instant of change, and Bergson's world is constituted by such heterogeneous simultaneities. Within the time-image, "the present instant is always in a state of self-distinction: it is both the moment that it is tending to become while also being the moment that is right on the verge of passing away" (Shores 2014: 222). The mode of being of the simple present, defined as Perfective Non-Past (PNP), is that of precarity. There is a constant danger of the present solidifying into false memory. Past and future durations are each mirrorings, refracting the borrowed light of the throbbing, enduring present. Beyond this effective simultaneity of present and future, the now that is given, there is no alternate elaboration. Novelty must be made in actuality, now or never. From our perspective, the instant is a stillness. Soon, physics shall have the tools to give a scientific, objective elaboration to abstract speculations relating to the crystallized nature of time. We cannot expect philosophy to be given much credit, but we can nonetheless content ourselves with the knowledge that the philosophy of the twentieth century has contributed, concepts which are of use to future generations. The positing and later empirical verification of the time crystal lends a retrospective plausibility to Deleuze's assertion, namely, that "the still life is time, for everything that changes is in time, but time does not itself change, it could itself change only in another time, indefinitely" (Deleuze 1997 [1989]: 17). The stillness of time ought no longer be held to stand in contradiction to the view of reality as the totality of dynamic processes! In light of our investigation and Shores's interpretation, no longer are we obliged to accept the mutual incompatibility of a process-based view and the concept of duration-as-instant. Time passes inside of the time crystal, conceived of as time-image. The time crystal is the actualized incompossibility of present and future, both of which are reverently situated inside the present, while the virtual's reality is illuminated by the PNP. Stillness is time as it endures. *Ecstatic is the stillness that prepares the next thrust of change.*

By enlisting the help of Deleuze and Shores, we hope to have to better elucidated our understanding of Bergson's actualism. Time as crystallization furnishes us with an account of time that incorporates the instant. We must dive down into time, if we are to recover the simplicity of impermanence. "The deeper the point we touch," Bergson remarks, "the stronger will be the thrust which sends us back to the surface" (CM: 147). What does intuition show us? Note the careful use of words. Interpretation delivers us over to the observation of an indivisible, simple *point*. Not a stream, not a tendency, not a river of becoming, but a point. The "when" of thought coincides inside the time-image with the "where" that unites both accord and discord. This point, however, is not the static point known by geometry. Rather, it is a qualitative multiplicity: intuition perceives "the continuous fluidity of real time *which flows along, indivisible*" (emphasis ours, CM: 150). The impassivity of the present endures, forking into the past of indiscernibility and the future of uncertainty. Between two extremes, there is a pointlike mobility, an infinitely compressed chaotic turbulence refracted onto itself. We are compelled to differentiate stillness from inertness, the former denoting a constancy of repetition, a dependability of movement rather than an infertile remnant like the latter. Time crystals are Protean entities, systems that integrate permanent instability into their operations, creating a type of "movement" which "would require no external force" (Gibney 2017: 165). Were time as such definitively proven to be a repetitive movement of movements, as Bergson supposed, we could say good riddance to immobility while *also* maintaining the primacy of the moment and continuity. The stillness of duration is the state of change in both its indivisibility and continuity. "No more inert states," Bergson exclaims, "no more dead things; nothing but the mobility of which the stability of life is made" (CM: 151). The discovery of the time crystal and its acceptance by science sheds new light upon the Bergsonian doctrine, for the latter conceives of duration in terms that resemble (if only by analogy) the characteristics of the time crystal. We wish to draw attention to the unexpected productivity of metaphors that, originating in philosophy, traverse and transgress the boundaries of speculative, theoretical, and empirical science. A mobile point, this is the destination toward which intuition seeks to extend itself. Bergson's purpose is to make thought more elastic. We are change, we are the formlessness underlying the synthesis of past and future: "Let us (. . .) grasp ourselves afresh as we are, *in a present* which is thick, and furthermore, elastic, which we can stretch indefinitely backward by pushing the screen which masks us from ourselves farther and farther away" (emphasis ours, CM: 152). It is within the present that we find the maximal intensity of elasticity. No prefiguration of the mobile can be given before the universal givenness of the instant. *The joy of impermanence is the wonder of passage.* The goal of Bergson's doctrine is a conversion of our perception into an awareness of simultaneity and duration. This alteration or makeover of the soul is an invocation, or rather, an initiation into the immediate perception of duration. The

1911 lecture, "The Perception of Change," contains what amounts to the penultimate Bergsonian representation of simultaneity, namely, the "panorama" supposedly seen by those who suffer near-death experiences. Because of its special relevance to our investigation of Bergson's simultaneity, we are obliged to quote this selection in full:

> In people who see the threat of sudden death unexpectedly before them, in the mountain climber falling down a precipice, in drowning men, in men being hanged, it seems that a sharp conversion of the attention can take place,—something like a change of orientation of the consciousness which, up until then turned toward the future and absorbed by the necessities of action, suddenly loses all interest in them. That is enough to call to mind a thousand different "forgotten" details and to unroll the whole history of the person before him in a moving panorama. (CM: 180)

Prior to this exposition, as if to dispel or foreclose any misunderstandings, Bergson also states that, in duration, "what we have is a present which endures" (ibid.). The near-death experience, the moment preceding, or even penetrating into, the mystery of death, affords a direct empirical proof of the simultaneity of past and future within the present. Alfred North Whitehead, the great philosopher of the twentieth century who was explicitly influenced by Bergson's outlook, notes that "the present moment is constituted by the influx of the other into that self-identity which is the continued life of the immediate past *within the immediacy of the present*" (emphasis ours, Whitehead 1933: 233). This is in accordance with Bergson's view. It is always a present that composes the habitat of any change. Ordinarily, consciousness is oriented toward preparing future bodily movements, coordinating our expectations with possibilities. The moment preceding death changes everything. The attention of the organism, having been fatally opened by an experience of its own finitude, normally directed toward life and action, is diverted toward the past. As one ages, the haunting, ghostly pale light of what is no more looms ever brighter, blackening out the light of the present, until the point wherein memory is all that is left. Dementia lies at one extremity, the point in which a subject has become entirely virtualized. Those suffering from the advanced stages of dementia live in a disordered world of memory; their sole mode of being is entrapment in a virtual existence, an interiority unamenable to change, a prospect that is a nightmarish prospect for the young soul,. The limit-experience related above by Bergson, however, differs from dementia significantly, since the former provides an immediate lucidity. The past invades the present in both cases, but in the near-death experience the present transforms into an overloaded singularity. Just prior to losing themselves in the panoramic moment of death, some are able to recover times formerly thought lost. The arrival of the end relaxes intelligence, yet this negation is, for Georges Poulet, a positive reinvigoration, a resurrection of the past in a saturated present: "This deficiency and absence, this wholly negative event (. . .) stimulates the bursting forth in the spirit of the most positive thing possible, total memory—that is, the self embracing itself" (Poulet 2011: 492). In Poulet's interpretation, Bergson's exposition of near-death "panoramic vision," in "The Perception of Change" (as well as, in an earlier iteration, *Matter and Memory*), represents a positive potential. Memory has a positive being. Is this truly the case? Melting back into total memory—must we view this as a necessarily "positive" datum? As distinct from Poulet, we hold that this going-under of intelligence shows the simultaneity of past and future, but only since the occurrence takes place in a person living in the moment. *It is in the present that duration endures.* One must be unfortunate enough to undergo a near-death experience so as to access the panorama of the dying, and also fortunate enough to live to tell the tale. Otherwise, the present would go extinct along with the dying person. However illuminating it may seem, Poulet's formulation regarding the concept of near-death panoramic vision actually obscures things. "By losing all and by losing ourselves, we gain everything and we gain ourselves" (Poulet 2011: 493). Unfortunately, the case is far different. *By collapsing into the virtuality that is memory, we regain a past at the cost of losing ourselves.* Poulet's characterization of the panoramic vision in dualistic terms—positive access to "total memory"—obscures the immediacy of intuition. Equivalently to the intuition of time, the necrotic panorama takes place in the present, an intensified moment. Attaching positive aspects to this abnegation confusingly leads to a valorization of memory that has no foundation within the Bergsonian method, the goal of which is not the recovery of a lost time, in the style of Marcel Proust (cf. Megay 1976). Our goal is not some reactionary return to a mummified past. The spiritual intent driving Bergsonism is *the redemption of the present*, even at the expense of abandoning past and future. By no means do we mean to imply that the past is nothing whatsoever, or that antiquity has zero bearing whatsoever upon posterity. Neither do we wish to suggest that the future cannot impact our own time. What we do affirm is the necessity, the absolutely imperative duty, to save the present. Salvation cannot be deferred any longer, it must happen now or never. Urgency is now, duration is our instant, our time, our shared burden.

The panoramic vision of the dying is proof not of the precedence of memory, as Poulet's interpretation would have it, but rather of the inherent presentism of real, creative time. Impending death reconnects the organism to a more intensive present, resulting in a reawakened simultaneity of past, present, and future. As we have stated, Bergsonian philosophy is a reversal of consciousness, the goal of which is initiation into the discordant yet continuous multiplicity of things. Through this submergence, we "plunge into" change, "for the purpose of deepening and widening" perception (CM: 158). Roused to action, perception is reattached to a discordant present filled with chance. The contours of actual things are, in the Bergsonian view, situated halfway between abstract ideas and pure continuity. A surface, a segment, a line, is a construct informed by our body. If this dependence moves along a boundary or border, this movement in turn is recoded along a dream plane into an apparent necessity. "Changed" perception is a socioecologically oriented application of thought to immediacy. Intuition is the pure perception of change, untinged by separation. Despite its momentariness, enlightenment is preceded by real intellectual effort. Immediate duration, in other words, does not preclude the need for practice on the part of the Bergsonian meditator. At its most essential level, Bergsonism is about *changing* perception in a performative sense. To *change* something in this second, perfective sense means something different from inducing a modification of mental states. Rather, the goal of *changing* is to open up and broaden perception. This can be achieved solely in the present. As Miliĉ Ĉapek explains, "if duration is real, future as future is unreal, i.e. not yet present, and its coming into being is a genuine and not merely apparent novelty" (Ĉapek 1971: 108). Vague or abstract language cannot, in itself, decompose the presuppositions and prejudices preventing us from coming into coincidence with change. The cognitive field is, in the case of intuition, inverted. Were we to remove the "in" from "intuition," we would be left with the phrase "tuition." Those who achieve such an intellectual effort in effect invert the normal process of education. The phrase "tuition" originates from the Latin verb *tuitio*, denoting a "caring for," "watching over." On this etymological basis, we may say that not only is intuition a vision of the base of consciousness or a merely aesthetic representation of the becoming of reality; it is also an ethically grounded act of self-care, in the fullest sense of the term. To care for one's self may seem a daunting task, especially if combined with a demand for immediacy. Introspection cannot be gleaned from external sense data or any other agent's cognitive field. Intuition is blind to mere facts, just as pragmatic intelligence is unwilling or unable to incorporate elements it lacks an organic interest in. The introspective moment of self-care is a hard-earned immediate enlightenment, both in the sense of penetrating to a condition lacking in mediation, and of temporal suddenness. Fortunately, the Bergsonian meditator is not constrained to begin from nothing. Others have preceded us in this process of opening, great personalities, avatars that have gone through ripening. For centuries, artists have achieved a revelation of that which is hidden to normal perception (CM: 159). The artist is a revealing agent, an indicator solution added to experience from without. Artistic creation is, for Bergson, a mode of seeing differently. As Guerlac writes, "Art place us in a kind of dream state. It elicits a sympathetic response on our part, a virtual participation in the feeling or idea" (Guerlac 2006: 52). What the mere existence of art as a separate discipline or way of life proves is that perception can be extended beyond the confines of a pragmatic instrumentality. If intuition is the reversal of care, rechanneled into the self's interiority, it also constitutes an intensification of perception. Interiority hangs together with breadth. The maximal level of self-care is the perception that has been extended inward. Intending to become large and all-encompassing, experience triggers a non-relational mental state. Guerlac's use of the phrase "virtual participation" need not detain us excessively, for the virtuality of the artist's participation only refers to the external realm of causality. Our artist refrains from action, while in reality actualizing herself through her creative movements. An artist is not unlike a brain: both serve to actualize memories (CM: 162). According to Stephen E. Robbins's helpful summarization, in the Bergsonian system of philosophy "perception is *virtual action*" (Robbins 2000: 28). Intuition is a non-relational mental state that gives itself in the mode of disclosure. Because of the discord or corporeal hesitation we as living beings experience, we are set into motion situated in a body which functions as a set of problems. Arriving at places of habitude, our consciousness is able to record the interdependent moments dotting the journeys we have taken. If perception is action in its nascent, schematic, virtual—memorized—state, then intuition as self-care denotes the actualization or, better yet, purification of perception. Far from separating the agent from action, intuition allows the meditator to plunge back into change with renewed force and commitment. Philosophy in general is a chain of traditions circling around images. In concepts made bright by reflection, signs are joyfully and astonishingly illuminated. For a moment, chaos has been fixed, remade into an apparent order. The origin itself is disordered, complexly stochastic motion. In Ovid's *Metamorphoses*, we find the following representation of primordial Chaos:

> Before there were sea, earth, and all-covering sky,
> the outlook of physical matter was identical everywhere,
> which they called Chaos, a shapeless, disorderly bulk,
> nothing but motionless mass and amassed within
> the warring seeds of entities ill-connected.
> (Translation altered by authors, quoted in: Gildenhard and Zissos 2017: 11)

Intriguingly, the intuition of the Roman poet reveals stillness within movement. The Chaos of Ovid is described as "shapeless," but the external absence of motion belies the permanence of change within the "warring" interstices of this qualitative multiplicity. In Ingo Gildenhard and Andrew Zissos's interpretation, the poet shows that "a condition of endless morphism that never results in the (renewed) stability of form required for a genuine metamorphosis" (ibid.). It is only compared with an outside that the same can attain coherence. As one of the moderns, longer does Bergson live in Ovid's well-ordered world. Ruth Lorand's contention that disorder is impossible in Bergson's philosophy strikes us as rather too conservative. "It is inescapable," she declares, "that we always find some kind of order" (Lorand 2000: 90). The organism has a regularity of its own that shapes its dynamic engagements with its world. Disorder is, for Lorand, "virtual," a construct of intellect that "denotes (. . .) not the absence of order, but the absence of the specific order one expected to find" (Lorand 2001: 91). In *Creative Evolution* Bergson, lacking as he did the semantic of later complexity theories, rejected the notion of "disorder," preferring to interpret evolutionary processes as manifestations of a liberating tendency somehow innate to matter. Lorand prefers to characterize evolution as constituting an example of an "unpredictable order," and even speaks of highly complex "lawless orders" such as aesthetics (Lorand 2001: 84, 116). But could the reverse also not hold? What if "order" in general is also an inadequate way of characterizing reality? The endless morphism of process never results in anything other than a transitory stability, for creativity is absolutely free and unpredictable. In this sense, a chaos can be discerned underlying free creativity. Ovid's gesture is a magical act. By describing it as at once motionless yet replete with "warring seeds," the Roman poet fixes chaos in place. The thought that becoming can yield genuine changes without finalizing any form was, for the Roman culture of the time, an unthinkable prospect. A diversion of apprehension is required if we are to give back to perception its broadness. Indeed, Bergson declares confidently that "conversion of the attention would be philosophy itself" (CM: 163). The philosophical project as such is a perpetual questioning, even breaking, of our inherited cultural, historical, and cognitive frames. As Al-Saji reminds us, "through repetition and habituation, a particular order of difference and meaning becomes the norm—a certain level becomes *the* level according to which we see, receding itself from visibility. But it is not only to individual bodies that habit belongs; social horizons themselves become habitual and institutional" (Al-Saji 2009: 384). Intuition, as a method of self-care, is the antidote to cognitive closure resulting from habitual adherence to a certain social horizon. Convention is invisible to those who follow it. Even those who critique a manner of seeing are themselves in turn liable to be coopted by a rival orthodoxy. The Bergsonian meditator strives to resist the inertial force of institutionalized modes of seeing. The necessities of social life, the mechanisms of normalization, and the regulation of perception all separate us from the apprehension of the original mobility of reality. A *changed* perception refuses restabilization, for it directly observes instability and impermanence in all things. Every institution is revealed as groundless from the changed standpoint. Immediacy of change breaks open the habituality of socially encoded repetition, as well shattering our subjective prejudices. Unfastened, habit, like memory, proves to be an empty shell, a cocoon of sorts that has developed to shield the growing organism. The transformative morphic force of intuition leads to immediacy, exercising its power upon a present given to access. In intuition, we are "immersed in the immediacy of durée" (Barnard 2011: 52). As interiority, self-care is a direct interest, a claim that forecloses, for a while at least, the operations of both memory and divination. Past and future, as distinct categories, are suspended; here, the intuitive moment prevails: "In it the past becomes identical with the present" (CM: 185). Integrated into the moment of self-care, the past becomes interesting. This resituated past is a destabilized antecedence, a dynamically qualitative change. Working our way back to interiority, we are able to "re-establish continuity in our knowledge" (CM: 167). The creative mind is a concern with self that blends into the impermanence of reality's permanent, original mobility. Ovid's characterization of Chaos as an immobile mass was most probably inspired by the poet's own experiences regarding objects. We can easily visualize things lacking in form, for instance, mud. Merely extrapolating the illusory formlessness of some things, however, never will give us the idea of Chaos in its purified state.[10] Nor does relative movement do justice to the "original mobility" of self-actualizing reality. "It is only in the immediacy of our own perceptions that we can come to know that movement cannot be divided into parts," writes G. William Barnard (2011: 75). Such an interpretation runs the risk of reducing Bergson's position to an idealist subjectivism, for the reader may easily gain the unfounded impression that knowledge relating to *our perceptions* is of central interest while, in truth, *access to immediacy* is the goal of intuition. As soon as our attention has been successfully converted, perception is transmogrified into immediacy, the latter denoting the coexistence of two instants or occasions. From a human perspective, consciousness is a requisite precondition, but other modes of access are also imaginable. Plunging into change is not an epistemological quest. It is the revelation of an actual present within a life. *Coexisting, a few may brighten thy life.*

The Second Lecture contained in "The Perception of Change" commences with a momentous pronouncement on Bergson's part. "*We shall think of all change, all movement, as being absolutely indivisible*" (CM: 167–8). Each and every moment is characterizable as a mobile stillness. Not unlike Ovid's Chaos, the Bergsonian moment is an apparent immobility truly pregnant with intra-mobilities. The moment stands out because of its indivisibility; "it is indecomposable" (CM: 169). Nothing moves, nothing is moving, for that which passes is inseparable from its changes. We cannot speak of an agent outside of its performances. Every causal situation, every occasion in the universe, is a sensual field, that is, a place of contact and metamorphosis. It is no accident that Bergson's philosophy has fundamentally influenced the "vital materialism" of Jane Bennett, to quote one contemporary example, for this philosophy contains the grains of a monumentally cosmic expansion of the idea of sensation, as well as a recognition of the inherent vibrancy and activity of reality (cf. Bennett 2010: 62–82). Since any object is, to a greater or lesser degree, accessible to other things, we may speak in every case of "elements that make up the sensual field," this latter concept denoting the area of interpenetration (Harman 2005: 161). As immediate data, actualized elements arrive at relationality. The key tenet of Bergsonism is the following: "Movement is reality itself" (CM: 169). Were we to experiment with colliding Bergsonism with Harman's speculative realism, we could obtain a highly interesting hybrid, one that excludes the possibility of an immobile object, while also emphasizing the objective nature of each process, or duration. In Harman's system, a complex web of cosmic sympathy, what the author terms "allure," is what holds the cosmos together. As Harman writes in a characteristically concise yet poetic manner, "allure is that furnace or steel mill of the world where notes are converted into objects. The engine of change within the world is the shifty ambivalence of notes, which both belong to objects and are capable of breaking free as objects in their own right" (Harman 2005: 179). A Bergsonian response to this position would be to emphasize the changing, shifting nature of both "notes" and their corresponding "objects." Durations are nested within larger durations to infinity. In a limited sense, one could say that a process can function, potentially, as the note of a more overbearing, broader process. Returning to the example of birds and their habitats, it would not be unimaginable to characterize an entire bird species as a "note" immanent to a particular morphospace or ecology. Each thing is decomposable into a set of correlating movements, whose sole common thread is their temporal correlation. This decomposability in no way contradicts Bergson's assertion that the moment (remember, each moment is already a duration!) in itself is incompossible. Attempting to segregate and ossify one set of movements into a definite object is doomed to certain failure. Beneath every object, new layers of mobility and impermanence may be unearthed. There is the unity of the shadow of a process within each actualized object, yet this seemingly concrete reality also betrays the incompleteness of each actualization. Everything is in the making, events in time produce indivisible moments, and this production is, in turn, indecomposable. "All change is an indivisible change," on the ultimate level of description there is nothing apart from internal continuity (CM: 171). Would this mean that we are obliged to describe the entirety of the universe as "a fluid continuity, from one end of the cosmos to the other," as Barnard suggests? (Barnard 2011: 87). Such holism, uncannily reminiscent of Hindu traditions relating to the unity of the world, seems not wholly unwarranted. Hindus are known to have "prostrated" themselves before Bergson's house near Geneva (Grogin 1988: 82). Does this colorful historical anecdote, taken together with the emphasis placed on continuity, therefore prove without a doubt that Bergsonism conceives of the world as a singular, unified entity, as the All-in-One? We should not rush to unwarranted conclusions, however seductive such a comparison may be.[11] The indivisibility of the moment, even the integration of every single duration into some global, universal "duration of durations," cannot in itself result in the complete erasure of multiplicity. Guidance is always to be sought from the intuitive experience of immediacy. At any single moment, countless other durations are in play, immeasurable and measurable temporalities alike, each of which exerts a different influence upon events. "There are changes," but beneath change "no things which change" (CM: 173). Even the greatest of durations, the most all-encompassing of temporalities, the time of the cosmos, lacks an underlying substance. Rephrasing Kant, Bergson affirms with the utmost audacity and radicalism that "change (. . .) is the thing itself" (CM: 174). However intently we would seek to discover anything differing from process, we find nothing apart from more processes. Each duration is a crystallized moment. It is precisely this stillness which allows for the possibility of time's flow. As we have seen, for the Deleuze of the 1980s (a most faithful disciple of Bergson's actualism indeed) time is given in its purest form in the still-life. A seemingly immobile object, like a bicycle at rest, refers, in its very inertness, to the immediately momentary nature of change: "A bicycle may also endure; that is, represent the unchanging form of that which moves, so long as it is at rest, motionless, stood against the wall" (Deleuze 1997 [1989]: 17). The moment is time, being the indivisible inhabitation of impermanence. Instead of spatialization, we have here an absolute resistance to juxtaposition. The absolute indivisibility of the moment is precisely what prevents it from being reordered into a serial category, and thereby degraded into a spatial concept. A bicycle stood against a wall, far from being an external presentation of time, is a living actuality, equal in ontological value to our own subjective experience. Each moment is indivisible, a unity of givenness, while externality is transitive.

PRIORITIZING THE ACTUAL

Why then would it appear that the "inner life" of the human subject appears to constitute a privileged locus of change? Reality is "only a movement of movements," however the status of substance as change is most acute in the experience of innerness (CM: 175). Let us remember that everything has duration. The immobile bicycle described by Deleuze also "endures," by virtue of its objectified stillness. Ultimately everything forms the same continuity, "the term duration" being "applicable to any type of being," as Eelen Fell points out (Fell 2012: 145). Be that as it may, we are nevertheless obliged to distinguish between the concrete level of explanation and the metaphysical level. On the one hand, "the sameness of everything," the equal status of each object—every existing thing being an instance of endurance—relates to ontology (i.e., metaphysics) (ibid.). On the other hand, concrete durations also compose unstable, irreducible multiplicities. Far from being subsumed into an all-encompassing sameness, difference is preserved inside a life which is the interiority of a certain duration, but our human subjective interiority is not a synonym for the *entirety* of *all* interiorities. No duration can be discerned apart from the instant, each instant being an indivisible moment, differing in size from other occasions. "Real duration is what we have always called time, but," Bergson adds, "time perceived as indivisible" (CM: 176). In other words, a duration cannot be anything other than a moment. It is for this reason that, writing decades later, Deleuze will characterize immobile objects as perfect encapsulations of temporality: "The bicycle, the vase and the still-lifes are the pure and direct images of time" (Deleuze 1997 [1989]: 17). Uninterrupted stillness is *where* time is. Recent studies have proven that the quality of sleep is determined in large measure by the lack of movement on the part of the sleeper's body. The less we move, the higher our level of rest, the greater the relaxation of our organism (Liu et al. 2020: 440–445; Wisden and Franks 2020: 366–367). Conversely, lack of sleep, influenced by stressfully excessive movements of the body, translates into poorer health, even premature death. Stillness, even on the relative, concrete plane, is the prerequisite for mobility further down the road. Each duration is a stillness building up into another mobility. In a way, Arthur Oncken Lovejoy is correct in his characterization of Bergson as being unwittingly "Eleatic" (Lovejoy 1913: 328–329). This depends on the way we define time. If time is thought of as a quantitative succession, then Bergson is not an Eleatic. As we shall see, Bergson affirms the reality of time against philosophers like Zeno of Elea, who seeks to divide temporal flow into spatialized slices or atoms of time. But Bergson simultaneously maintains the indivisibility *and* multiplicity of durations. Observed from the inside outwards, each duration persists in a condition of stillness. Observed from the outside inward, each duration blends into the permanence of mobility, or the immutability of impermanence. The instant is not a mathematical point. Rather, it is a flow replete with moments, functioning as the sub-durations or notes of greater temporalities. Potentially at least, intelligence could be extended backward to encompass the entirety of a person's history. For such an enhanced, majestically extended mind, the past would be transmogrified into "an undivided present" (CM: 180). Conforming to nothing but our own intuition of the interiority of time and, by extension, change, we discover a sincere multiplicity that is also a continuum. Each thing has its moments of solitude, for a duration is a moment of stillness preceding some action. Prior to the subdivisions constructed by analytical reason, continuity persists: "What we have is a present which endures" (ibid.). Remapping original mobility onto the self does not entail reducing change to the modifications of subjective experience. Similarly, drawing a map is akin to the creation of a system of reference, but surely cannot equate to the transformation of the territory represented by the map. A past is always preserved, and the most unpromising of moments contain new dynamic states in the making. "The preservation of the past in the present is nothing else than the indivisibility of change," Bergson assures his listeners (CM: 183). When we say of something that "it moves," this refers to a simultaneity extending across the whole terrain of a certain duration. Dynamism uplifts traces left by previous becomings, but the productive power of the present is what contains the endless supply of actualizations. Change is best understood as a "natural indivisibility" composing "the very substance of things" (CM: 184). Reality is what resists categorization. The very status of a "real" object depends upon the resistance it displays. Émile Meyerson, Bergson's contemporary, characterizes this resistance of the object as "the irrational" (*l'irrationnel)* (Meyerson 1962 [1908]). The problem with this term, as Capek points out, is that it introduces an epistemological and normative dimension to an ontological topic. We do not have any basis for characterizing material processes as "irrational," for this would assume that reality can be divided along the lines of segments amenable to our actions, and territories which recede from our knowledge. Contingency only appears as an irrational remainder or excess in the context of a cosmology that accepts necessity (Ĉapek 1971: 108). But the modern view of the world excludes necessitarianism. Nothing exists merely because it must exist. No object in our universe exists necessarily.

Qualitative changes are endowed with a performativity entirely of their own making. The sheer availability of perspective means that alteration matters. A still moment is a preparation leading up to flight. William James refers to "places of flight filled with thoughts of relations, static and dynamic, that for the most part obtain between the matters contemplated in the periods of comparative rest" (James 1950 [1890]: 150). For Bergson and James alike, the stillness of the moment is already pervaded by motion, just as Ovid's static, unchanging Chaos is filled with alteration. There is inside every birth an unreal unity (the self-identity of the "static") within the broader flow (the dynamic of the "flight"). Bergsonism treats metaphysics as a method of giving birth to movement inside of ourselves, serving as an introduction to flight. Already in James's work, which predates Bergson by some years, rest is treated as relative. Intuition has constructed a thought capable of accessing the self-elaboration of reality. If this is so, then this capability demands a philosophical register. From the relativity of rest, we strive toward the active time flow lacking differential, spatialized, segregated units. Deleuze's bicycle, the still-life, is the image of temporality. Like a peep show, Deleuze and Bergson show us a representation of time, but neither of them adheres to the pretense that movement cannot be entirely represented. As Ĉapek emphasizes, "The basic imageless heterogeneity of duration (. . .) can be grasped only if we abstract from the sensory diversity of its contingent content" (Ĉapek 1971: 173). The Bergsonian thought of duration is without image, being posterior to the Deleuzian "time-image." Heterogeneous continuity cannot really be captured by even the most creative of metaphors. In his "Introduction to Metaphysics," Bergson distinguishes between two modes of knowledge. The former pertains to the relative, while the latter "attains the absolute" (CM: 187). Witness here an intriguing turn of language! The first form of knowledge is analysis, while the latter is metaphysics. Instead of merely connecting or relating to the absolute, metaphysics reaches inside of it. The guiding differentiation in this foundational text will be one that is already familiar to us: the entire essay revolves around the difference between inside and outside (Lawlor 2010: 27). Analysis only reaches the relative, for it is a multiplication of perspectives. Bergson's metaphysics is altogether different from the analytical standpoint. It is not a perspectivism in the usual sense of the word, for it does not result in the institution of a multiplicity of movements in the place of unity. Rather, metaphysics is a practice of togetherness, a generality of alteration within individuality. The internality we strive after shall be a triumphant interiority, a complete refusal to distinguish ourselves from the unity that has been achieved through intuitive penetration. Far from de-emphasizing the real in the manner of an idealist, Bergson is adamant that this philosophy is a realist project. "The movement will not be grasped from without," as in the case of perspectivism, "and, as it were, *from where I am*, but from within, inside it, in what it is in itself. I shall have hold of an absolute" (emphasis ours, CM: 188). If we are to achieve a sympathy with becoming, we are compelled to separate ourselves from our own situation. Togetherness paradoxically demands a decoupling from one's self. The dynamic pathway of intuition corresponds to an organic unity that refuses the mask of subjectivity in favor of a permanently changing body of being/s. The whole is perturbed, all of the time, and all of time is this perturbation. As Ansell-Pearson highlights, going "beyond" one's own condition means transcending "accrued evolutionary habits of thought that prevent us from recognizing our own creative conditions of existence" (Ansell-Pearson 2018: 17). Intuition begins when we leave our own place. *The ripening cannot be reached where I was.* If we remain locked in our habitual states, only retrospectively rationalized intellectual equivalents and abstractions will be unearthed from the surface dimensions of change, extracted traces that cannot but do violence to pure immediacy. Think of the participants in a reality TV show, narrating events that have already occurred as if they were happening. This cannot but strike us as absurd, as even the most skillful editing cannot seamlessly integrate retrospective narration with either spontaneous or scripted events. Thinking *beyond* is a profession of movement, a fixation upon openness, a vision pronounced aloud, a creativity unbound from repetition. Intuition, Bergson informs us, is but a new term for a much older concept, namely, *sympathy*. If analysis is the multiplication of perspectives revolving around the object itself, then intuition is the force of thought, the flight that goes beyond one's place: "We call intuition here the *sympathy* by which one is transported into the interior of an object in order to coincide with what there is unique and consequently inexpressible in it" (CM: 190). At this point, we do not wish to burden the reader with a long-winded etymological investigation of the phrase "sympathy." Needless to say, Bergson's "Introduction to Metaphysics" is not, as Ĉapek would have it, a work of "epistemology" (Ĉapek 1971: 149). At heart, sympathy is a fundamentally ethical and ontological concept. It is definitely synonymous with what we describe as "dynamic togetherness." Abolishing the difference between I and Thou entails the breaking of all restrictive frames and cognitive windows. Sympathy, akin to the direct experience of light, the abolishing verticality and horizontality alike, is a thought that cannot content itself with remaining enframed. Unmediated directness is fixation upon the absence of fixity. *We are, it came.*

Sympathy is in its very essence a "moral feeling" (Lawlor 2010: 36). Witness the paradoxicality, even ineffability of Bergson's definition of intuition as the sympathy that gives access to what is unique about an object, allowing the meditator to "coincide" with what is "inexpressible" in reality. This latter quality is the individuality of a thing. In the sympathetic domain, we admit of no degrees or levels of access. Penetration is immediate, the quality of inexpressibility is already there, in its nakedness. In this duration outside of all dogmatism, a whole, an absolute even, is detected operating inside of the object of sympathy. To borrow from Harman's speculative realist idiom, it may be stated that sympathy is a form of speculation that gives affordance to the "hidden execution" of objects. As Harman writes evocatively, "The executancy of a thing is a dark and stormy essence that exceeds any (. . .) list of properties" (Harman 2005: 104). For Harman and other representatives of the contemporary speculative realist and object-oriented ontology movements, the weakness of phenomenology lies in its reduction of objects to phenomenality. Tom Sparrow's criticism of the phenomenological approach could just as well refer to any dogmatic mode of subjectivist thought: "Phenomenology does not get us to the noumenal, it keeps us chained to the phenomenal, where we have been all along" (Sparrow 2014:

1). This is precisely Bergson's problem with all forms of idealism. Closed, dogmatic thought keeps us where we are, locked within our current state, without allowing us to step out into any *other* interiority. Bergsonism affirms that thought can be something other than a frame. As distinct from the phenomenological tradition, the intuitive method strives to drill into the hidden executancy of things themselves. This form of sympathy is not a merely subjective feeling or an introspective datum of consciousness peeled back onto itself: the introspective power of intuition references an infinitely expandable register. Any object can serve as the object of our intuition, no differentiation is made between organic and inorganic, human and nonhuman. Going beyond is a flight without limits. Already in the Stoic School, we see a holistically oriented blending of the ethical and the ontological. For the Stoics, the body is a web of affects, while the concept of "sympathy brings out the physical interconnectedness of the world with itself and the entities in it" (Brouwer 2015: 27). The rhythm of awareness entails a tincture, a blending with both higher and lower temporal heterogeneities, each thing having an interior aspect resistant to any and all conceptual cocoons. Intuitive awareness hinges upon a recognition of the limitation of representation. Any visually oriented image necessarily fails to do justice to the hidden executancy of objects. Interiority is the splendor of limitless containment, of durations nested into durations. Not unlike the mystic, the person who reaches intuitive awareness, even actual coincidence, with the chosen object of meditation, must take a leap over conventional language, in favor of what Evelyn Underhill in her classic study of mysticism calls an "exalted language which induces in sensitive persons something of the languid ecstasy of dream" (Underhill 1912: 94). The hiddenness of the thing readily supplies the mountain our speculations must climb. Actuality is the excessive correspondence flooding into our overwhelmed consciousness, to the point where we feel ourselves as a table, a rock, an oyster, a plane in flight. "Metaphysics," Bergson states, "is the science which claims to dispense with symbols" (CM: 191). This is a bold statement, one that must be investigated thoroughly. Can any philosophy really do without metaphors, rhetoric, and symbols? This can only be the case if and when thought is translated into action. A true metaphysician would be somebody who not only writes of the moment of duration, but one who actually exceeds her condition. Coincidence with what was formerly hidden is conformity with the pure power of becoming, the power of things. Subdued, domesticated, regulated reason must be unbound, its frames shattered, if the power within is to be unmasked, relieved of its enchainment to the avatar of identity. If analysis carves up reality into a multitude of perspectives, then intuition reunites us with intimacy. A Buddhist meditation manual, the *Visuddhimagga*, describes concentration as "the centring of consciousness and consciousness-concomitants evenly and rightly on a single object" (Buddhaghosa 2010: 82). Intuition would be a not dissimilar practice of reorienting our attitude and attention toward the present moment, the unfragmented singularity of an object in particular. There is no limit to the recentering of consciousness, for the reversal of consciousness into our self, the peeling back of our thoughts into the recesses of our hiddenness is but a preliminary to the broad-based expansion of concentration in all directions.

By no means must we say that intuition is completely, irrevocably ineffable. Even though it composes an excess that cannot be represented by an image, a variety of images are nevertheless capable, "through the convergence of their action," of redirecting consciousness toward the intuition of pure change (CM: 195). Provided that our use of metaphors does not fossilize becoming into a static solidity, we are entirely warranted in utilizing words that enlighten the imagination and guide the mind into the interior alterity of objective execution. Intuition is a cosmic sympathy, for it rests upon a recognition that "my essence or 'whatness' is connected to all the events throughout the past in general" (Lawlor 2010: 35). Instead of circling around a potentially juicy and flavorable fruit, intuition is the actualized experience of the sweetness of fructification. The absolute, purified of all division yet still remaining multiple, is receptive of immediate observation. Continuity does not mean that everything melts into an undifferentiated Pre-Socratic *apeiron*. The fear of object-oriented philosophy regarding process is that a reduction of being to becoming could potentially jeopardize the individuality of objects. Harman in conversation with Deleuzian philosopher Manuel DeLanda emphasizes that process philosophy rejects the individuality of objects. In a world of pure becoming, there can be no space for discrete, individual things. The individuality of a thing is its essence, and this guarantees that each object has an "innate consistency" and "a certain durability" of its own (DeLanda and Harman 2017: 57). Would Bergson be guilty of the crime of affirming change at the expense of uniqueness? As we have seen, in Bergsonism all times definitely do not melt into a single, undifferentiated universal, cosmic time. Instead of duration, there is an infinite variety of durations nested into one another. Changes are all there is, movements of movements. This by no means implies that we must collapse the individuality and singularity of a particular duration into an undifferentiated mass of Becoming with a capital "B." As Pete A. Y. Gunter observes astutely, in works such as the "Introduction to Metaphysics," any "doctrine of sheer uninterrupted becoming is rejected. Becoming, Bergson insists, is rhythmic" (Gunter 2005: 153). There is a limited place for interruption, hesitation, and divergence within the flow, otherwise how else could a novelty come into being? Without gaps inside of becoming, nothing could emerge. On the plane of relativity, fractal structures, mythical elements, and splendid epithets, to name a few random examples, are manifesting, without ever being entirely completed. What, after all, is a formation? In vain do we attempt to infuse it with a substantial, unchanging existence. Interconnectedness implies impermanence. Without gaps between the various durations, without an in-betweenness worming itself into change, there could be neither connectivity nor inclination. The error of conventional thinking consists in the equation of the gap pertaining between two durations and spatial discreteness. Durations are heterogeneous multiplicities, resonant flows. Thus, the present may be envisioned as contaminating the past and future, while the latter two terms project themselves back into the moment. When time flows beyond itself, it becomes non-perfective, a present unavailable to itself. Any view which would collapse entities and durations into an all-encompassing, pantheistic world-unity is excluded by the example given below:

> Even with an infinity of sketches as exact as you like, even with the word "Paris" to indicate that they must bear close connection, it is impossible to travel back to an intuition one has not had, and gain the impression of Paris if one has never seen Paris. The point is that we are not dealing here with parts of the whole, but with *notes taken on the thing as a whole*. (emphasis ours, CM: 201–2)

To return to Harman's verbiage, a thing is always more than the sum of its notes. Decades later, the realism of the twenty-first century accords remarkably well with Bergson's processual realism. True, the note-taking described in the quote above is intended in a literal sense. Somebody who sketches Paris will undoubtedly take notes of the most remarkable buildings, characteristics, and landmarks of the city. But in a more ontological sense, each thing already displays a set of attributes that belong to its reality. For Harman, each object can be conceived of as a "volcanic core" uniting around itself a variety of characteristics "articulated into countless facets, notes, qualities, accidents, and other halos, echoes, and shadows" (Harman 2005: 193). To wit, let us note here again the eerie similarity in language between Bergson and Harman. The former uses the phrase "halo," for instance, on several different occasions, at times influenced by his contemporary, the American pragmatist James. There is not much here in Harman's sentence that Bergson would find issue with. The sovereign reality of the object withdraws from contact, just as an alien duration withdraws from the advance of analytical reason. No set of perspectives comes close to exhausting the manifold nature of each duration. If we are to break the shell and truly enter into another duration, a stupendous intellectual effort is requisite. Wildly speculative gambits, expeditions, and ambitious research agendas are necessary. We are encouraged to take a gamble by claiming that immediate perception of the object's otherness through sympathy is, indeed, possible. Bergson issues a challenge addressed to all future philosophy, indeed, to the humanity of tomorrow: "Philosophy should be an effort to go beyond the human state" (CM: 228). The success of transcending our condition, defined as the entirety of our habits, is far from guaranteed. By no means are we assured that this striving will ever be successful. The imperative is directed not at the achievement of a state, but the making of an effort to begin with. Going "beyond" is "an attempt at an enlarged perception," not a finished state but a process of growth which is always open, always in the making (Ansell-Pearson 2018: 98). An enlargement or enhancement is never closed, for it is relative to a previous state of closure. After intuitive awareness, we remain as we were, yet something has changed. An estrangement has occurred. No longer does the meditator identify entirely with herself. By getting inside of another duration, we are impassioned in a manner we could not have foreseen. *Influenced, I am underneath the magnetism.* The doctrine of cosmic sympathy by no means disappeared with the passing of the Stoic School. Through various eras, this concept has periodically resurfaced in both literature and philosophy. The seventeenth-century English poet Henry Vaughn gives expression to the hermetic doctrine of cosmic sympathy in the following manner:

> 'Tis a kind Soul in *Magnets*, that attones
> Such two hard things as *Iron* are and *[. . .]*,
> And in their dumb *compliance* we learn more
> Of Love, than ever Books could speak before.
> For though *attraction* hath got all the name,
> As if that *power* but from one side came,
> Which both unites; yet, where there is no *sence*,
> There is no *Passion*, nor *Intelligence:*
> And so by consequence we cannot state
> A Commerce, unless both we animate.
> For senseless things, though ne'r so call'd upon,
> Are deaf, and feel no Invitation;
> But such as at the last day shall be shed
> By the great Lord of Life into the Dead.
> 'Tis then no *Heresie* to end the strife
> With such rare Doctrine as gives *Iron* life. (Vaughan 1678: 9–10)

Alan Rudrum differentiates three different forms of sympathy: the threefold "commerce" pertaining between God, the Creator, and created beings; the "influence of macrocosm on microcosm"; and, thirdly, "an influence operating through the senses" (Rudrum 1974: 133). While two of these types of sympathy relates specifically to subjectivity, even in these cases it is a rather broad category, encompassing all objects in the first case, and all living things in the third. This teaching bears a resemblance to Bergsonism, for in the latter too, sympathy is not an exclusively sensual or subjective experience restricted to consciousness. Feeling and sense have a role to play, but intuitive awareness reveals far more than anything that can be seamlessly integrated into everyday perceptions. Intuition is an impassioned *personification* of the inorganic. "True empiricism" for Bergson is rooted in a recognition of irreducibility on the level of ontology, an attitude which keeps "as close to the original as possible, to probe more deeply into its life, and by a kind of *spiritual auscultation*, to feel its soul palpitate" (CM: 206). Far from reinstating a misguided phenomenological idealism, true empiricism listens to the objects themselves. A world full of sense would be an enlivened cosmos, redounding with invitations, affordances, and wonders. Each landscape would be at once an existential heartland, a differential space, and a chaotic Babel. Rejecting the primacy of the visual faculty, Bergson enjoins us to *listen*. The notes of an other duration make themselves heard before being seen. Authentic empiricism is capable of stating problems in a novel way, and of going beyond conventional experience (Deleuze 1997 [1966]: 30). Considered as imageless thought, intuition rejects the priority of vision. Bergson's metaphysical empiricism is directed toward an astute listening, an auscultation of the moment. Referring to ambient composer Pauline Oliveros, scholar Julia Steinmetz writes that "the practice of Deep Listening is not limited to the hearing" (Steinmetz 2019: 128). A complete, unmediated intuition is one that mobilizes the entirety of the body, transforming corporeality into a register of rhythm and reverberation. Intuition is best understood as "a kind of intense listening" (Lawlor 2003: 66). Philosophy, defined as a concretion or, better yet, an occasion of intuitive penetration, is turned toward the ground zero of becoming. This turning is also a tuning into the reverberations of the thing. Instead of ready-made concepts that have attained a state of completion, intuition supplies the deep listener with "conceptual directions" (CM: 209). Intelligence itself is a continuous flux, in which provisional presuppositions and uncertain intentionalities are far more frequent than ordered genres, localized objects, and discrete concepts. Besides psychological and physical time, there is a "physiological time" as well (Gunter 2005: 151). There is a time of the surface and a time of the depth, a time of the unspoiled, intact, soft skin, and a time of the wound, a time of large, macrocosmic things, and a time of microcosmic molecular processes. A truly empirical philosophy must necessarily focus upon the community of all these temporalities. "Philosophizing," after all, "consists in placing oneself within the object itself" (CM: 210). The effort of intuition, this manifestly sympathetic action, consists in a speculative gambit. Bergson's things are made of time. We must focus on the transient community between past, present, and future if we are to understand Bergson's philosophy, the concept of duration in particular. Instead of asserting the survival of the past, or even worse, the subordination of the present to either past or future, we must conceive of the present as constituting the bedrock of the synthetic actualization of all pasts and futures. Without a present, there could be no memory, while the future would dissolve into unintelligible dreamscapes lacking any meaning. We must emphasize the dominance of the present in the Bergsonian framework. It is, in fact, no exaggeration to claim, as Matyaš Moravec has done, that Bergson, despite all appearances to the contrary, is actually a presentist! (Moravec 2019: 197–224). Let us read the following sentence attentively: "Inner duration is the continuous life of a memory which prolongs the past into the present" (CM: 211). Here Bergson is not saying that the actuality of the present is a mere aspect of a larger "virtual" past. It is not the present which is prolonged, inescapably, into a past that has already reached a state of completion. Neither does Bergson say that the creativity of the present moment depends on a prior past. Just the opposite is true: the past is fatally dependent upon the vibrancy of a present. Everything depends on the actual, the virtual being but a mirrored present; the latter is the burden dragged along by the former into a Future-Presentive. No spatial juxtaposition need be implied by all this. An example may be of help in this regard. Paleontologist Brian DePalma's team uncovered a late Cretaceous fossil site in North Dakota where, remarkably, the tektites and rocks created by the meteor impact that wiped out the dinosaurs are preserved. This find was as close as anybody can get to what amounts to a geological "photograph" of the K-T extinction event, the tektites presumably having rained down from the sky "within 45 minutes to an hour of the impact" (DePalma et al. 2019: 8190–8199). Would this not be an example of a memory, a rearticulated geological temporality unearthed and resurrected from the very bowels of the Earth? Yes and no. As memory, the find, replete with "fish stacked one atop another and mixed in with burned tree trunks, conifer branches, dead mammals, mosasaur bones, insects, the partial carcass of a *Triceratops*," as well as "marine microorganisms," represents a set of entities no longer present. Hauntingly, a fossil is a present absence, the trace of something not here, yet still eerily reminiscent of what formerly was.[12] Yet discovery is nonetheless present, in the form of a trace. A certain past is extended, through the efforts of paleontologists, into our own present. It is only the existence of this present, this now, which makes the act of uncovering possible. Had the K-T impact exterminated not 75 percent of terrestrial life but 100 percent, then certainly no curious animals would be left to unearth the remnants of previous life forms on Earth (barring, of course, the subsequent and somewhat unlikely arrival of an alien civilization with any scientific interest in the geological history of this planet). Even in the absence of scientific observers, we can imagine a variety of geological and climate processes which could promiscuously mix their own durations with the geological memory. On occasion, rivers are known to reveal fossils.

But had the remarkable North Dakota fossils remained completely hidden to all agents, with the exception of the bedrock they had been incorporated into, then we would be warranted in describing them as exclusively "geological" memories. An object is a memory inasmuch as it is unrevealed, inaccessible. The virtual is a past present that has yet to be resurrected by a future actuality. Without any actual, human or otherwise, the virtual, as memory, has no presence. To actualize, *the past must extend into the present*—and not vice versa! Paradoxically, a memory can only become recalled if it ceases to remain a past.

Without the actual, the virtual is both impotent and powerless. Instead of actualization as such or becoming in general, Bergson enjoins us to always think of becoming as inherently individual in its nature (CM: 212). While states and conditions are mobile, and things are decomposable into movements, this does not imply that all processes can be melted down into a generalized changeability. Bergson's "absolute" is a dynamic, particular, agonistic milieu, a unity in diversity. Instead of absolutizing change, authentic empiricism recognizes that nothing exists apart from the qualitative multiplicity of changes. The difference of a duration does not translate into a complete discreteness. We can maintain unbroken continuity without needing to abandon difference altogether. Change is not a movement between separate spaces. Never may the mobile be discerned inside any point in particular, for "the passing, which is a movement, has nothing in common with a halt, which is immobility" (CM: 213). The real is not a collection of points, undivided movement being impossible to reconstruct from a set of stoppages. Intuitive awareness is submergence within the moment which also represents an ascent of the spirit. The latter is, in fact, this very process of accessing another, alien duration, intuiting the time of something other than ourselves. According to the standpoint of magic, the soul of a magician is capable of connecting with the resonations of the universe. In an occult work, the poet Vaughn writes that, by a union with the force of cosmic sympathy, the soul can "infuse and communicate her thoughts to the absent, be the distance never so great" (quoted in: Rudrum 1974: 135). Would the self-projection of the past into the living present not constitute, similarly, an action-at-a-distance, a haunting communication of absence within presence? Memory would, on the presentist view, be the introduction of a disappeared former present into the objective simplicity of the present moment. Time is its own passing, a perpetually absconding continuity. Rejecting Bergson's positive outlook, Bachelard prefers to emphasize "the negating value of all knowledge that is truly actual" (Bachelard 2000 [1950]: 34). Temporality itself is not immune from the desire to negate. Besides the relatively well-ordered time of subjective succession, we find "ill-made rhythms" that "have been tossed into delirium, forced ones given peace and listless ones aroused"

(Bachelard 2000 [1950]: 21). The present, as a fragile synthesis, is far from equilibrium. The ambition to possess all durations, however impossible, is nonetheless our guiding light when it comes to the practice of metaphysics, illuminating the eternal creation of different concepts and concepts of difference. A real object consists of both indivisibility and endurance. Reality is duration, which we define as the disappearance of the past through the synthetic action of a present moment, tending toward a not yet apparent future. Each object is a "real indivisibility" (CM: 216). Bergsonian time is evental and processual simultaneously. Commerce is allure, attraction, magnetism, while intuition is the super-perceptual pathway the meditator traverses to regain this vibrance. Without a present, lacking, in other words, the power of an actual, the virtual would remain in a condition of stagnation. Without the miracle of the moment, our world would swiftly degenerate into a nightmarish assortment of distended times, shattered lines, diseconomies of semblance, archetypes without particularities, decayed entities lacking purpose, entangled molecules collapsing into disordered gray goo, and, the very worst, a state wherein past, present, and future could not refer to each other. The present is everything, neither past nor future have any independent existence. This is why Bergson states that "duration will be the 'synthesis'" of both "unity and multiplicity" (CM: 218). Instead of cutting out segments of time, or extracting slabs of solidified time, intuition shows the true nature of time as actualizing crystallization. Never do we find a final condition, a time-crystal that has hardened into a perfectly repetitive state. Each repetition is a creative act, an aimless expenditure of energy. Everywhere always, pasts and futures are derivative of vibrant, authentic actual moments. The effort of intuition installs us within the object itself, inside of the being of another time. Or, differently put, sympathy reinstalls us inside of a body of becoming. In Barnard's characterization, intuition may be described as a "moment-to-moment immersion into the flux of life" (Barnard 2011: 212–213). Sympathy is a breaking of the shell. Dying to one's self, the corners of the window are shattered. We discover that until now we have been blinded, numbed, boarded up. Immersion in time is a unity, an astonishing recomposition of our ownness within a mode of expanded togetherness. Starting from "self-sympathy we can even sympathize with the entire universe" (Lawlor 2003: 66). The interconnection of each duration with the other is an immediacy conducive to receptivity. An active contemplation would be a consciousness of change observed within the attractive depths of the changeless. Metaphysics is incessantly caught between the "pure repetition" of materiality and the "eternity" of the living moment (CM: 221). Matter too has a duration, in the form of pure repetition. In Bergson's philosophy "the universe itself is in duration" (Barthélemy-Madaule 1970: 121). The time of life, however, is that of eternity, eternal creation. In a way, this is a reproduction of the dualism of matter and memory outlined in Bergson's work of the same title, *Matter and Memory*. But this duality is only apparent. The time of life differs from the time of the inanimate only in terms of its qualitative intensity. Living time, defined as the time of the actual, is "a duration which stretches, tightens, and becomes more and more intensified" (ibid.). But memory here actually corresponds to the repetitious aspect, the materiality of time! The sinuosity of the present, on the other hand, is the actual, intensive moment, the becoming which has not yet become. Undoubtedly, all this talk of presentism on our part can strike the reader as somewhat counterintuitive. But Bergson himself defines philosophy as a thoroughgoing reversal of our normal, habitual thinking: "To philosophize means to reverse the normal direction of the workings of thought" (CM: 224). Instead of a chasm separating the duration of living creatures from the merely repetitive duration of matter, Bergson prefers to posit a difference of degree. Each duration includes something of the other durations. As we shall see regarding Bergson's evolutionism, life could not exist without being wedded to matter. Some durations are installed, mounted as it were, upon other times. Of course, such spatialized language must be conceived of as nothing more than an ill-fitting, necessarily clumsy use of metaphor. The reversal of thought is the ultimate intensity, the supervening and transcendence of reflection, the productive synthesis through which thinking breaks open its chrysalis. Through the intuitive method, intelligence transforms itself into a strange species of immaterial eagle, capable of soaring in the sky and probing into microbial depths. Extending infinitely, the effort of intellect going beyond its own condition shows us "the Magnets which so strongly move And work all night upon thy light and love" (quoted in: Rudrum 1974: 136). Immediately perceived, duration is the immeasurable excess, the play of chaos destructuring itself. The intuitive method is a reorientation and a conversion, a diversion of consciousness away from practical exteriority toward the heterogeneous continuity of interiority. A thing is an "urge to movement," and metaphysics would be a practice that not so much seizes the existence of alterity, as lets it flow (CM: 236). Concentration is neither disinterestedness nor interest, neither the fixation of attention, nor its absolute relaxation. Rather, to concentrate means to intensify our thinking, to the point wherein the alterations of the flow are all that remains. Inside of alteration, we capture nothing. This is the still-life, emerging from the duration encapsulated by the intuitive act, that possesses and remakes us, completely destroying both desire and the absence of desire. Neutral is the creative validation of the passing. Indetermination marks neutrality with its own trace. The latter is what fructifies the understanding, yielding a sweet bliss, a taste of the beyond within the immanence of *this* now. A now like no other, that is the closest thing to enlightenment here.

NOTES

1. James Burton's excellent study has pointed out a marked similarity between many of Dick's and Bergson's concepts. For the most part Burton concentrates though on Bergson's idea of cosmically "open" religion, as outlined in the latter's *Two Sources of Morality and Religion* (Burton 2015). The topic of Bergson's social and moral philosophy shall form the focus of a later volume of ours and will not be discussed here.

2. This is absolutely not the manner in which musicians themselves would view things!

3. The examples of incoherent serial retcons provided by Proctor include *Coronation Street* and *Dallas*. In the latter case, the retconning was so extreme that in Season Eight the viewers are informed that none of the episodes in Season Seven actually happened. This latter would surely classify as one of the most drastic instances of forced retroactive continuity in the history of television!

4. In a recent piece on the philosophy of time in Object-Oriented Ontology/Speculative Realism, Arjen Kleinherenbrink has argued for a reading of Gilles Deleuze's philosophy which advocates a doubling of time along the lines of the distinction between Aion and Chronos in Deleuze's *Logic of Sense*. As Kleinherenbrink, following critics such as Peter Gratton, sees things, presently the temporal status of objects is unclear in Harman's speculative realist philosophy. Objects in speculative realism seem to persist in a permanent present (Gratton 2014: 99; Kleinherenbrink 2019: 543). As a solution, the author proposes that we differentiate between "the time of incorporeal events" and "the time of what happens to [. . .] bodies themselves" (Kleinherenbrink 2019: 546). The former would correspond to virtuality, while the latter is the actual. In Kleinherenbrink's view, it is really actual, that is, *corporeal* time, which is the sole real mode of temporality ("Aion constitutes a present." Kleinherenbrink 2019: 547). In this book, we shall seek to advocate for a multilevel theory of purely immanent duration/s which avoids positing a virtual time lacking in corporeality.

5. In this regard, the Deleuzian interpretation, as given in *Difference and Repetition*, which conceives of duration as relating to "internal difference" is very pertinent, and one we are generally agreement with (see: Bryant 2008: 49–73).

6. This assertion is also emphasized by Agner Fog in an influential study of cultural evolution, who argues that cultural selection is characterized by an accelerated tempo of adaptability (Fog 1999: 77). As of late, this view has been the object of heated debate. Some recent historical studies of cultural evolution suggest that the rate of social change is a great deal slower than Dawkins and others influenced by meme theory had thought (Lambert et al. 2020: 352–360).

7. https://ontology.ihmc.us/kaos.html.

8. When Martin Heidegger and Graham Harman alike claim that each thinker has "one great thought," and the whole of a philosopher's teaching can be distilled into a single intuition, they are both following in Bergson's footsteps.

9. In a private communication, Keith Ansell-Pearson called my attention to the significant divergence between Deleuze's early works and his *Cinema* books. An alteration in Deleuze's interpretation of Bergson becomes evident from comparing

his statements from the late 1950s and early 1960s and the 1980s. While the former years are characterized by a stringent commitment to the primacy of the virtual, the picture becomes more complex in the *Cinema* books and Deleuze arrives at something approximating an actualist position. My primary point is that Deleuze's earlier works fail to grasp how Bergson does *not* privilege either memory or virtuality at the expense of percption and actuality.

10. Lorand's denial of disorder rests on a confusion between authentically ontological chaos (the irreducibility of reality to consciousness) and what we have termed "KAoS." This second type of disorder is the irritating absence of an expected order within a certain ecological niche or contexture, actually constituting an aspect of a certain broader order.

11. Bergson has had a profound and provocative influence on modern Indian philosophy. Two particularly well-documented examples are the Hindu traditionalist philosopher Sri Aurobindo and the process philosopher Vaddera Chandidas. (On the former's relationship with Bergsonism, see Maitra 1988 and Bhattarcharya 1972. On Chandidas's reading of Bergson, see Raghuramaraju 2019: 70–90.) Although Chandidas does not cite Bergson explicitly, A. Raghuramaraju maintains that the former's 1975 work, *Desire and Liberation*, can be read as a reaction against the excessively positive outlook advocated by the French philosopher. As Raghuramaraju explains, "While Bergson relates non-being with negation, Chandidas relates negation to permanence. Permanence, not non-being, is based on or is sustained by negation" (Raghuramaraju 2019: 85).

12. On our notion of "absentology," see (Horvath and Lovasz 2016, 2019: 46–47; Horvath 2016: 23–73; Lovasz 2016). This phrase was first coined by sociologist Jennifer Croissant in a landmark 2014 text: The projects of agnotology and absence (should that be absentology?) require a great deal of taxonomic work, and this paper is meant to add to the conversation, not as a matter of lexical policing, but as a necessary step in theory-building and developing the capacity for cross-case comparisons in studies of ignorance, as well as to articulate a possible framework for studies of other kinds of absences (Croissant 2014: 18).

Chapter 2

Completing Relativity

In a recent book, the renowned Italian physicist and popularizer of theoretical physics Carlo Rovelli differentiates between a scientifically grounded ("physical") and a supposedly emotionally charged ("philosophical") outlook on time. The "veneration of time" on the part of philosophers such as "Heraclitus or Bergson" has generated a range of philosophies "without getting us any nearer to understanding what time is" (Rovelli 2018: 109). In a single sentence, the physicist disqualifies an entire tradition of thinking about time. The feeling of time passing is radically distinct from time in its objectivity. There can be, for contemporary physics, as interpreted by Rovelli of course, no reconciliation between the feeling of time's flow and time in itself: "Perhaps the emotion of time is precisely what time is for us. I don't think there is much more than this to be understood" (Rovelli 2018: 110). Is such a sweeping assertion correct? Furthermore, may we simply disqualify Bergson and all other philosophers from the debates relating to the ontological status of time? When writing a book on Bergson, should we simply exclude *Duration and Simultaneity* from our investigations altogether, leaving temporality to the physicists, ignoring this embarrassingly "unscientific" book? To all three of these questions, we are compelled to answer in the negative. It would have been inexcusable for Bergson to not attempt a confrontation with the celebrated physicist Albert Einstein and the theory of relativity. Our question is the following: is the Bergsonian view compatible with relativity? The amplitude transition philosophy strives for would be a veneration of time, a near-deification which is almost more than love yet something less than a complete coincidence with the object of striving. In this Rovelli is correct: an element of emotion is involved in all authentic philosophies. The dominant viewpoint on the Bergson-Einstein debate is that of Alan Sokal and Jean Bricmont, who argue in a chapter of their notorious

Intellectual Impostures—rather unfortunately excluded from the English translation of their work—that Bergson flatly misunderstood Einsteinian relativity and, furthermore, Bergsonian philosophy is largely to blame for postmodern "abuses" and misappropriations of scientific concepts (Sokal and Bricmont 1997). Bergson simply did not understand the import of Einstein's theories, and that is that. As a consequence, the former's interpretation is *per definitionem* invalid, simply inadmissible in the court of science because of its factual inaccuracies. If the entire work is the rotten fruit begotten from a mutual misunderstanding, then what is the point of including *Duration and Simultaneity* in our own book?

Nothing prevents us, even on the level of interpretation, from freely choosing our coordinate systems. Even the most controversial or discredited of works can be of interest. In particular, the Science Wars of the 1990s and their unfortunate aftermath which reverberates to this very day point toward the necessity of reasserting the legitimacy of philosophy. The Sokal 2.0 Affair, for instance, has only served to further underline the deep divisions between the human sciences and natural science (Mounk 2018). From their mutually segregated ghettoes, researchers working in separate fields hurl invectives at one another, often without much common interdisciplinary commonality. Any middle ground appears to be a No Man's Land, with very few having the bravery to confront deeply entrenched divisions. Bergson's philosophy we believe can be of help here, for its creator was animated by a fundamentally interdisciplinary ambition, seeking to synthesize the various results of different scientific disciplines within an all-encompassing metaphysical vision. Far from being unscientific, Bergsonism is an eminently intellectual and scientific enterprise. Against Bertrand Russell's infamous charge of irrationalism, founded on an astoundingly superficial reading of Bergson, Pete A. Y. Gunter argues that the Bergsonian method consists of both an "analytical empiricism, close to the methods and assumptions of science, and a dynamic empiricism, the complement of the first" (Gunter 1969: 29). In the wake of its many scandals, the philosophical outlook, defined as both the respectful veneration of time and a dynamically enhanced, liquified empiricism, awaits rehabilitation. We follow the spirit of relativity in choosing *Duration and Simultaneity* as our own coordinate system. Different works and texts age at contrasting speeds. The immense popularity of physics has contributed to an acceleration in the aging of Bergson's text. However controversial such an undertaking may be, it is necessary to delve into the details of the critique of Einsteinian relativity this book contains if we are to rescue Bergson from the charge of science denial. By consequence, an understanding of the Einsteinian theory is also required. In this, a reading of Einstein's *Relativity: The Special and General Theory*, intended for a lay audience, will be more than sufficient. Additionally, we shall also engage in a close examination of Paul Langevin's

1911 text, "The Evolution of Space and Time," in which the author—a close ally of Einstein—works out the details of what later became known as the "Twin Paradox," one of the most intensely debated thought experiments in the history of relativity and, for that matter, theoretical physics in general. After these preliminary steps, we shall be sufficiently equipped to examine Bergson's arguments, as presented to the reader on the pages of *Durée et simultanéité*.

EINSTEIN'S REVOLUTION

Relativity theory is an invitation to think in terms of particularities rather than universality. Never do we encounter time as such or truth in general. Already in the Euclidean system, the "straight lines" we speak of are not having zero truth status (Einstein 1921 [1917]: 2). It is a habit of thought to relate a geometric representation back to a spatially discrete object. We are in the habit of viewing the correlation between a geometrically depicted line and the "real" object as some empirical proof of truth. It is purely the particular system, in this very moment, which counts, but not as an absolutely true existent. Correlation, Albert Einstein argues, has nothing to do with any incontestable truth. The correlation between a line we arbitrarily draw upon a map and, say, our walk, has only an apparent verity. For us, the coincidence between our real walk and the line decorating the map is certainly empirically verifiable. Yet this description remains only that, a representation, a convention lacking a stable ontological basis. Already prior to Einstein, the French scientist and polymath Henri Poincaré advocated for a view of geometry and even scientific concepts in general as constituting conventions, convenient constructs designed to assist scientific work but not genuinely referring to reality. In Jimena Canales's summarization, "a conventionalist perspective, rather than aiming to describe how things really were (as a realist would), maintained instead that scientific descriptions arose from the particular needs of different professions and the individuals who espoused them" (Canales 2015: 76). The truth status of the line is a matter reducible to practical convenience. As Einstein states, without citing Poincaré but following the latter's spirit, "by the 'truth' of a geometrical proposition (. . .) we understand its validity for a construction" (Einstein 1921 [1917]: 4). The power of practically oriented thinking demands a corresponding conceptual production, directed against the resistance of emotions, passions, and prejudices which becloud judgments relating to both time and truth. To avoid misunderstandings, we must clarify what we mean by the term "conventionalism." According to Poincaré's position, "although none of the geometries are true or false," this by no means implies "complete neutrality on our part. A particular geometry can still be more convenient for our purposes, given our experiences, and given the nature of the world that produces these experiences" (Ben-Menahem 2009: 52). Experience has nothing whatsoever to say about truth. Empirical data provides us with an always precarious access to the resistance of the constant modification of our environment to our own designs. The experience of the resistance of reality "tells us not what is the truest, but what is the most convenient geometry" (Poincaré 1905: 70–71). Underneath frontally oriented thought, there is a decision which, in turn, is always already effected by the contours of an inhospitable, hostile environment.

Which geometry must we follow, if we are to find what we are searching for? The choice of coordinates is filled, one could say, contaminated by the contingency of the environs of the agent. Every act of measurement necessitates the postulation of a point which "we employ as a standard measure" (Einstein 1921 [1917]: 5). The coincidence of a potentially infinite range of historical, cultural, and natural accidents has been required for us to reach the point whereby we posit a point of reference. Einstein chooses Trafalgar Square as an example, being a well-defined place. But immediately, we are confronted with the nebulousness of reality. What if a cloud hovers above the otherwise clear-cut, identifiable square of worldwide fame? We could, Einstein argues, construct a pole that reaches the cloud, measuring its distance from the ground. But the nebulosity of the cloud prevents any definite positing of distance! The complexity of reality means that, in actuality, measurement is always subject to uncertainty. It is possible to fix a point in one place, but the contours, shapes, and sizes of real objects are perpetually changing: "In practice, the rigid surfaces which constitute the system of coordinates are generally not available" (Einstein 1921 [1917]: 8). If Euclidean geometry operates with straight lines, then a geometry wishing to penetrate into real, objective time must habituate itself to the perpetual modifications of real objects. Without imagining rigid, fixed objects, geometry cannot even begin to proceed with its task, which is, above all else, the practical orientation of our actions. The significance of the beclouded square lies in its mediation of arbitrary features and nebulosities within geometry and physics, as well as the consequent thoroughgoing rejection of anything resembling absolute Newtonian movement. In the Einsteinian view, the generality of movement is a hallucination of consciousness. We are informed that "there is no such thing as an independently existent trajectory (. . .) but only a trajectory relative to a particular body of reference" (Einstein 1921 [1917]: 10). In the special theory of relativity, the possibility of any absolute has been destroyed. The relativist revolution initiated by Einstein abolishes any absolute privileging of a coordinate system, André Metz's protestation to the contrary notwithstanding ("Einstein's theory of relativity does not abolish the absolute") (Metz 1969 [1924]: 159). Indeed, the body of reference itself dissolves into a dynamic, qualitative multiplicity of differential coordinates. No longer may we talk of absolute movement, mobility having no independent existence in the Einsteinian view. It is always a particular, relative development we talk of when we speak of change. At this point, Einstein introduces the idea of a "Galileian system of co-ordinates," to which the classic laws of mechanics apply. Uniform motion, or movement in a straight line, would pertain to Galilean systems only (Einstein 1921 [1917]: 12). The straight, Euclidean, normative line is relative to systems governed by the law of inertia. A real simultaneity exists in the world, one erroneously equated by Metz with a new "absolute," as grievous a category error as any. This is the simultaneity of Galilean and non-Galilean systems of coordinates. At every moment, both classical and relativistic physics pertains, as the dual aspect of one situation. From a general point of view, the same event can be viewed in two opposite ways. That which appears, upon first inspection, as a straight line, can also be observed to constitute a parabola, depending on the situation of the observer. If the observer is moving, the straight line traversed by another moving object is deformed into a parabola (Einstein 1921 [1917]: 10). No longer must physics obey the Galilean imperative to think in terms of Euclidean geometry.

The Earth may be compared to "a railway carriage travelling with a velocity of about 30 kilometres per second" (Einstein 1921 [1917]: 17). More complex laws apply to a body in motion than a stationary object. This recognition represented a revolution in physics, contributing in no small measure to an exponential dynamization of cosmology. In a daring gambit, Einstein staked his scientific credibility upon empirical predictions relating to the deflection of light from the Sun. The expedition led by Arthur Stanley Eddington, observing the solar eclipse of May 29, 1919, on the island of Principe, empirically corroborated Einstein's mathematical formula on the deformation of light with empirical measurements of the eclipse (Dyson et al. 1920: 291–333). Light was indeed bent by the gravitational force of the Sun to the degree Einstein predicted. The findings were presented at a sensational meeting of the Royal Society in November of 1919, almost immediately catapulting Einstein to international fame. The headline of the *Times* on November 7, 1919, read "Revolution in Science/New Theory of the Universe/Newtonian Ideas Overthrown" (quoted in: Canales 2015: 17). Quite a spectacular headline indeed! Instead of a straight line, the dissemination of light follows a curved pathway. An important ramification of the Eddington's eclipse expedition was that the Earth too was found to be always in motion. While the classically formulated law of light propagation still held, from November of 1919 onwards it did so in a deformed, distorted manner. If an empty space would exist, then light could theoretically travel in a straight line, but the inherently contorted, misshapen, curved nature of reality makes this impossible. After the Eddington Event of May 29, 1919, the swerve displaces, or rather, mangles the straightness of the line. The observer becomes interchangeable with the ray of light. Situated relative to a railway embankment, we have a luminosity traveling almost at light speed. Gravitational distortion accounts for the difference between its actual velocity and the theoretical speed of light propagation in empty space (Einstein 1921 [1917]: 22). Actualization inherently does not fulfill the entire potentiality of movement in a vacuum. In emptiness replete with zero gravity, light could travel uninterrupted, but real movement is always characterized by deflection and aberration. To think the emergence of real simultaneities and durations, it is necessary to abandon the fixity of reference. Einstein enjoins us to imagine two bolts of lightning striking the train tracks. We attach clocks to several sections of the track. Will the two lightning strikes be simultaneous? Einstein answers in the negative. Time will necessarily pass differently for passengers on the train than for clocks situated along the tracks. Two events "simultaneous with reference to the railway embankment" will not be "also simultaneous relatively to the train" (Einstein 1921 [1917]: 30–31). Einstein's relativity decidedly affirms the absence of both absolute movement and absolute time. The relativity of simultaneity means that "*every reference-body* (co-ordinate system) *has its own particular time*" (emphasis ours, Einstein 1921 [1917]: 32). To think in terms of simultaneity means to think in terms of relations. Bergson, in a response to Metz—more on that controversy later!—is entirely correct in his characterization, according to which "Einstein's universe is simply a nexus of relations" (Metz and Bergson 1969: 177). Time itself is a heterogeneous multiplicity of temporal interrelations and mutual causalities. Does this not in itself resemble the Bergsonian affirmation of multiple durations? Real simultaneity is distorted by gravitational effects. Time has no relevance outside of a particular body of reference. This circumstance, as proven by the Eddington Event, is a nonarbitrary, empirically proven feature of reality. Inseparably, measurement implies the positing of a point of reference, but this positionality is nevertheless always arbitrary. It is a nonarbitrary fact that real bodies refer to each other, and that time is this very process of mutual referentiality. Nonetheless, exactly how positionality is enacted in each situation is a contingent circumstance. No measurement is possible without already presupposing something in comparison to which the measurement shall be conducted. In a letter addressed to Otto Naumann, Einstein summarizes relativity by stating clearly that "no physical reality can be lent to time and space" (quoted in: Crelinsten 2006: 90). Returning to the example of the train and the railway embankment, everything depends on the initial position of the observer in relation to the observed facts. A distance measured from the train is not necessarily equal to that measured from the embankment or any other, more static place (Einstein 1921 [1917]: 35). If there is such a huge rift separating two different registers of observation, how may the laws of classical mechanics nonetheless coexist with relativity?

It is at this juncture that Einstein introduces the Lorentz transformation, based on Hendrick Lorentz's work. What this mathematical construct allows Einstein to do is differentiate between measurements conducted in a mobile and immobile context. Every component of the train-human-railway embankment assemblage is, of course, in motion, if for no other reason than their positionality upon a planet orbiting a star at a huge velocity in space. But for the sake of argument, Einstein characterizes the embankment as a static system. If we compare a measuring rod in motion with one that is not presently moving, we find that "the rigid rod is (. . .) shorter when in motion than when at rest, and the more quickly it is moving, the shorter the rod" (Einstein 1921 [1917]: 42–43). The Lorentz transformations are supposed to prove in a formal manner that "as a consequence of its motion the clock goes more slowly than when at rest" (Einstein 1921 [1917]: 44). This assertion forms the basis of what will become known as Langevin's "twin paradox." Velocity results in the retardation or contraction of temporal flow. On this basis, time may be described as an elastic rubber band. Instead of a container, time is a multiplicity of durations, the progression of which depends upon their movement. Time does not differ from movement. The velocity of the flow influences the propagation of light as well (Einstein 1921 [1917]: 48). Einstein does not stop here though. Not only is the length of a body dependent on its speed, but even mass is structured so as to be inseparable from energy. Variation is introduced into the very heart of extension. Momentously, relativity constitutes an upheaval that liquifies all constants by paradoxically utilizing a constant value—the speed of light—to decompose a previous cosmology. This is a deconstruction of classical physics that does not completely eliminate the previous Galileian-Newtonian outlook. The latter has a place, but only on a certain plane. Mass too is variable, "the inertial mass of a body (. . .) varies according to the change in the energy of the body. The inertial mass of a system of bodies can even be regarded as a measure of its energy" (Einstein 1921 [1917]: 55–56). Once we transition from one type of situation or position to another, a Lorentz transformation is required to prevent us from losing our train of thought: movement itself is translation. Lines are not normal, curvature being an empirical feature of reality. Transmitting displacements, Einsteinian relativity undoes any privileged positionality. Einstein himself is unambiguous when he proclaims boldly that "there is no such thing as a 'specially favoured' (unique) co-ordinate system" (Einstein 1921 [1917]: 63). This audacity is occasioned by the failure of the Michelson-Morley experiment to discover anything like the "ether" which had preoccupied physicists for generations. The ether was, for a long period in the history of physics, a convenient convention, needed to stabilize cosmology, functioning as a stable external point of reference, its introduction coinciding with the Copernican abandonment of the geocentric worldview. What relativity made possible, with its jettisoning of the ether, is an authentically *acentered cosmology*. No longer is there an absolute motion or a singular point of reference. Instead, we ought to refer always to "motion with respect to the body of reference chosen in the particular case in point" (Einstein 1921 [1917]: 64). Instead of being a defect, aberration is the norm when it comes to the propagation of light. The straightness of the line is, in actual fact, the imperfect, virtual, powerless aspect of mobility, while deformation, as confirmed by the Eddington expedition, is a perfection, an actualization of the line. *A swerve is a completed line.*[1] Shockingly, the dependence of size upon velocity entails that we live not in a three-dimensional world, but a Minkowski-type "four-dimensional space-time continuum" (Einstein 1921 [1917]: 65). The time it takes for a duration to endure is equivalent, in Einstein's eyes, to its extension. There is no difference between a spatial extension and a temporal duration in the theory of relativity. Time is but one of the four basic coordinates of a system. Each object has three spatial coordinates, x, y, and z, in addition to its time coordinate. The world is therefore a continuum in which time is always already fragmented into a heterogeneous multiplicity of different becomings. "Time is robbed of its independence," for temporality has been placed back into the individual entity (Einstein 1921 [1917]: 66). Without delving into the production of excessively formal proofs, we may take the following summation as exemplary: "Minkowski spacetime can be thought of as composed of a stack of instants of time" (Belot 2013: 191). The spacetime continuum comprises an accumulation of durations, embedded within one another. From the perspective of interiority, each duration appears to be nothing more than a pointlike instant but traversing the "worldlines" hypothesized by Minkowski, we observe real change. What is important from the perspective of special relativity—remember, everything heretofore related here is in connection with the special theory of relativity—Minkowski spacetime allows us to extrapolate and deduce ontological conclusions on the status of time in its objectivity. The relativity of simultaneity implies a spatialization of the transient: "Temporal separation behaves (. . .) like distance in Euclidean geometry" (Belot 2013: 190). For the sake of convenience, we differentiate between "when" and "where." From the relative point of view, these two are one and the same. Einstein himself is explicit when he states that "the time co-ordinate" in special relativity "plays exactly the same role as the three space co-ordinates" (Einstein 1921 [1917]: 65). It is just as ill-advised to speak of an extension without time as a time without extension. As we shall see, in spite of all misunderstandings to the contrary, Bergsonian duration similarly does not lack extension. The extensive and the intensive, the spatial and the temporal are locked in a *chiasmus*. It is only when situated within the zero energy state of virtuality that time lacks extension. Without a body that registers its passing, temporality cannot have efficacy. Aging, decay, rot, decomposition, these too can serve as coordinates, the clocks, so to speak, of a particular biotemporal flow. Outside of a particular coordinate system, time is nothing.

Each movement is accompanied by a contraction in length. There is, in the relativist universe, nothing apart from an infinity of worldlines. As Hermann Minkowski himself puts it, the entirety of the cosmos is considered to be none other than the sum of movements: "We (. . .) get an image, so to say, of the eternal course of life of the substantial point, a curve in the world, a worldline [*sic*], whose points can be clearly related to the parameter t from -∞ to +∞. The whole world presents itself as resolved into such worldlines" (Minkowski 2012 [1907]: 40). In relativity, the rigid point dissolves into a curve situated between negative and positive infinity. This is not so very different from Bergson's assertion in numerous instances that the new physics will, sooner rather than later, dissolve the world into a mass of insubstantial movements, vibrations, and dynamisms.[2] From special relativity, we must proceed to the general theory of relativity. Rather than historiographic authenticity, our commentary is an attempt to grasp the essence of the relative view. Only after the intuition of relativity has been mastered can the Einstein-Bergson controversy be addressed directly. Specifically, the relativist intuition is a necessary preliminary to the completion of relativity through Bergsonism. Provocatively, we hope to prove that Bergson, far from missing the point, actually extends, even explodes the rigidity of the mathematical point in a manner akin to Einstein and Minkowski. If anything, Bergson goes further than many rivals, because duration as qualitative multiplicity does not depend on the positing of the speed of light as an absolute value. Viewed from this philosophical perspective, it is Einsteinian relativity which is incomplete, because of is fatal reliance upon a supposedly absolute measure. Since relativity ties itself to the speed of light as an absolute footing for measurements, it remains committed to an Occidental metaphysics that privileges light at the expense of darkness. We posit that worldlines pass through both light cones and cones composed of dark matter. In the world as we know it, based on contemporary cosmological knowledge, the probability of worldlines traveling through dark matter is far greater than a traversal through enlightened realms. A light cone can only be a frightfully rare exception and not the rule. The world line and its accompanying cones will be reintroduced at the point wherein we synthesize Bergson's intuition of duration with Einsteinian and Minkowskian relativity. How then does general relativity differ from special relativity? As we have seen, in Einstein's view "every motion must only be considered as a relative motion" (Einstein 1921 [1917]: 69). In the second decade of the twentieth century, Newtonian absolute movement is dethroned, overthrown, shattered into an infinite variety of pieces. No longer is there any general motion, and we may choose any object we please as a reference-body for our calculations. Nothing commits us to privileging any single perspective. Einstein in effect liberates the thought of the subject from enchainment to its own, phenomenal, familiar realm. We have here something again remarkably similar to Bergson's intuition. Both the physicist and the philosopher are committed to a transcendence of the human condition. No reference-body is unique, not even our intrinsic position. The relativist can posit any number of positionalities: all are equal without exception (Einstein 1921 [1917]: 70). Every body comes equipped with a field, a medium that exhibits a force of attraction and repulsion upon surrounding objects. When speaking of objects, we must therefore also describe the fields in which they appear. There is no body that fails to display the presence of some type of field. While there is, in Einstein's view at least, no immediate causality—objects can only travel at the speed of light, a position which clashes with Bergson's affirmation of immediacy—we can nonetheless contend that any object "produces a field in its immediate neighbourhood directly" (Einstein 1921 [1917]: 75). From an ontological angle, this would imply that objects are inseparable from their corresponding fields. New Realist philosopher Markus Gabriel argues analogously that the existence of an object is a direct correlate of its positionality within a field. As Gabriel states, in full accordance with the spirit of Einsteinian relativity, "for physical objects to appear in the field of sense of the universe is for them to be physical, where this might be defined in terms of what can ideally be observed or claimed to exist by some idealised future physicist. For unicorns to appear in the colouring book *Unicorns are Jerks* is for them to be associated with our standard fantasy of unicorns" (Gabriel 2015: 158). On every level, causal relations are deformed by the interventions of gravitational fields. Einstein proved that the acceleration of bodies depends not on their material or physical states, but rather on influences radiating from their fields. "Gravitation," Einstein proclaims, "is the cause of acceleration" (Einstein 1921 [1917]: 76). Of course, it would be difficult indeed for us to attribute a definite quantified physical value to the gravitational distortion exerted upon Gabriel's unicorns by their particular fields of sense. Any such hypothetical value would be the unmeasurable product of speculation. The denial of existence, however, by no means implies the reduction of the ontological nature of an object to a "merely" imaginary status. Everything exists, inasmuch as it connects to a certain field. As Gabriel points out later on, "absolute non-existence is impossible. Everything exists, but in different fields of sense. It does not co-exist. There is no all-encompassing field in which surprisingly there somehow are unicorns and there are no unicorns. There are unicorns (for instance, in *The Last Unicorn*), and there are no unicorns (for instance, in Milwaukee)" (Gabriel 2015: 178). It cannot be said that there is a universal gravitational field encompassing merely fictional or intentional objects and real values, objectivities and actualized causalities, respectively. Everything is actual, even the most improbable of rainbow-colored unicorns.

Because acceleration is independent of the nature or condition of a body, we may state that even fictional entities such as unicorns are affected by the intensities of their respective gravitational fields: "The ratio of the gravitational to the inertial mass must (. . .) be the same for all bodies" (Einstein 1921 [1917]: 77). Naturally, Einstein refers here to the entirety of real physical bodies, say celestial objects, planets, and other bodies in motion. The physicist would find any extension of this principle to absurd or "unreal" (from a commonsense perspective) entities a travesty and a misreading. A unicorn, for Einstein, lacks any force, for it lacks gravitational mass. But we can easily construe an *imaginary force* by positing the unicorn's gravitational field. Any object, even one without apparent mass, has, if we accept Gabriel's view, a field of sense. This latter furnishes us, in theory, with imaginary values from which we could, if we so wished, derive the inertial mass of a unicorn. Even in the case of imaginary objects, it can be stated that the Einsteinian position holds: "The gravitational mass of a body is equal to its inertial mass" (ibid.). Would the extension of relativity to all objects, including virtual ones, be such a travesty as it appears upon first impressions? The relativism inherent within relativity, as Gabriel points out, is far from absolute. We must always be attentive to *which* gravitational field we are speaking of when interpreting a certain modality. The selection among various fields is itself nonarbitrary: "The structure of the field containing the object is as objective as the object itself, and it defines what it is for the object to appear within it" (Gabriel 2015: 276). The motion of geometric space has been sliced into pieces by the Einsteinian Event in physics. Temporal flow is a heterogeneity of temporal details, given momentary coherence by their respective fields. The positing of gravitational distortion is, for Einstein, of key importance, because this allows for the denial of the constancy of the propagation of light in a vacuum. Gravity deforms rays of light, as proven empirically by Eddington's expedition (Einstein 1921 [1917]: 90). In hindsight, it would be all too simple to discount the various obstacles which Eddington and colleagues had to surmount to capture adequate photographic plates of May 29, 1919, eclipse. The worldwide success resulting from the announcement of the findings, corroborating Einstein, retrospectively obscured the very real difficulties faced by the expedition. As Eddington wrote in a notebook at the time,

> On May 29 a tremendous rainstorm came on. The rain stopped about noon and about 1:30 when the partial phase was well advanced, we began to get a glimpse of the sun. We had to carry out our programme of photographs in faith. I did not see the eclipse, being too busy changing plates, except for one glance

> to make sure it had begun and another half-way through to see how much cloud there was. We took 16 photographs. They are all good of the Sun . . . but the cloud has interfered with the star images. The last six photographs show a few images which I hope will give us what we need. (Quoted in: Crelinsten 2006: 131)

What this example shows is that any discovery is situated within a real field of sense. The simplicity of the Lorentz equations and other formalizations introduced by Einstein belie the complexity of the particular field of sense which corresponded to the eclipse and its photographic reproduction. As a matter of fact, only one of the sixteen photographs taken by Eddington's crew actually proved to be of sufficient quality to prove anything whatsoever. One could experiment with enumerating the various components of the field of sense depicted above, producing what amounts to a "Latour-litany," a list of significant objects.[3] The description commences with a "tremendous rainstorm," noontime, the various phases of said eclipse, the clocks, the cameras, plates, clouds, the Sun and Moon, the members of the expedition, and, of course, the all-important images. Eddington refers to the interference of a cloud with celestial images. Could Einstein's introduction of a fictive cloud hovering above Trafalgar Square be an unintentional homage of sorts to this cloud ominously situated above the island of Principe on that fateful day? How does Einstein's imaginary cloud relate to the very real clouds which do, on certain dreary English mornings, becloud Trafalgar Square? Presumably, this meteorological event is not infrequent in England. Yet there is only a single cloud that corresponds to Einstein's own intentional field of sense. All this would seem a mere embellishment, but it has far more than entertainment value. The moment described in the Eddington's notebook is the instant of discovery, a duration replete with a complex field of its own. In retrospect, the discovery corroborated Einstein's equations relating to the curvature of light rays. This corroboration overcodes, so to speak, the event of discovery. The verification of Einstein's theory is, to borrow Levi R. Bryant's term, a "bright object." Interestingly, Bryant in turn utilizes the concept of gravity in defining this mysterious class of entities: "A bright object is a machine that gravitationally overcodes the local manifestations, movements, and becomings of other machines" (Bryant 2014: 202). No gravitational manifestation can claim universal validity. In the latest strands of new realist thought, we find an integration of Einsteinian insights. None of the gravitational fields has a unique privilege or centrality, even though some are certainly stronger than others. In the above example, the Sun is overcoded for a limited space of time by the Moon, and the entire expedition is dominated by an eclipse. It is not paradoxical in the least to say that the eclipse functions as a bright object, in the sense that it dominates and overcodes its own field of sense.

Relative to every moving object, we posit a gravitational field. The last unicorn leaping into an unknown future is also affected by its own imaginary ecology. The character of a field, after all, depends entirely "on the motion chosen" for the object by its ontological modality (Einstein 1921 [1917]: 91). A gravitational field gives its components what they need, furnishing them with inertial mass and force. Temporal details have proliferated into a gravitationally mediated flow of multiple geometries. The nature of a body is an in-betweenness, situated within the object's field. There is, of course, the possibility of equivalence among bodies, in which two things would coincide completely. Each reference-body is interchangeable. The slicing of the world into separate fields has important temporal ramifications. General relativity necessitates a diversity of times: "In every gravitational field, a clock will go more or less quickly, according to the position in which the clock is situated" (Einstein 1921 [1917]: 95). The diversities of fields imply a non-coincidence of temporalities. General insights are impossible in a world of unequal flow speeds. No simultaneity can be absolute, and this is true of the coordinates as well. Tragically, Einstein does not realize that through an absolutization of the speed of light, he remains trapped within a basically metaphysical Western logistics of illumination. Without the positing of this final remnant of the Occidental Absolute, Einstein could have passed through to a theoretically coherent completion of relativity. Apart from positionality, relativity should not posit anything else. The universe cannot be a domain of domains either, for "it would have to exist in a more-encompassing domain that could not in turn be investigated by physics" (Gabriel 2015: 141). There is no duration of durations or time of times, just as there is no singular world line containing all other lines. Instead, there is an infinite or near-infinite heterogeneity of movements. From the circumstance of gravitational distortion, it follows that "the idea of a straight line also loses its meaning" (Einstein 1921 [1917]: 97). After Einstein, we cannot remain committed to a Euclidean geometry or a belief in straight lines. The right angle has lost all traces of its former legitimacy. Interestingly, of late studies of visual perception have abandoned the fixity of the term "item." Instead of searching for items, implying rigid, discrete, separated components of a perceptual flow, the literature on perception is moving toward the primacy of "eye fixations." As the authors of a recent paper claim, "adopting fixations as the conceptual unit allows all kinds of displays into the visual search fold, including real world scenes and X-rays, rather than only those with clearly defined items" (Hulleman and Olivers 2015: 24). Perception as such would not constitute so much a search for atomistically segregated items, but rather a transitory fixation relative to an ecology or visual field. *Contexture occasions texture*. Each surface is, for Einstein, a continuum, not a set of different items (Einstein 1921 [1917]: 98). We can move from one object to another, from surface to surface, traversing the fields. Even "the division of the field into electric and magnetic forces is a relative one with respect to the time-axis assumed," notes Minkowski in the 1908 Göttingen lecture, "Space and Time" (Minkowski 1920 [1908]: 86). The partial attraction between events is a function of both their temporalities and distortions. Relativity is a reconceptualization of time along the lines of heterogeneous deformation. Each line is a swerve, and straight lines are nothing but convenient fictions.

Relativity explodes the homogeneous. As Minkowski highlights, the world-postulate is a liberal one, allowing projection in time and space to be "handled with liberty." No longer is there a single space, but "many spaces" (Minkowski 1920 [1908]: 74). Einstein enacts a similarly liberal maneuver through the introduction of Gaussian coordinates into the discussion of general relativity. This is necessary for resolving the following thought experiment. Let us try to completely overlay the surface of a table with squares composed of rods. All goes as planned if the temperature of the table remains constant. But the introduction of a heat differential, mimicking the inequalities in temperature levels which may be found in the real world, complicates matters greatly. Our rods will deform, and the marble table cannot be covered successfully. By heating the center of the table, we discover, through the deformation of our measuring rods, that "the marble slab is no longer a Euclidean continuum" (Einstein 1921 [1917]: 100). This pressing issue can only be resolved by abandoning the construct of the square, and introducing "a system of arbitrary curves" (Einstein 1921 [1917]: 103). These "Gaussian co-ordinates" are infinitely dense, and we are therefore able to cover the entire surface of the table. What Einstein is saying here is that reality behaves similarly. Our concepts must become more arbitrary and exhibit weirder behavior, if we are to truly access the arbitrariness of reality itself. The arbitrary Gaussian continuum, Einstein makes clear, "can be applied also to a continuum of three, four or more dimensions" (Einstein 1921 [1917]: 105). Einstein's overheated hypercomplex marble table is a metaphor for an accelerating universe of uneven temporalities. Light propagation cannot be constant. The immediate presence of arbitrariness in existents is a circumstance that physics must account for. Absolute simultaneity is made impossible by the arbitrary curvature of real, existing lines. Straightness is imaginary, virtual, whereas the curve coincides with the actual. The velocity of light cannot be thought of as a constant when we transition to general relativity, for this velocity "must always depend on the coordinates when a gravitational field is present" (Einstein 1921 [1917]: 111). Force is a transmission between coordinate systems, reproducing the absolute relativity of simultaneity. The Einsteinian revolution rehabilitates the arbitrary in the context of cosmology. We are witness here to the liquefaction of both bodies and clocks. Systems of notation too are in movement, flowing along with a field which distorts their effects: in general relativity "it is impossible to build up a system (reference-body) from rigid bodies and clocks" (Einstein 1921 [1917]: 112). Spacetime definitively consigns the mathematical point to oblivion. As Minkowski already made clear, there is no point in clinging to the idea of a static point, because every axis consists of nothing but quickly vibrating molecules. However microscopic it may seem, the point we draw on our paper or render upon our computer screen is also a nebulous, vibrating arbitrary motion. The possibility of rendering a point is beclouded, smudged as it were, by the triumph of relativity.

All is in motion, but each motion pertains to a particular directionality. There is no general motion, just as there cannot be any general space. Every point of the continuum can be assigned four numbers, but which four these shall be is a matter of convenience (witness here again the Poincaréan element within the Einsteinian system). What is focal is a moment of bursting kinetic energy, made possible by an intensive journey. The point explodes, leaving glimmerings of inertial energy and liquified times. Concerning physical points, "the only statements" which "can claim a physical existence are in reality the statements about their encounters" (Einstein 1921 [1917]: 113). The focal is nothing apart from a collection of situations and relations. The violent, explosive moment of kinetic energy is a dynamic reference, desynchronizing time into an infinity of spacetime continua. Finally, cosmological thinking is delivered from the tyranny of the straight line, the oppression of the point, and the dictatorship of Newtonian absolute motion. The challenge, from here on out, will be to think in terms of relativities and relations, abandoning both absolutism and substance. "Encounters," reiterates Einstein, "constitute the only actual evidence of a time-space nature with which we meet in physical statements" (Einstein 1921 [1917]: 113–114). Instead of discrete, fixed points coming into contact, we observe the mutual encounters between motions. A dot is an alternation—this is the point! Every relation imaginable is also a translation, understood as the delivery of a novel moment in a spacetime. Synchronizing in a flash of color, the new moment is a uniform translation if and when it occurs among Galilean systems. But in the real world, what we have are irregular, arbitrary Gaussian coordinates. Time is an encounter between a clock, or observing system, and a body. Outside of a conjunction between two objects, time has no existence (Einstein 1921 [1917]: 114). Every statement relating to the physical, objective nature of objects is resolvable into the representation of a coincidence between two existents. The encounter is an agreement among arbitrarily emergent coordinates. In real spacetime descriptions, the use of rigid bodies of reference cannot be maintained.

The curvature of light discovered by Einstein results in a revolution in physics which replaces rigid "reference-bodies" with the arbitrary "Gauss coordinate system" (Einstein 1921 [1917]: 115). Implicit in this replacement is a realist ontological premise. According to Einstein, reality discloses a curved nature, an assertion borne out by empirical measurement of the angle of the Sun's rays during May 29, 1919, Principe eclipse. Directionality is the dominant principle inherent within Einsteinian relativity. To obtain the nature of an object, it is sufficient for us to know its direction. Knowledge of an object is awareness of its tendency. Every transformation is a translation, formalizable as the transition of "one Gauss co-ordinate system into another" (Einstein 1921 [1917]: 116). Radically, Einstein rejects the idea of a rigid body. As he writes, "*in gravitational fields* there are no such things as rigid bodies with Euclidean properties" (emphasis ours, Einstein 1921 [1917]: 117). With characteristic soft-spokenness, Einstein downplays the momentousness of this discovery. What is the physicist really saying here? We must carefully note the phrase "*in* gravitational fields." Never is there an object outside of a field. Therefore, in actuality the above statement must be amended in the following manner by subtracting its—for our purposes, unnecessary qualification: *there are no such things as rigid bodies*. Again, this stands in complete accordance with Bergson's view, expressed years earlier. In the latter's 1904 "Introduction to Metaphysics," we read "reality is mobility" and "all reality is (. . .) tendency (. . .) if we agree to call tendency a nascent change of direction" (CM: 222). Impermanence is introduced into the very essence of the reference-body. Einstein advances the idea of a "reference-mollusk," with its attendant tactile associations of softness. A soft clock, while selected arbitrarily, is nonetheless a far from accidental choice of metaphor. All mollusks, not unlike all clocks, "can be used as reference-bodies with equal right and equal success" (Einstein 1921 [1917]: 118). However improbable it may seem, even the mollusk, so much less complex than a human being, can also serve as a reference-body, for it too has a gravitational field. Later on, Einstein stresses that general relativity allows us to "free ourselves from the distasteful conception that the material universe ought to possess something of the nature of a centre" (Einstein 1921 [1917]: 127). Mollusks have open circulatory systems and lack anything like a central point. Their blood mixes with their effluence. They are closed organisms which nonetheless lack many of the borderlines and boundaries that characterize more complex animals. Analogously, a three-dimensional spherical space is "finite," yet "has no bounds" (Einstein 1921 [1917]: 133). The intuition of "world-radius," conceived of as a sphere, points toward the possibility of infinity within closure. General relativity, however, does not exactly posit some type of "world sphere." Rather, because of the deviations and curves of its individual parts, Einstein posits the existence of a "quasi-spherical" universe (Einstein 1921 [1917]: 136). Relativity obtains a deformed, desynchronized world. It is the multiplicity of gravitational fields that prevents the finalization and perfection of the sphere. Einstein however does not take the ultimate step and deny the very existence of a universe. Intriguingly, before the triumphantly positive closure of his book, Einstein imagines the possibility of an absolutely deformed, contorted universe:

> We might imagine that, as regards geometry, our universe behaves analogously to a surface which is irregularly curved in its individual parts, but which nowhere departs appreciably from a plane: something like the rippled surface of a lake. Such a universe might fittingly be called a quasi-Euclidean universe. As regards its space it would be infinite. But calculation shows that in a quasi-Euclidean universe the average density of matter would necessarily be nil. (Ibid.)

Since Einstein's time, theoretical physics has warmed up to the possibility of a truly "quasi-Euclidean" universe, an irregularly curved cosmos. In 1984, a review article authored by several physicists hypothesized the existence of cold dark matter (Blumenthal et al. 1984: 517–525). The concept of dark matter has proven of great convenience to researchers ever since. Presently, the standard cosmology model states that 28 percent of the universe's estimated mass is composed of cold dark matter, 68 percent of "repulsive dark energy," and 4 percent of Ordinary Matter (Gribov and Trigger 2016: 1–10). From the perspective of human access, it can be said that the vast majority of matter (96 percent) is almost completely inaccessible, being all but equivalent to nothing whatsoever. For Einstein, such a cosmology would surely be repulsive. The darkened world we inhabit is characterized by a surfeit of repulsion and a deficit of attraction. Recently, the hypothesis of a "periodically waveguided multiverse" (PWM) has been put forward as a possible explanation for the emergence of dark matter and antimatter: "Ordinary matter (electron and proton), antimatter (positron and antiproton) and dark matter (electron and proton) in the PWM are intrinsically identical" (Gribov 2019: 21). Were this conjecture ever to be empirically proven, the plausibility of an infinitely recurring multiplicity of durations and fields could be verified. In the words of W. Berteval, "it is not (. . .) the curve which is a complication of a straight line, but the straight line which is the extreme limit of a curve" (Berteval 1969 [1943]: 220). The real world is composed of curves, directions, tendencies, and changes. All is becoming, but there is no generalized "all" or a becoming in itself.

DECOMPOSING THE TWIN PARADOX

After drawing out the consequences of Einstein's theories of special and general relativity, we proceed to an elucidation of Paul Langevin's "Twin Paradox." After these two constructive elements—the Einsteinian theory itself and Langevin's article—are integrated, we shall be able to deduce what, if anything of Bergson's criticism is justified. In a lecture originally published in 1911, Langevin posits the relevance of evolution on the level of both concepts and physical entities. Physical concepts themselves are not immune to the effects of evolutionary selection. A "progressive adaptation" among the various physical theories may be identified (Langevin 1911: 32). Metatextually, the text of the lecture ingeniously embeds itself in the evolutionary process. Echoing both Einstein and Minkowski, Langevin recapitulates the absence of a generalized, universal space or time. There is no such thing as time in itself. Neither can any physical theory be said to stand the test of time, including that which issues this pronouncement. Impermanence pertains even to the most fundamental of our concepts of reality. At every moment, change is occurring. The Darwinian struggle for survival characterizes not only the lives of organisms but also physical theories. Concepts too have a "rhythm of their own," a life which is subject to selective pressures (Langevin 1911: 33). Decades prior to either meme theory or the neopositivist movements, Langevin already operates with an evolutionary epistemology.[4] To function efficiently, a new theory must be more economical with the facts. Evolutionary success describes here the ability of a concept to furnish explanations in a timely and economic manner. A successful concept mines reality, as it were, using "more or less completely the body of facts already established by the theories over which it triumphed" (ibid.).[5] Curvature, irregularity, and impermanence are pervasive in the scientific realm too. What the Michelson-Morley experiments show is that ether is nowhere to be found. There is no such thing as "absolute translational motion" because there is no absolute environment to which measurements can be adjusted (Langevin 1911: 35). When moving from one system of coordinates to another, various measures of magnitude and time are also subject to change. Entirely consequentially, Langevin is essentially enhancing or, to use an expression from Einstein, *dilating* the implications of relativity theory. A world is not a block of definite extension. Rather, it is an "ensemble of events" (Langevin 1911: 38). Systems no longer correspond to our everyday, commonsense perceptions. For the sake of convenience, we act as if the apparent, microlevel correspondences we encounter can be extrapolated into a general simultaneity. We suppose everything to occur inside the same time. This time we inhabit is the sole time which exists for us. Inside of a duration, such a view is not wholly erroneous. But a world is *an ensemble* of events. This means that not only is our world *not* the sole set of events—there can very well be occurrences outside of our own domain, at least if we accept the ontological possibility of multiverses and multiple Gabrielian fields of sense—but also that a world is nothing other than an ensemble of events. In itself, *a world is not*. Following Minkowski's terminology, Langevin holds that "the shape of a body at a given moment is determined by the set of simultaneous positions on the worldlines of various material points which constitute this body" (ibid.). In Minkowski's model, to each body there corresponds a certain light cone. All systems are dependent on their particular light cones, but each body of reference is also decomposable into a set of segments or miniature lines. Mobility is the usual, even original state of things. "Regularity," writes Berteval, "is a simplification (one, moreover, never completely realized) of the complexity of things" (Berteval 1969 [1943]: 221). Simultaneity is the product of an encounter, being the togetherness of at least two separate durations. Relativity tends toward a full symmetry, even coincidence, between space and time. It is truly correct that in relativity, space and time blend into a type of coalesced positionality, for motion is the symmetry of all states within an event. The actualized manifestation of surfaces is a unified time, when viewed from within a duration. We observe a fiery chariot, racing to the finish line, unfurling a vast pageantry of colors, signs, smells, flags, and clouds of dust in its wake. All of these heterogeneous elements are simultaneous with the chariot race itself. Something is there of an infinite concept, but equations are like strangers, or friends who have grown coldly distant, if they are not related to the concrete situation, to a duration in particular. The time of the chariot with flaming wheels is inseparable from the length of the racetrack and the velocity with which it wins the race. Its time is the period it takes to traverse its path. The traversal is the time. Of course, the horses yoked to the chariot have their own vital temporalities. So too does the chariot racer and the various wooden and metal components which make up the chariot as vehicle. These diverse durations are more or less united by an intentionality, namely, the time it takes to win the race. Even the speck of dust contained by the airborne cloud has a duration of its own. Perhaps, having fallen to the ground, it shall remain there for millennia, long after the racetrack has been demolished and subsequent civilizations have lost all interest in holding chariot races.

All these considerations seem to indicate that lengths of time are inseparable from spatial categories. As Langevin puts it, "the interval in time, as well as the distance in space will become," as a consequence of relativity theory, "variable with the reference system, that is, with the motion of the observers" (Langevin 1911: 41–42). Absolute coincidence is a dangerous game, however. Some of the most memorable events in races occur when the members of opposing teams come into too close proximity. In the case of racecars, shockingly lurid accidents can happen. One study has shown that non-scripted violent sports are significantly more enjoyable for spectators than scripted incidents. Contingency has an enjoyment value which is yet to be sufficiently understood by psychology (Westerman and Tamborini 2010: 321–337). We crave surprises, collisions, accidents. Is this a product of our fatalistic, perhaps even masochistic subservience to Lady Luck? Or rather, viewed from another perspective, is the enjoyment of violence not rather the sympathetic feeling of a subjective coincidence with causality? The accident or the nonscripted act of violence or the spontaneous sexual act, are all these not instances of what Langevin calls a "double coincidence"? In such cases, the encounter goes too far, resulting in the shattering of integrity. Langevin introduces a thought experiment, in which a car is fitted with a hole into which we may throw multiple objects. The car and the objects thrown are all in motion, compared to both each other and the road. If we throw the two things into the hole simultaneously, "their endpoints coincide in both space and time, it may result in a shock and the breaking of the objects" (Langevin 1911: 42). Langevin maintains that the shock event, as excessive coincidence, is an absolute fact for all observers. Everybody near the scene, stationary observers and the passengers of other cars in the vicinity included, will agree that the collision of the two objects has resulted in their mutual destruction. When things come together at the same point, annihilation is produced. Nevertheless, the time it takes for the breaking to occur will differ for the various perspectives. Viewed from exteriority, duration is a distance mediated by light. The speed of light is "the only velocity that is preserved when passing from one reference system to another and plays in the electromagnetic world the role which is played by the infinite velocity in the mechanical world," observes Langevin (1911: 43). The positing of the speed of light as an absolute reference represents one of the last remainders of Newtonian physics. Replacing both ether and absolute movement, this represents an arbitrary selection, introduced into relativity as a matter of convenience. The result is that every point, world line, and cause is enchained to a certain physical occurrence, namely, the presence or absence of light. A world line is unintelligible without a corresponding light cone. But what of the various occluded and occulted distances? The cosmos of the prevailing Standard Model contains more darkness than light. For our intents and purposes, we may bracket the Einsteinian privileging of light, seeing in it—as Langevin himself suggests—a remainder, an archeological trace, of Galilean-Newtonian mechanics. What is important is that we realize that "the order of succession of two events in time has no absolute sense" (ibid.). The order of causal relations can be reversed, provided the observer changes direction. While objects cannot be brought into complete spatial adequation in space without destroying one another, a temporal coincidence is achievable (Langevin 1911: 45). There is an imaginable minimum spatial distance, the transgression of which results in a shock event. A speed greater than the speed of light could bend causal succession, reversing the progression of events, allowing us to communicate with a past that has already occurred. Each portion of matter or world line has a *proper time* of its own. This latter term we take to be synonymous with Bergson's multilevel concept of duration/s. Precisely which time must be considered "proper" depends on the type of process in question. Time elapses differently for observers situated within a "non-uniform" motion from those situated within a uniform movement. Here we arrive at the crux of the matter. In the case of a uniform reference system compared with a nonuniform reference system,

> The two events considered may be taking place at the same point, in relation to which a portion of matter has traveled a closed cycle and has come back to its starting point thanks to its non-uniform motion. *And we can say that for observers related to that portion of matter, the time period elapsed between the departure and return, i.e. the proper time of the portion of matter will be shorter than for observers who would have stayed connected to the reference system in uniform motion.* That portion of matter will have aged less between its departure and its return than if it had not been accelerating, *i.e.* as if it had remained stationary relative to a reference system in uniform translation. (Langevin 1911: 49–50)

If one observer accelerates at enormous velocity, departing from another observer who remains within uniform motion, then for the former time will dilate. Like a rubber band, temporality is extended by acceleration. The distance traversed by the accelerated observer translates into an overproduction of temporality. Differently put, there pertains *a conversion of spatial acceleration into temporal accumulation.* For the accelerator, the temporal measurements of the observer remaining locked to Earth's regular orbit become inaccessible. Similarly, the latter cannot gain admittance to the time measurements of the accelerated agent. The maximum accumulation of time is achieved by those who are subjected to the greatest increase in speed: "Those have aged the least, for whom the motion during separation was most distant from uniform, and who have suffered the greatest accelerations" (Langevin 1911: 50–51). Aging is a matter of movement. Time measurements are only ever snapshots of the velocities. Two events can become, at least in theory, mutually desynchronized. Taking a space traveler accelerating to something approximating the speed of light. Langevin reckons that an enormous gulf will open between her time and the time of her companion who stayed on Earth. A significant difference will become apparent upon the former's return to Earth. The exact size of the temporal disparity (two years for the space traveler equal two hundred "Earth years") is a concrete prediction on Langevin's part. If a spaceship were to approximate a speed "one twenty-thousandth less than the velocity of light," such would be the size of the temporal disparity (Langevin 1911: 51). Readers are even provided with a calculation relating the amount of energy required to build such a spaceship! However hare-brained all this may sound, Langevin's thought experiment nonetheless provoked a great deal of controversy. How does the phrase "paradox" come into the picture though? The paradoxical nature of Langevin's example stems from the necessity of choosing the place of observation. Conceiving of two observers as twins, one could say that the space voyager sees Earth as relatively stationary, whereas the sister who stays on Earth can similarly conceive of herself as in movement. As we shall see, Bergson cites the interchangeability of perspectives in relativity theory to attack Langevin's paradox of the twins. If each perspective can be switched with the other, there is no way for us to decide which of the twins is accelerating and which remains locked in uniform motion! Langevin maintains an absolute asymmetry between the accelerating and nonaccelerating twins. The return journey results in a change of reference: "Conditions are reversed: for each of them sees the life of the other as remarkably fast, two hundred times faster than usual, during the year for him to return the Explorer sees the Earth perform the actions of two centuries" (Langevin 1911: 52). Does the seemingly arbitrary introduction of a reversal in the space travel narrative resolve the paradox?

Everything depends on the concept of time we choose, this choice being both physical and philosophical. The Einstein-Bergson controversy hinges upon an issue of continued relevance to scientific debates in the late twentieth and early twenty-first centuries, namely, the topic of the ever more frail borderline between the humanities and the natural sciences. Bergson advocated in favor of an active role for philosophy in scientific debates. This is evidenced by his strenuous critique of both Einstein's relativity and Langevin's thought experiment. On April 6, 1922, in the College de France, Bergson famously confronted Einstein. After the latter had delivered a lecture to the cream of the French intellectual and scientific community, Bergson expressed a few reservations regarding the theory of relativity. Bergson's original train of thought revolves around the interiority of duration. As he states, "each of us feels himself endure: this duration is the flowing, continuous and indivisible, of our inner life" (Bergson et al. 1969 [1922]: 128). Simultaneity pertains not between different places, but different durations. Up to this point, the Einsteinian and Bergsonian positions are in full accord. What distinguishes them is the divergent status they give to the interior aspect of time. Bergson and Einstein alike reject "impersonal time," conceived of as a container of processes, or time of times. Real duration, however, differs from "measurable time" as "the multiple times of relativity theory" are "all far from being able to pretend to the same degree of reality" (Bergson et al. 1969 [1922]: 129). There must be a significant difference between an experience of time and the temporality of a fictional abstraction. It is of significance, often ignored by impatient critics of Bergson, that the philosopher never questioned the import of relativity. This is not a case of a philosopher flatly denying a physical theory. Rather, Bergson is attempting to construct a philosophically rigorous foundation for relativity. "The relativistic point of view," he states in remarks addressed to Einstein, "does not exclude the intuitive point of view, and even necessarily implies it" (Bergson et al. 1969 [1922]: 130). Simultaneity necessarily implies an instantaneous perception, as well as indivisibility. In cases of causal coincidence, the connection between observer and observed we can speak of both an indivisible act of observation, and a recognition of multiplicity. The scene of a casual relation is "given intuitively," as a simultaneity available to access. Because of our status as intermediate-level observers, the entirety of the world cannot be simultaneous. An "absolute simultaneity" is a projection of consciousness, a fiction obtained by extrapolating our localized experiences of simultaneity. Were we to be somehow equipped with a "superhuman consciousness, coextensive with the totality of things," an absolute simultaneity could be realizable (Bergson et al. 1969 [1922]: 131).[6] Our scale prevents us from ever reaching such a state of complete "coextension" with all times and spaces. The philosopher's agreement with the Einsteinian outlook is evinced at several instances, for instance: "Simultaneity contains nothing absolute" (ibid.). In a fundamentally realist vein, Bergson argues for a non-conventionalist view of time. It is not the correlation between two time measurements which really matters when we talk of time, but that "between an indication of a clock and the moment at which one finds oneself, the event taking place—something, finally, which is not the indication of a clock" (Bergson et al. 1969 [1922]: 132). It would be all too simple to misinterpret this statement as implying some type of subjectivism or idealism. The introduction of "microbe clocks" on Bergson's part into the discussion allows for the absence of any consciousness. Events and coincidences are still in the making in the absence of a conscious, reflexive observer. The point is not that time is manufactured by organisms, as if experience had any primacy over measurement. What Bergson is attempting to remind listeners of is the important difference between real time (the interior time of duration/s) and abstract, formalized, in a word, measured time. The moment is irreducible to any measurement or convention. Specifically, no clock can exhaust the contents of a duration. Against the clock, Bergson posits the irreducibility of time to measurement. The object of contention here is not a completely unfounded separation between subjective and objective times, so much as a recognition of the double-sidedness of temporality in its *objectivity*. In a brief reply, Einstein pathetically misses the point of what Bergson is communicating. The physicist conceives that the philosopher is trying to defend the primacy of lived, experienced time. Denying the ontological validity of "psychological time," Einstein asserts that "there is no philosopher's time; there is only a psychological time different from the time of the physicist" (Bergson, Einstein and Piéron 1969 [1922]: 133). Erroneously, Einstein seems to think that the "interiority" of duration refers exclusively to the interiority of a subjectivity. As David Scott observes, commenting the April 6 encounter, "Einstein (. . .) misread Bergson's intention" by "assigning the being of duration primarily to be the psychological" (Scott 2006: 191). If anything, this grave misunderstanding stems from Einstein's equation of inner duration with the subjective experience of time. Bergson's radically realist relativism was lost in translation, with disastrous consequences for the relationship between science and philosophy.

Ilya Prigogine expresses a widely held consensus in the following remarks: "Einstein gave a presentation of his theory of special relativity, and Bergson expressed some doubts about it. It is true that Bergson had not understood Einstein. But it is also true that Einstein had not understood Bergson" (Prigogine 1999 quoted in Canales 2015: 350). However provocative it may seem, we assert that the misunderstanding was not actually mutual. Bergson understood full well the momentous import of relativity theory. The misunderstanding lay solely on the side of Einstein. Our hypothesis is that the philosopher's relativism surpasses the physicist's, in both its scope and radicalism. A change of movement implies a change in time, for each movement is situated within a situation or positionality: "It is not because clocks go more slowly that time has lengthened; it is because time has lengthened that clocks, remaining as they are, are found to run more slowly" (DS: 15). The presence of a clock does not affect the feeling of temporal flow. While velocities relate to one another, the interiority of time cannot be amenable to quantification. Inside of each duration, there is a lack, an absence which lies outside of the coordinate systems. "All that science can tell us of the relativity of the motion perceived by our eyes and measured by rulers and clocks," explains Bergson, "will leave untouched our deep-seated feeling of going through motions and exerting efforts whose dispensers we are" (DS: 34). Each process is a dispenser of efforts, going through movements it itself is producing. None of the coordinates may be counted as being crucial, special, or central. When it comes to causal relations, no independent center is present. A clock need not be a physical instrument. It is, at the very least, a potential access to the surface of a duration, "an ideal recording of time" (DS: 42). The ideality of the "clock" is a synonym for its virtuality. One can even say that measured, evaluated, and observed times all compose the virtual, that is, ineffectual aspect of temporality. Both the moment and the flow are one and the same, two aspects of the same time phenomenon. This double aspect or duplicitousness of time is akin to the alternation between wave and particle. The "simultaneity of the instant" and the "simultaneity of flow" are "distinct but complementary things" (DS: 54). The inner life of a duration, a motion stemming from within that duration and extending out of it, as well as the continuity of the undivided motion in its external effects, all these form a single, indivisible elasticity. As a discrete category, the instant is a construct, useful for computing the simultaneity of a physical event with a time measurement. The continuous however is synonymous with the interiority of duration. In itself, the innerness of time is the flow of things. To each occasion of corporeality, there corresponds a gravity and an attraction. Without the crucial distinction between real duration and logico-scientific construct, Bergson's realism would dissolve into yet another conventionalism. Bergson emphatically does not state that all durations are one. "The universe" only "*seems* to us to form a single whole," for we as finite creatures extend our own durations through acts of sympathy to other, alien temporalities and macroscales (emphasis ours, DS: 45). Without a distinction between real, that is, interior duration, and conventional time, we would be left without any way of identifying the sovereign reality of temporalities, as given to themselves. There is no question of privileging the human subject here. An inner duration refers to the inaccessible aspect of *a* time, "inaccessibility" here denoting the absence of connection among objects and not a subject-object relationship. What we have in the Bergsonian universe is a continuity of qualitatively heterogeneous processes. Time and being alike are "a flow or passage, but a self-sufficient flow or passage, the flow not implying a thing that flows" (DS: 44). The self-sufficiency of the flow not only suggests the absence of substance but also the substance of time as a *firewalled transition*. A real event streams differently, and this difference cannot be translated unproblematically into the strokes of a clock. Truly, the future is also present in a field of sense, as an unbounded yet also timeless intensity. Each motion is a performance from within. To flow is to endure, and contemporaneity means a connectivity with at least one other duration. As Bergson states, "if I draw my finger across a sheet of paper without looking at it, the motion I perform is, perceived from within, a continuity of consciousness, something of my own flow, in a word, duration" (DS: 50). A superhuman awareness would be able to penetrate into inorganic temporalities, experiencing all progressions.

Instead of setting out to legitimate some type of subjectivist idealism, Bergson's intention is to prove the objectivity of movement. In both relativity and Bergsonism, change is understood as the primary component of worlds. "Absolute change" is present throughout the worlds inhabited by objects, and concerns "our knowledge of the interior of things," in the form of "a psychology that reaches into metaphysics" (DS: 36–7). In other words, a psychological intuition of interiority is an instrument and not the end-goal of the philosopher's interrogation of relativity theory. Most of the implications of relativity are, for that matter, fully accepted in *Duration and Simultaneity*. The "reciprocity of displacement" on the part of objects coming into contact in causal relations "is actually given" (DS: 33). There is no recourse here to some vague, virtual power of the past. Everything pertains to the actual, to which corresponds past and future. Scott describes "clock time" as "the deficient form which modernism gives itself as temporality. Looking at the clock and orienting oneself toward time reduces time to the 'now,' our always awaiting something (to come presently) in the present" (Scott 2006: 191). The goal of Bergson is to uncover the being of time as a simultaneity of qualitative multiplicities. We cannot content ourselves with defining temporality in a conventional or quantitative manner. Enumeration, culminating in a reduction of time's flow to mere succession, fails to exhaust the being of time. Besides the formalized simultaneity of two instants, we have the spontaneous "simultaneity of two flows" (DS: 52). The entire argument hinges around the definition of simultaneity. What does the coincidence between two durations mean? Never can the being of time be taken as given. None of the conventional theories and formalizations are capable of giving us time, for the latter is a real velocity. Compared with the complexity of reality, measurements are motionless. The interior aspect of time does not display any segmentation. Within their monadistic marrows, durations are thick, indivisible unities, sufficient unto themselves. "If a duration has not instants, a line terminates in points," implying that the propagation of movement from a standpoint of reference is not retraceable to any single instant, the latter being "something that does not exist actually, but virtually" (DS: 53). The real enjoys a primacy in regard to the time of the clock. If both past and future are parasitical upon the present, then the same must be said for clock time as compared with the duration of reality. Bergson's argument amounts to an appeal to the incalculability of duration. However close the correlation of the clock and the reality it purportedly measures, two durations will never entirely concur, for the clock has a duration of its own which is distinct from the objects whose times it translates into conventionally constructed units of measurement. The clock itself, or whatever else functions as a time measurement instrument, is just an object among objects. Objective simultaneity "is absolute and has nothing to do with the synchronizing of clocks" (DS: 55). In a 1924 lecture on the ontology of time, Martin Heidegger follows precisely the same train of thought when he remarks that "the clock that one has, indeed, every clock, shows time but only the time of being-with-one-another-in-the-world" (Heidegger 1992 [1924]: 17). Without a togetherness in the world, the clock could never function as a mechanism for representing time. This does not imply that the presence of a subject is necessary for time to "happen." Such a conclusion would be preposterous. Rather, this is a fundamentally relativist recognition that the diversity of objects is the prerequisite of time. Without a heterogeneity, absenting at least two things or processes that bear comparison, we cannot identify any temporal flow. There is no time-in-general waiting to be discovered. If Einstein deformed the straight line into a curvature, then Bergson resolves the line into a set of evanescent smudged points. The postulation of an all-encompassing time originates in an immediate experience of the correlation between two neighboring events. Universal time is a projection of consciousness, erroneously derived from the observation of closely related correlative events. However powerful the hold of such a construct upon our minds, the primacy of real, localized enduring simultaneities ought to be recognized. Such a recognition of the local nature of simultaneity is a relativity which is "more Einsteinian than Einstein" (DS: 56). A stationary point will not do. Neither will an absolutization of time or space be permitted. The movement of a moment ago returns again: time is the place within which the temporal operation differentiates into a "real" flowing and a propagation domesticated, translated, tamed by quantification. Wild operativity, composed of relative molecular motions, is molarized into the being of the parasitical clock. When segregated from its neighbors, it appears that there is no change in the moment, yet intuition shows that within the instant there is unveiled a pure, constant modification. Stationary yet rounded by neighboring qualitative instants, actualization is made of relative motions, transformations which are never entirely equivalent. Each process is fully capable of functioning as a "spectator" whose "completely qualitative duration admits of a more or a less without being thereby accessible to measurement" (DS: 58). The velocity of the actual is superior to that of the virtual, for the latter lacks efficiency of any kind. Relative motions come to a result, only for this end-product to dissolve into either a memory or a retroactively construed futurity.

Underneath the quantitative enumeration of artificial successions, Bergsonism uncovers real time, the radical gravitational acceleration of relations, untangled by intuition into "the simultaneity of flows that leads us back to inner, real duration" (DS: 61). Spatialization will not do, because it threatens to undermine the sovereignty of change in the life of objects. What we seek is a revival of timescales without falling prey to the falsity of locking time inside cumbersome mathematical formalisms. The supposed "incorrectness" of Bergson's mathematical speculations must be viewed in light of the author's entire project as a whole. For instance, the aforementioned Prigogine describes "Bergson's struggle with the Lorentz transformation in *Duration and Simultaneity*" as "pathetic" because the philosopher "completely misses the point" (quoted in: Canales 2015: 350). To this uncharitable dismissal, we must retort that it is Prigogine who misunderstands the Bergsonian project. The point is not the precise reproduction of a mathematical–physical train of thought—although, in our estimation, Bergson passes the test of precise reconstruction too, displaying a keen understanding of what is at stake in relativity theory. Rather, the aim of *Duration and Simultaneity* is a fundamental critique of the monopolization of the being of time by science, as well as a revolt against what anarchist thinker George Woodcock called "the tyranny of the clock" (Woodcock 1944). In modernity the rhythm of life comes to be regulated by an all-encompassing societal quantification of time. As opposed to this reduction, Heidegger at one point in *Being and Time* identifies temporality with the solar principle: "The sun dates the time which is interpreted in concern" (Heidegger 2001 [1927]: 465). Physics is overwhelmingly concerned with an objective definition of time. Ironically, such a striving to get a handle on the physical reality of time drives Einsteinian relativity into a forgetfulness of time's indivisible, enduring being. The accelerations and transformations of real processes cannot remain characterized by their relationships with clocks. Measurement invariably tends to decompose duration into a set of spatialized instances. The instant itself is legitimate, when conceived of on its own terms, as a quality. Actualization is not a lag, but a constant parthenogenesis. However many additional dimensions we may add to our enumeration of discretized instants, theory will never transcend to the significance of durational local simultaneity. Even were we to fill up every single existing particle with "microbial clocks," never could this measurement correspond to the spontaneity of duration. When Bergson says that "it is impossible to distinguish between the duration, however short it may be, that separates two instances and a memory that connects them, because duration is essentially a continuation of what no longer exists into what does exist," he is seeking after a revival of the primacy of the moment, and most definitely is not attempting to reduce the actual to the virtual (DS: 49). Memory enjoys no primacy when it comes to the self-elaboration of process. Consciousness is but one of the many forms duration can take. In the case of human beings, to be sure, it is consciousness that composes the type of presence which "infuses living duration into a time dried up as space" (DS: 60). Space, in this conception, is like a dry riverbed, lacking the cascading, ebbing discharges of movement. The spatial is the virtual reality of actualization. Bergson would have no quarrel whatsoever with Einstein, had the latter not folded time back into spacetime. In Minkowski's original construct, however, the spatial too is nothing but a chaotic medley of worldlines, traversing a diversity of light cones. Physics cannot remain oblivious to the implications of time's own mode of being. Durational interiority resembles a melody or refrain. If each duration were to be treated as an interchangeable, freely convertible quantity, then we would lose the indivisibility which guarantees, on a metaphysical and ontological level, the individuality of each duration. Einstein's gravitational field in turn gifts individuality to endurance on every single scale. Anything whatsoever which takes place occurs in both an individual and continuous manner. Relativity shows the equality, indeed, solidarity, of fields. Occurrence need not be thought of as denoting some correlation with a brain. The error of psychologization lies solely on Einstein's side of the fence, as the physicist has been caught red-handed spuriously attributing a subjectivism to Bergson. The sole content of "immediate experience" is "a continual flow" (DS: 63). Through an accident of evolution, it has transpired that living things are capable of accessing temporal flow through a consciousness of time. Nowhere does Bergson claim though that the being of time is necessarily drawn out or wholly exhausted by the presence of a consciousness. This would be to gravely misunderstand the import of Bergson's realism concerning time. The positing of a time consciousness is a merely provisional formulation on the philosopher's part. What matters is the self-actualization of time itself, and not how it gains registration. No type of clock time, including the biological timekeeping mechanisms of organisms, directly accesses the flow. We have therefore an internal differentiation within duration, between the gliding movement of the duration's sovereign individuality, and a spatio-temporalized instantaneous discreteness. Nothing restricts the self-identity of local simultaneity to time consciousness. Bergson's realism is an affirmation of the inexhaustibility of reality, the resistance of real things which cannot be translated unproblematically back into the language of abstraction: "The theory of relativity cannot express all of reality" (DS: 64). Nothing is capable of giving expression to all there is, for each existent is enchained to a particular set of fields of sense. This observation applies even to this theory itself. To state things in a metaphilosophical vein, the process of interpretation is capable of rescuing a formerly dormant, forgotten philosophy from its consignment to the dreary hallways of memory. Our rereading of Bergson's work is intended as a reinvigoration, a striving to breathe new life into Bergsonism.

AGAINST THE TYRANNY OF THE CLOCK

Temporal processes cannot be traced back to a clock set at a precisely delineated moment. But do not Bergson's own proclamations not contradict our own realist reading? The following sentences, which arguably lend themselves to a phenomenological interpretation, do seem to privilege the phenomenal, experienced realm: "where there is not some memory, some consciousness, real or virtual, established or imagined, actually present or ideally introduced, there cannot be a before and an after; there is one or the other, not both; and both are needed to constitute time" (DS: 65). Surely the above smacks of what Quentin Meillassoux has labeled "correlationsim." Under this term, we should understand "the idea according to which we only ever have access to the correlation between thinking and being, and never to either term considered apart from the other" (Meillassoux 2008 [2006]: 5). A superficial reading of Bergson's statements would indicate an almost caricaturistic correlationism. Is there really no such thing as time outside of consciousness?—asks the realist incredulously. Let us decompose the above sentences into a set of components. We have, first of all, the *absence* of a memory. This posited nothingness is an absurdity on Bergson's own terms, for there is no such thing as nothing. Something is always in the making. Hence, memory, as the parasitical virtuality of the actual, is ever present, in the form of a *derivative* presence. Just as a financial derivative, say, a futures contract, has no independent being of its own, its valuation being tied to what is known as the *underlying* asset, so memory cannot have a substance independent of underlying actual duration. If we have memory, Bergson asserts, we have a *before* and an *after*. Past and future are therefore second-order parasites of a mechanism (memory) which in turn is a derivative of the vibrating present. Consciousness, as a symbol, is but another term here for the actual. This does not translate into the monopoly of consciousness. Without a human consciousness, there cannot be a human time, but other durations flow in different ways. Times in the plural are nothing more than the eternal self-actualizations of moments. All events are equal, in a qualitative and not a scalar sense. Between the underlying, instantaneity of the moment, and the lived experience, there is a spatialization, identified as a practical tendency or instrumentality, of agents. The plurality of times in relativity is, in Bergson's judgment, derived from the heterogeneity of real durations and their causally effective—that is, actualized—interrelations: "a time lived and recorded by a consciousness is real by definition" (DS: 70). From the perspective of radical relativity, this entails that the reality of *a* time depends on the *almost* coincidental *correlation of durations*. In the case of humans, our time certainly is the product of our encounter with a clock and real processes located in our vicinities. Underlying our rhythms, however, are multitudes of smaller and larger durations, extending toward infinity. These considerations cannot be obviated by the charge of correlationism. Force is particular to a certain system composed of at least two durations. A human being can be sliced up into a consciousness of time and a corporeal biological duration, a difference between the experience of time and the feeling of aging. Looking in a mirror, we observe worrying lines traversing our face. We are confronted, suddenly, with evidence of age. Skin has become worn out through overuse. Coinciding with this particular moment, the instant of awareness, there are other torrential durations. On a macroscopic scale, a variety of times are always in motion. Considerations of relativity pertain to all reference frames, including those temporalities which abscond from direct, immediate access. Real time is recorded time, but even the "real" in this sense does not fill the entirety of the actual, the latter being immeasurably more than the sum of virtual instants. Passing through these breakups, we understand that the line dissolves into points, and the points melt into smudges. Transversal lines, defined as fuzzy sets, bear the signature of impermanence. Subjective time experience does not have a monopoly upon duration. Were this so, the charge of correlationism would stand before the court of philosophy. But Bergson fully accepts the "hypothesis of relativity," which consists in "rejecting the privileged system" (DS: 81). It is based on this consideration that Bergson will proceed to deconstruct Langevin's thought experiment. If no systems are privileged, this also means that all positions are interchangeable. Every duration can, at least potentially, when emplaced on the virtual plane, be translated into any other duration. Times intermingle with each other, the breakups reproduce themselves in our conscious constructs, subdivisions proliferating into an artificially generalized, universalized, and axiomatized Second Universe. Gradually, the intuition of indivisibility nonetheless returns duration to its private endurance. Working outwards from within relativity, Bergson will, in essence, attempt a demolition of Langevin's paradox of the twins by means of a radicalization. Real time is the subterranean fountain underlying time measurements, including the estimations manufactured by lived, experienced vital time experience. Even were we to accept a privileging of consciousness, as a correlationist reading of Bergson would suggest, this would at most imply a correlation between a consciousness and its world. Nowhere in *Duration and Simultaneity* is consciousness restricted to an anthropomorphic category. Of late, concepts such as subjectivity have been extended far beyond the human realm. A recent review article posits the presence of something resembling subjective experience in insects (Klein and Barron 2016: 1–19). Even were Bergson to be judged, in retrospect, to have been guilty of a certain strand of correlationism, by no means would it follow that the philosopher is also guilty of anthropocentrism, or any undue species chauvinist privileging of the human dimension. We nonetheless maintain that Bergson, far from being a correlationist, is a realist when it comes to time. Duration is not exclusive to the presence of a consciousness. Even if the past and future were to depend on memory, as Bergson seems to suggest at times, this does not mean that the present can be subordinated to past or future. Rather, these latter two virtualities freeload upon the vitality of the actual moment, defined in the previous chapter as "togetherness."

How then does Bergson dispose of the Langevin thought experiment? Furthermore, does the former's treatment of the twin paradox do justice to the problem? In a sense, Bergson's entire book hinges on the success or failure of his interpretation. As we have seen, Langevin claims that the acceleration of an observer and their corresponding clock will result in a temporal desynchronization. The space traveler, after spending two years heading at immense speed, will return to an Earth upon which two centuries have elapsed. Time, because of the voyager's acceleration, will have become dilated, that is, extended in space. In Bergson's opinion, expressed in the first Appendix to *Duration and Simultaneity*, all this talk of time dilation is a non sequitur. If the interchangeability of perspectives truly holds, as relativity says, then we cannot decide which of the two observers actually accelerated in the first place: "The physicist, going from system to system, will always find the real time in the systems in which he installs himself and which, by that very fact, he immobilizes, but will always (. . .) have to attribute more or less slowed times to the systems which he vacates" (DS: 163). Time dilation is an abstraction that does not correspond to physical reality. It is not unlike mistaking the distancing of a person from us with a real reduction in stature. As Bergson maintains, "My distantly-removed individuals are real enough and, as real, retain their size; it is as midgets that they are phantasmal" (ibid.). The physicist who attributes a slowing down to the clock of the space traveler confuses an abstraction with a reality. Inasmuch as the clock of the explorer is in motion, it is real, but it must also necessarily operate in the same manner as any other clock. Inside of the accelerator's duration, we observe the same phenomenon of temporal flow. Peter and Paul are real, normal persons. If Paul traverses a certain distance, he does not become transformed magically into a person of reduced stature. Relativity, according to the Bergsonian interpretation, would mean the "infinite multiplicity of imaginary times and a single, real time" (DS: 164). The singularity of a real duration corresponds to the intensity of its remoteness. As opposed to the positing of an imaginary Paul, hurtling across space, we must assume the duration of a Peter, the physicist doing the actual imagining. Paul Langevin extrapolates various constructs and prejudgments into the shape of an imaginary Paul, capable of slowing down the process of aging through traveling at huge speeds. An important truth has been forgotten here, namely, that imagination is exterior to the actualization of real movement. What Bergson's exhortation is directed toward is a recognition of the objective status of durations, as opposed to proceeding as if each abstract, virtual body were equal in effectiveness to actual bodies of reference. *It is a grave mistake to confuse the virtual with the actual.* As a possibility, becoming more youthful through space travel exists. Paul, as an imaginary protagonist constructed by the physicist's imagination, occupies, to use Gabriel's expression once more, a different field of sense from the *real* Peter (who in turn, is also a fiction of the philosopher Henri!). In and of themselves, both the space traveler Paul and the terrestrial Peter occupy the same type of relative immobility: "One cannot move with respect to oneself and, consequently, the physicist-builder of Science, is motionless by definition, once the theory of relativity is accepted" (DS: 167). Both protagonists of this tale are, when compared to themselves, immobile. Therefore, Paul's clock too will measure time in the same manner as Peter's clock. There is, furthermore, a symmetry and interchangeability between the terrestrial and the celestial perspectives. As Bergson underlines, "If I am with Peter, who then chooses himself as system of reference, it is Peter who is motionless; and I explain the gradual widening of the gap by saying that the projectile is leaving the cannon, and the gradual narrowing, by saying that the projectile is returning to it. If I am with Paul, now adopting himself as system of reference, I explain the widening and narrowing by saying that it is Peter, together with the cannon and the earth, who is leaving and then returning to Paul. The symmetry is perfect" (DS: 168). Langevin's thought experiment runs into an insurmountable paradox. We are bound to choose between either Peter or Paul's perspective. There is no way of knowing which of the observers is really accelerating, exactly because inertial movements are interchangeable. Metaphilosophically and metafictionally, we as readers are also compelled to decide whether Paul or Henri, the physicist or the philosopher, is correct. One of the two characters, Peter or Paul, is unreal, or rather, virtual, while the other is real as an actuality. In truth, of course, both of them are abstractions inasmuch as they are the protagonists of a thought experiment. The Bergsonian argument could be undermined by denying the existence of both protagonists. Like an amusing reproduction of a comic book image in which the validity of all characters and events represented therein is obviated by the addition of a sentence ironically denying their existence, we could claim that no lived duration corresponds to either Peter or Paul. Therefore, the experience of the terrestrialized fiction does not actually invalidate the fictionality of the space voyager. In sum, the fictionality of abstraction, identified with virtuality, cannot ever disqualify a real duration. As a science fiction, Paul the space explorer has a type of weakened, spectral temporality. If Peter is the point of reference, Paul degrades into a fictional entity, forever condemned to remain "incapable of reading anything at all; for, insofar as (. . .) he is in motion with respect to motionless Peter (. . .) he is nothing more than a blank image, a mental view" (DS: 170). Never can a temporal construct, a mere convention, abolish the reality of the interiority of a real duration. The innerness of time is something that transcends the validity of a thought experiment. None of the subdivisions of distance introduced by Langevin do justice to the heartlands, the hidden cores of times.

As Miliĉ Ĉapek has observed, because of the philosopher's acceptance of the basic premises of relativity theory—the validity of the Lorentz transformation and the elimination of ether, absolute (i.e., universal) movement and any absolute frame of reference, "it is impossible to see Bergson as an 'adversary' of Einstein" (Ĉapek 1990: 297). Is the highly negative evaluation of Bergson's critique of relativity by certain physicists justified? Ĉapek strongly doubts that this view, dominant in the Bergson- and Einstein-literature until the past couple of decades, is justified, and for good reason. After the publication of the second, 1923 edition of *Duration and Simultaneity*, containing the appendix outlining Bergson's reasons for disqualifying the twin paradox, the physicist André Metz wrote a scathing critique of Bergson's reasoning. The first issue is that the entire book deals with special relativity exclusively, without really engaging the general theory. Relativity is built upon the recognition that, in the case of any object or being, "the flow of time is bound up with the notion of movements" (Metz and Bergson 1969 [1924]: 138). Against the Bergsonian idea of a gulf between clock time and objective, real time, Metz maintains—in a rather positivist fashion—that time measurement does correlate with physical processes. We have the empirical means of knowing the proper times of objects, through such physically grounded modes of timekeeping as the half-life of cadmium ions. These are facts revealed and verified empirically through the methods of physics and chemistry. To dispute this would be to fall prey to an excessive relativism, a dire prospect Metz resolutely opposes: "Not only do we know what equal durations are, but also we possess the practical means of measuring these durations" (Metz and Bergson 1969 [1924]: 140). Any suggestion to the contrary risks reintroducing unfounded philosophical speculation into science. Each object has a proper time of its own. This is the time of the object itself, which is independent of consciousness. Relativity has made possible a subdivision of time into a near-infinite multiplicity of durations. Metz describes the progressive aspect of relativity theory as being constituted exactly by this decomposition of Time into several times: through its positing of "proper time," relativity "introduces reality inherent in beings and things themselves" (ibid.). At the risk of sounding repetitive, where, we ask, is the contradiction here between Einstein and Bergson? After all, already in *Matter and Memory*, written decades prior to the Einstein affair, Bergson states unequivocally that materiality too contains duration! (MM: 247–8). Reducing Bergson to a self-referentially onanistic idealism is nothing more than a grotesquely false parody. Relativity has recognized the "dislocation of simultaneity" which "occurs (. . .) for events which take place at points distant from each other" (Metz and Bergson 1969 [1924]: 143). Ordinarily, we assume time to be a container for all events. The reality is that time, in itself, depends upon the effects of gravitational fields and acceleration. Flow is distorted, as it were, by movement. Or, more radically, one could say that the length of time corresponds to the length and speed of a trajectory. The times of two clocks, the relatively immobile one and that of the space traveler returning to Earth, will run at different speeds, claims Metz. Time is "really physical time, ordinary time, and not a scientific monstrosity created by more or less arbitrary conventions," Metz notes caustically (Metz and Bergson 1969 [1924]: 145). Whatever we may think of Metz's own thought experiment of the moving sidewalk extending to infinity, the takeaway is that proper time is, indeed, proper to things themselves. We cannot brush away the time dilation displayed by the astronaut Paul's clock by relegating it to the status of mere fancy or baseless abstraction. Dilation has an objective status. The length of an object is its "proper dimension," being "an intrinsic property of" the "physical object, just as the proper time (. . .)" is "an intrinsic property of clocks, or of moving objects carrying clocks with them" (Metz and Bergson 1969 [1924]: 149). Again, we discover yet another physicist missing the point, for it is the very contention of *Duration and Simultaneity* that the slowed down time of Paul's accelerated clock does *not* correspond to the proper, objective time of said clock! Not even the computation of time is really at issue. What Bergson is talking about is that computation is never *entirely* adequate to proper time in its absolute mobility, and categorically does not deny the practical utility of time measurement. A proper time is that of a real object which evades complete, unmediated detection. If anything, Bergson rehabilitates the idea of proper time, rather than denying its ontological status, as Metz erroneously implies. There is some truth to the latter's criticism to the effect that Bergson commits a mistake when equating the motions of the two travelers. Coordinates in relativity are more than mere conventions, "they are always the dimensions themselves, as they are actually measured" (Metz and Bergson 1969 [1924]: 151). The retardation of Paul's clock is an "absolute event, independent of the observers" (Metz and Bergson 1969 [1924]: 152). Even were Paul to be absent, the spaceship being left to run on autopilot, none of this would detract from the objectivity of time dilation. An acceleration and reversal of sufficient size would actually result in a deceleration of time flow. The clock aboard the space vessel is objectively slower than terrestrial clocks. The Langevin thought experiment is therefore more than an abstraction, being an accurate representation of a physical truth, as revealed by relativity theory. Movement and time as such have been dethroned, and it smacks of reactionism on the part of Bergson to pretend otherwise. Metz seems to believe, based on a misreading of the philosopher's intentions, that Bergson is seeking to revive the Newtonian idea of absolute cosmic motion. The absolute priority of movement in ontology does not, however, entail the positing of a global, all-encompassing mobility. This seems completely lost on Metz, who persists in reading Bergson as a Newtonian reactionary hell-bent on restoring the Galilean-Newtonian universe. Ironically, it is Metz who argues that relativity is fated to return us to an Aristotelian concept of time! (Metz and Bergson 1969 [1924]: 140–141).

What we can accept from Metz's critique is that Bergson does commit a grave error. No master is perfect. Defending Langevin against the allegation that he ignored the reciprocity of durations, Metz states—from the standpoint of relativity theory, correctly—that "reciprocity is not applicable" to the example of Peter and Paul if we consider Paul's return trip as well, for reciprocity only pertains to inertial movement (Metz and Bergson 1969 [1924]: 155). Until this reversal, Paul is truly engaged in an inertial movement. But the interchangeability of Peter and Paul is eliminated by changing course at the midway point of the space voyage. The swerve eliminates the inertial straight line once more. Viewed in this light, Bergson's error seems an almost fatal one, with dire consequences for the credibility of the entire Bergsonian project, at least when it comes to the status of *Duration and Multiplicity* as a philosophically and scientifically grounded critique of scientific misuses of philosophical concepts. But from the viewpoint of philosophy, the situation appears much less drastic. The only way of knowing whether Langevin's thought experiment holds would be to accelerate, say, a radioactive metal and compare its half-life with that of a relatively immobile piece of the same matter. Luckily for us, an empirical test concerning the phenomenon of time dilation was conducted in 1971 using four cesium atomic clocks transported aboard commercial flights. Their values were tested against one another, returning values that appeared consonant with the theory of relativity. Empirically speaking, Einstein and Langevin's ideas were vindicated with pinpoint precision. But when the traveling clocks were measured against the stationary clocks located at the Naval Observatory in Washington, a very different picture emerged. Even relatively small accelerations were supposed by researchers to effect the cesium. While the speed of an airplane is much less than that of a spaceship, the causal relation can be hypothesized to be analogous to that of Peter and Paul in the thought experiment. What researchers found was that actually it was the time of the immobile cesium clock which was dilated, as compared to the moving clocks. In other words, Paul returned older than Peter! (Unnikrishnan 2020: 1–63. See also: Hafele and Keating 1972a: 166–168, 1972b: 168–170).[7] This empirical proof appears to contradict Bergson, Langevin, and Metz alike. In truth, time dilation, at least in the case of atomic clocks transported by planes as compared with a stationary clock, operates oppositely to that predicted in the Langevin thought experiment. Going faster actually makes our clocks run faster too. Reality is stranger than theory. While acceleration is not reciprocal, as Bergson would have it, neither does acceleration dilate time in the same manner as standard relativity theory states. Neither is Metz's contention that Bergson "is anxious to save universal time" justified, because, as we have had occasion to state repeatedly, absolute Bergsonian mobility itself is relative, a characteristic of particular durations (Metz and Bergson 1969 [1924]: 158). Mobility cannot be globalized in the context of Bergsonism. In all fairness, neither Metz nor other colleagues could have known that Langevin's twin thought experiment can be tested through any empirical means. The first cesium clock, which became the primary frequency standard, was built by Louis Essen in 1955, long after the Bergson-Einstein controversy had been relegated to a mere historical curiosity in the history of ideas. In retrospect, the philosopher's suspicions regarding the twin paradox were borne out, if not precisely in the manner Bergson would have anticipated. In another sense, however, the unpredictability of the evolution of science is something which fails to surprise neither the Bergsonian disciple nor any of Einstein's true adherents.

Metz's critique does not impact the philosophical import of Bergson's teaching, consisting in large measure of begging the question. "The physicist," Metz exclaims, "cannot prevent himself from believing that intrinsic properties of things do exist," falsely insinuating that Bergson somehow denies or forgets the objective, inner nature of time (Metz and Bergson 1969 [1924]: 160). There is more to movement than inertia, this much we grant Metz. None of the reference systems can be conceived of as truly static. Relativity destroys the possibility of immobility, for all is change and transformation, alteration being the metamorphosis of a duration. Bergson himself affirms this when he writes in the conclusion of *Duration and Simultaneity* that "the details discovered and described by physics belong to the thing and no longer to a mental view of it" (DS: 160). Far from constituting an idealist lamentation about the relegation by natural science of subjective experience to an unfounded, illusory epiphenomenon, Bergson accepts the ontologically valid status of objective durations. More than just "real," a duration is also actual. Real space and time, composing an interpenetrating qualitative multiplicity, evade all direct contact, not being "the space and time of any physicist, real or conceived as such" (DS: 155). Time is an excess that escapes experience. If anything, Bergson is denying any privileged access to the being of time in itself, a par excellence realist move! Not even consciousness or the conscious experience of time's passing can bring the entirety of time to the surface of manifestation. Against the primacy of both the clock and the act of observation, Bergson, as a truly radical relativist, is seeking to preserve the interiority of duration. Rarely has the intent of a philosopher been subject to a misunderstanding of such gravity. The time of the real Paul differs fundamentally from that of the abstracted, constructed, cutout version of Paul. Similarly, the duration of the spaceship also differs from any representation we may have of its temporal endurance. As inexhaustible differences, durations are richer than even the broadest, most far-reaching scientific observations or speculations. In the case of the fictionalized observer, Bergson graphically compares the act of abstraction with disembowelment: "He has been surreptitiously drained of his content, in any case, of his consciousness; from observer he has become simply observed" (DS: 103). The precise reason we must transcend mere representation is that the act of reproduction invariably reduces the content of duration to a caricature. If we are to avoid this mistake, we must learn to observe diversity within continuity. All of Bergsonism revolves around the problem of discovering the qualitative heterogeneities within continuity.

Apparent is not the same as intrinsic. Appearance to a scientific register cannot be equated as an absolute fact. To do so would be tantamount to committing what we may call Metz's Fallacy, the confusion of a scientific theory with a concrete reality. The "apparent" and the "unobservable" are neither antonyms nor synonyms, as Ĉapek makes clear (Ĉapek 1990: 308).

Bergson's mistake in the polemic with Einstein lay in occasionally ignoring the objective nature of encounters, and writing at times in a seemingly subjectivist manner. The encounter between two frames of reference is a real causal event. This resistance of the event to philosophical treatment extends to scientific observation as well. Relativity, in the form critiqued by Bergson, retains "of time nothing more than the light-line"; in spite of this reduction to the speed of light, "real duration continues to haunt us" (DS: 123). The real accordance between two reference bodies is not to be confused with the concepts of scientific inquiry. Theoretical physics remains a theoretical application of abstractions, nothing more. Bergsonism, on the other hand, registers the immeasurability and inexhaustibility of real time. As a fundamentally agonistic practice, this philosophy is built on the recognition that none of our modes of representation, not even intuition, can bring every content of the actual to the surface. An actuality is not a surface but a depth. How then can time be posited as continuous while maintaining a plurality of times? Ĉapek argues that the term "unity of time" must be understood not a "metrical sense" but a "broader, topological sense" (Ĉapek 1990: 312). A heterogeneous multiplicity contains many mutually incommensurable elements. Nonetheless, they all comprise a disharmonious whole. Ĉapek proposes bridging the gulf separating Bergson from Einstein through abandoning Bergson's distinction between space and time. Absolute simultaneity is impossible within relativity, because there is no such thing as an instantaneous cut in four-dimensional spacetime. We can only slice the elements of the world at the speed of light, provided we accept the latter as a stable point of reference. According to Ĉapek's estimation, "In *Duration and Simultaneity* Bergson came close to an explicit negation of absolute simultaneity" (Ĉapek 1990: 313). Bergson states explicitly that the only objective simultaneity to be found in the material realm is "the simultaneity of flows" (DS: 61). *Not* flow in the singular, but *flows in the plural.* What prevented Bergson from realizing that this amounts to what Ĉapek calls "the simultaneity of intervals" was the philosopher's fixation upon keeping space and time segregated (Ĉapek 1990: 313). If we excise this from the Bergsonian project, we can achieve a theory of time that merges qualitative change with heterogeneity. In Minkowski's original construction of the spacetime continuum, alternation is not disposed of. What Ĉapek highlights is that conceiving of space and time as a continuity does not imply the elimination of temporal flow or the continuity of duration. The three spatial coordinates, x, y, and z, as well as the temporal coordinate t, are all contained within a world, the latter being "the manifoldness of all possible values" (Minkowski 1920 [1908]: 71). Rather than eliminating or flattening time, what Minkowski actually does is internalize temporal flow. Far from constituting an entrapment of duration within space, the Minkowskian spacetime continuum constitutes the most radical extension of time into the very cores of objects. As Ĉapek notes intuitively, "The union of space and time, proposed by Minkowski, works in favor of time rather than that of space, in other words, that it can be characterized as a *dynamization of space* rather than a *spatialization of time*. This is why the term 'time-space' is probably far more appropriate than 'space-time'" (Ĉapek 1990: 314). However counterintuitive it may appear, the Bergsonian project is compatible with a radical reading of relativity theory. Timespace is the interiorization of modification, the introduction of change into the heart of each thing. Instead of rendering being static, Minkowski introduces us to the *timespace*, defined as the dynamic self-actualization of processes. Both the positivist dogma, which supposes the issue of time to have been settled once and for all by the natural sciences, as exhibited blatantly by the Metz Fallacy, and Bergson's own dogmatic blind spot regarding the absolute separation of time from space, can be excised from Bergsonism without threatening the metaphysical system's integrity. The primacy of the vibrating present, the status of duration/s in the making, these are not endangered by Minkowskian spacetime.[8]

Relativity also affirms the primacy of actuality, as does Bergsonism. Each cone, if separated from its grounding in the metaphysico-logistical privileging of light, becomes an actualization, a line of movement in a world. None of the worldlines can be said to be all-encompassing. Rather, durations are relative, being flows that can be separated topologically. If it is to be rescued, we must perform a transformation of Bergson's philosophy of duration. According to this reinterpretation, the Bergsonian time philosophy can be conceived of as an epochal theory of time. As Elie During emphasizes, "Bergson never considered restoring an absolute, universal Time in Newton's sense" (During 2007: 88). It is not so much a case of universalizing time as retrieving the interior unity of times. During recognizes a far from trivial detail which eluded Metz's attention, namely, that the "absolute mobility" of Bergson corresponds to the unity of *a* temporality, and not *of* temporality as such. In Alfred North Whitehead's philosophy of time, During explains, "duration (. . .) is an intrinsically spatio-temporal concept; simultaneity is indissolubly local *and* distant" (During 2007: 91). The Whiteheadian resolution of Langevin's twin paradox lies in the recognition of duration's inherently distant nature. A duration is never entirely accessed by another level or element of the world. Simultaneity itself is a distant duration. What then becomes of time? What happens to Paul in his spaceship? On the Whiteheadian view, as interpreted by During, "the speed associated with his path through spacetime was diverting motion through time into motion through space, enabling him to save time" (During 2007: 97). Whitehead, in a 1923 philosophical symposium dedicated to unpacking the relation of simultaneity to relativity, proposes a resolution of the twin paradox which builds on the positing of a simultaneously holistic and plural universe. We must begin, argues Whitehead, from unity. "The real diversity of relations of" the twin's bodies "to the universe is the cause of their discordance in time-reckoning" (Carr et al. 1923: 35). There is no such thing as a duration empty of content. Both Peter and Paul are composed of a multitude of other durations, the lengths of which can vary greatly. Both acceleration and deceleration disclose fundamental truths concerning the becomings of actual things. As Whitehead notes, "Acceleration and deceleration (. . .) express an essential fact of the life history of any body, and is not merely an accidental outcome of the arbitrary choice of co-ordinates" (Carr et al. 1923: 40). In a realist vein, we can affirm that it is indeed Paul who is accelerating, growing ever more distant from the Earth, only to turn around and return to this planet. Similarly to Bergson, Whitehead too rejects the ontological validity of time dilation, contending that time will, on the whole, pass in exactly the same manner for the astronaut as for the terrestrial observer. Nevertheless, this in no wise implies the rehabilitation of some "universal" time. With a radical gesture, Whitehead decomposes both twins into sets of diverse, heterogeneous durations: "It is the problem of science to conjecture the characters in the three-way spreads of the past which shall express the dependence of the three-way spread of my present experience upon the past history of Nature. These characters are collections of molecules (called the real star), some hundreds of years ago, and light-waves in subsequent years up to the present time, and finally disturbances in my body. With this point of view simultaneity is the foundation of science" (Carr et al. 1923: 41). What Whitehead suggests here is that the various interpreters of Langevin's thought experiment have been searching for simultaneity in the wrong places. Rather than just thinking of duration as the ipseity of different, spatially distant entities, Whitehead enjoins us to discover the heterogeneity of every duration. Simultaneity is already inside of every duration! This does not mean a breakdown of time into quantifiable, uniform, spatial instants. Durations are components of larger durations through and through, all the way along the *timespace continuum*. Bergson, not unlike Whitehead, speaks throughout his philosophy of "moments" and "states" which are not external to one another. David A. Sipfle's unorthodox interpretation of Bergson as an "epochal" theorist of time is, in our view, entirely justified, for duration is never just singular (Sipfle 1969: 280). There exists, in the processual view, both separation and interpenetration. To use Whitehead's terminology, the time of processes displays both a prehensive and a modal character. Unity/prehension and separation/modality both pertain. In Bergsonism too, there is to be found a recognition of the fundamental import of the qualitative instant. This is pure modification, prior to any formalization or abstraction (Sipfle 1969: 285). The momentum of the qualitative forms a continuity with its components, a prehensive application of reference, irreducible to any inertial frames. Worldlines form a multiplicity of cones already, within the dark interiors of microdurations. Simultaneity persists the cores of becomings. To quote

Ĉapek, "The proper time is (. . .) the causal core of all times" (Ĉapek 1990: 303). Through the intuition of the duration of a moment, Bergsonism returns us to the molten core of time, in the here and now.

NOTES

1. In the latter's affirmation of the idea of the "swerve," we may identify an unmistakable affinity between the Roman Epicurean poet Lucretius and the Einsteinian theory of relativity. The Lucretian world view influenced Einstein's outlook to a certain degree, and the physicist did indeed write a preface to the 1924 German translation of Lucretius's *De Rerum Natura* (Kuznetsov 1987: 67–108).
2. For example, Bergson describes the material realm in *Matter and Memory* in the following way: "Matter (. . .) resolves itself into numberless vibrations" (MM: 209). More on this topic can be read in chapter 4 of this book.
3. We borrow this phrase from Ian Bogost. It refers to the favored method of the great sociologist of science and philosopher Bruno Latour, who regularly maps networks based on exhaustive enumerations of even deceptively inconsequential objects. Every existent is for Latour an "actant" or agent, insofar as it causes changes in the world (Bogost 2012).
4. Although the rather forgotten yet momentously important French sociologist Gabriel Tarde, one of the pioneers of memetic theories, was a contemporary of Langevin. Of more than incidental interest is Latour's vehement advocacy of Tarde against Emile Durkheim. Rather than treating "social facts as a thing" in the manner of the Durkheimians, a Tardean approach, one that Latour approximates with his own Actor-Network-Theory perspective, would consist in recognizing that "all things are society" (Latour 2002: 120).
5. Paula Marchesini has argued that cosmologies informed by physics alone have proven incapable of overcoming inconsistencies. Julian Barbour's relativistic negation of time is incapable of accounting for the observed modifications and becomings of objects. At the other extreme, Lee Smolin conceives of a multiverse of many different universes constantly destroyed and reborn. Existing worlds survive a Darwinian competition for survival, yet Smolin fails to account for why an existing universe must adhere to any stability. In Marchesini's view, Bergson's philosophy can incorporate both absolute change and relativity. A multilevel view of duration allows for the simultaneous existence of moments and alternations. Physics depends upon a rigorous metaphysics (Marchesini 2018: 140–152).
6. An interesting example of such an experiment is provided by a YouTube channel ("V101 Science") that specializes in providing High Definition computer graphics animations of what a human observer would see, were they to descend from a spacecraft onto the surfaces of various planets otherwise inaccessible to human observers, equipped with an unbelievably and improbably resistant spacesuit. We are provided with a vista of horizons that "no living human" could possibly ever see.
7. "Richard Keating was surprised in 1982 that two atomic clocks traveling in opposite directions around the world, when compared with a third that stayed at

home, showed slowing that depended on their absolute speed through space—the vector sum of the Earth's rotation and airplane speeds—rather than on the relative velocities of the clocks. But he quickly accepted that astronomers always use the Earth's frame for local phenomena, and the solar system barycentric frame for other planetary system phenomena, to get results that agreed with the predictions of Relativity" (Van Flandern 1998: 87–88). It is a far from trivial circumstance that the inventor of Cesium atomic clocks (and incidentally a critic of Einstein and Langevin) physicist Louis Essen expressed misgivings as to whether atomic clocks are a suitable tool for proving relativity at all (cf. Essen 1971).

8. Our intuitions on this point are borne out by Michael Hinchliff's brief investigation regarding the relationship of Minkowskian relativity to presentism. Upon first blush, it would seem that presentism is endangered by the absence of absolute simultaneity in special relativity. If an event is distant from us, but accessible to another observer, and we have access to that observer, whom we experience as real, then we are trapped into having to maintain the reality of our fellow observer while simultaneously denying and accepting the reality of the event they are observing. But the presentist is not obliged to accept the transitivity of observations nor the relativity of presentism as a positionality. As Hinchliff writes, "The present can be identified with the here-now or it can be identified with the surface of the past light cone" (Hinchliff 1996: 131). Furthermore, one could counter anti-presentists by saying that it is the theory of special relativity which fits into a broader presentist framework (ibid.). Echoing Minkowski's original presentation, Hinchliff notes in passing the possibility that "an eliminativist answer about the nature of time" could very prove the most fruitful approach to thinking about the metaphysics of temporality. Surprisingly, this accords fully with the Bergsonian project, which hinges on an eliminativism applied to universal categories. There is no "time," no "duration of durations" in Bergsonism. There are durations, which are all, in reference to their ipseities, present. cf. Hinchliff, Mark. "The puzzle of change." *Philosophical perspectives* 10 (1996): 119–136, in particular 129–132.

Chapter 3

Being Becoming

The interpretation of emergents requires a corresponding plasticity in our own conceptual determinations. Intellectual work is a labor of connectivity. The entire project of this book is at once a reconstruction and a recommencement, an updating of the original author's doctrine. We begin with immobility and proceed in the direction of an ever more complex elaboration of movement. Even the smallest elements contain a diversity of perturbations. This applies to philosophical texts as well, and the myriad commentaries they occasion. Intriguingly, interpreters of Bergson have routinely suggested alternative titles more suggestive of the contents of his books. Hence, we here follow the suggestion of J. W. Scott, an otherwise less than sympathetic commentator of Bergson. In a 1913 article, the author recommends two titles of his own for *Creative Evolution* which would better reflect its contents: "Reality as Self-Creative" and "Being Becoming" (Scott 1913: 345). Following Scott, we opt for the latter title, as the essential thesis of both our own work and Bergson's own, the latter being arguably the centerpiece of the entire Bergsonian philosophy, is the following: *Being is becoming*. Change is the sole substance available. Upon reflection, even the most crystalline of diamonds dissolves into a set of processes. Impermanence is the substance of beings. Solidity is but an arrested movement; the unchanging is a mere projection of deluded consciousness, intent upon securing a sure grasp on surfaces in its environment. In truth, the unchanging is what we may call a mind-dependent reality. It is not that it does not exist in any sense whatsoever. Rather, the absence of change is a necessity of practically oriented consciousness. When viewing things as they are, the diversity of perturbations is all that remains. In this chapter, our goal shall be the reconstruction of Bergson's philosophy of organic temporality. Before proceeding, a few qualifications are in order. The question of whether changes in the science of biology over the past century have substantiated Bergson's philosophy or not is a thorny topic that has been broached by several commentators, including numerous authors working in the field of theoretical biology. On our part, we are in agreement with Messay Kebede, who argues that Bergsonism implies an "epistemological diversity," inclusive of both intellect, as informed by the natural scientific outlook—whatever that happens to be at the moment—and the philosophically grounded intuitive perspective. There is no need to reduce "philosophy to science," nor are we compelled to relativize or subjectivize "both science and philosophy" (Kebede 2019: 33). Even if the transformations of the life sciences over the past decades have, in a scientific sense, failed to throw up any definitive empirical proof of a vital principle at work in the universe, this does not invalidate the philosophical significance of Bergson's vital philosophy. We are agnostic regarding the scientific validity of speculative philosophical concepts such as the *élan vital* or "life force." Science has progressed much since Bergson's day. Most biologists today reject any type of teleology on the level of both organisms and their environments. Nonetheless, as Michael Ruse points out in a recent article, some key contemporary representatives of theoretical biology have maintained some limited form of teleology.[1] Although nobody in contemporary biology today takes goal-orientedness, conceived of as an all-encompassing global process, too seriously, because we are here engaged in philosophical commentary, we have greater liberty to diverge from the state of scientifically coded legitimate knowledge. Concepts themselves can remain intact, awaiting new utilizations, adaptations, and applications, long after their scientific plausibility fades. What *Creative Evolution* shows, besides the absence of a permanent existent, is the perpetuity of change. Impermanence, in other words, is the sole invariant aspect of existence. Being is a future-directed intentional becoming. Peculiarly, this yields the result that ideas constitutively open to change can, at times, effectively immunize themselves to alternation. Bergsonism is more than a *Zeitgeist*, for it composes an entire manner of viewing the endurance of a set of worlds. In this regard, no amount of scientific innovation can entirely relegate the sense of change. The permanence of change yields a moment of duration, a convention of heterogeneous instants. In speaking of disorder, we perceive a strangely perfect activity of adaptation. Every form which has been maintained through evolution is a differential presence, referring to both an environment and other vital presences in the vicinity of the organism.

TELEOLOGIES OF LIFE

To say the least, the ambition of *Creative Evolution* is not a minor one. The goal of this classic text is the presentation of a philosophically grounded metaphysical vision of the entire process of evolution. It is no exaggeration to call it a philosophy of life. Several billion years of history have been condensed into a compact book. When reading this tract, we must always keep in mind the limitations inherent to the very nature of such a project of condensation. A great deal of simplification is required if the presentation is not to get bogged down into an excess of detail. At times precariously, Bergson balances between delving into the details of the evolution of certain organisms and the broad strokes of metaphysical presentation. One key issue which does interest us is whether this work can be judged a success from a philosophical standpoint. An issue raised by some commentators has been whether Bergson's philosophy is monist or dualist. As the philosopher (and, incidentally, later the second president of India) Sarvepalli Radhakrishnan maintains, a key question regarding the entire philosophy of Bergson relates to "the ultimate unity or duality of life and matter" (Radhakrishnan 1917: 330). It would seem there is no getting around this dilemma. Matter and life are the same, or not. If they differ, the two must be distinguished in some way from one another. Unless it can be proven that the distinction between dualism and monism is itself a false problem, we are faced with a genuine issue that demands addressing. Therefore a key question of this chapter will be the following: does *Creative Evolution* imposes a dualism of matter versus life, or, alternately, is this apparent dualism a mere appearance? In what direction does Bergsonism enact this transcendence of dualism, if at all? Even were Bergsonism's view of life were ultimately to be found defective in the end, it could nonetheless be posited that it can form a base from which we can ascend once more toward a future philosophical-scientific synthesis, in the manner of a trampoline. We begin then from an empirical experience of change. As Bergson's contemporary, the esteemed biologist Julian Huxley wrote in his 1912 book, *The Individual in the Animal Kingdom*, "the progressive change or evolution (. . .) is one of the fundamental things of Life" (Huxley 1912: 3). It is no exaggeration to say that during the early twentieth century, the impression of a general direction in the evolution of life was treated as a self-evident doctrine. It was thought that there exists a type of continuity among animals and humans, a connection of all living things with one another. The common thread uniting all organisms was thought to be progress itself. In Huxley's words, the discerning mind can differentiate a "progress of continuance on this earth up towards the unattainable maximum of the undying" (Huxley 1912: 25). The metaphors used in scientific communication are never just superficial nods to cultural or social standards or mores. Rather, they inform scientific debates in important ways. At the time, it was a widely held consensus that life is heading toward an ever-greater differentiation. Echoing Bergson, Huxley speaks of "the upward progress of terrestrial life towards individuality" (Huxley 1912: 26). On this view, evolution can be conceived of as a progression, tending to the ever-greater complexity of individual organisms. The mental life of human beings is a great deal more complicated than that of previous animals, and the whole of evolution can be identified as a movement tending toward the creation of ever more individuated persons. The whole story of life boils down in Huxley's speculation to an almost Gnostic process of liberation from entrapment within matter. First, the most primitive organisms are composed of "substance without individuality." Second, multicellular creatures display "an individuality co-extensive with" their "substance." Third, in humans, consciousness becomes unmoored from its enchainment to matter, leading to "an individuality still tied to substance but transcending it in all directions." And finally, there can be posited spiritual beings that have completely separated themselves from embodiment, having "become an individuality without substance, free and untrammelled" (Huxley 1912: 31). This latter stage would be something outside what we call "human." Perfected individuality would, according to this teleology, correspond to the complete absence of a material body. At the height of its development, the body discards itself. Progress is a tendency heading toward the utter wasting of materiality. In hindsight, we can see clearly that the mistake made by Huxley and many other biologists at the time was the projection of contemporary cultural values onto scientific facts. Indeed, all of Huxley's works are dedicated to fitting empirical facts to his proposed equation of evolution with progress. To change means, in the long run, to improve. As Ruse has commented, Huxley and most other biologists in the late nineteenth and early twentieth centuries "saw no essential difference between epistemic norms and cultural norms" (Ruse 2009: 351). If reality is a constancy of change, then this ought to be for the better, otherwise the cultural expectations relating to perpetual improvement would be cruelly dashed. From the second half of the nineteenth century onwards, in the Western world "progress became the philosophy of the day" (Ruse 2009: 244). It is not without some irony that precisely the single factor which contributed to a broad-based decline in the belief in progress, namely, war, is referenced by the progressive evolutionist Huxley. In a key passage, the biologist compares life to war in a more manner reminiscent of a Heraclitus than a scientist. Referring to what the literature now refers to as "evolutionary arms races," Huxley compares the competition between predator and prey species to the military buildup of rival nations, a process that was already in progress during these years. If "one species happens to vary in the direction of greater independence," gaining an excessively large advantage over others, the interspecies "equilibrium" is "upset," leading to the spontaneous evolution of modes of defense (Huxley 1912: 116). Although the guns of naval vessels have been increasing in size and destructive capability, the armor of battleships has also been getting a great deal thicker. It is as if the *Stegosaurus*' scales and spikes were a reaction to the lengthening of the *Tyrannosaurus*' teeth. Through this dialectical relationship between predator offense and prey defense, the progress of evolution is driven forward: "Each advance in attack has brought forth, as if by magic, a corresponding advance in defence" (Huxley 1912: 115). This optimism was, to a large extent, vitiated by the events which transpired during the course of World War I. Far from leading to more progress, the technological advancement of weaponry led to a state of static equilibrium. Instead of the fast-moving warfare visualized by strategists prior to the conflict, combat was, for the most part, restricted to the stasis of trench warfare.[2]

Does Bergson commit the same error of projecting culture onto nature? In other words, is his concept of evolution teleological? This question can only be addressed by a thorough reading of *Creative Evolution*. Each commentary is itself invariably a work of novel creation, the manufacture of novelty from a momentary situation of correlation. We correspond with a temporally distant text and, building from this contact, enrich our own postulations and distinctions. We commence from the circumstance of change, as felt from our situation within the interiority of a subject. "I change (. . .) without ceasing" but does this entail a shifting between states? (CE: 3). Bergson does not believe this to be so. "Change," he writes, "is far more radical than we are at first inclined to suppose" (ibid.). We are compelled to always go further, questioning the integrity of all conditions, states, positions, and situations. Modification means something more radical than a mere transposition between two happenings or scenes, a transition from one immobility to another. If we observe a state in itself, we find plasticity everywhere. As we have seen in the previous chapter, the point is always already a smudge. Were we to stop progression—we refrain from writing "progress" here—then we could not gain access to anything remotely resembling duration. As Idella Gallagher observes, "Duration is all of one piece," in which "the whole of the past is accumulated and preserved and borne along with the present moment" (Gallagher 1970: 19). That is, the whole of *a* past, we hasten to add. Each state is a ripening, a swelling, or accumulation within a certain present. The past is given inside of its corresponding moment. Just as the point is inseparable from its smudges, so the actualized duration cannot be segregated from its broader history. Memory is endogeneous to duration, just as virtuality is hosted by actuality. The privileging of the durational moment by no means excludes change. *The mark is already a swerving, the fruit is a fruition.* Each state, be it conscious or unconscious, organic or inorganic, "is continually swelling with the duration which it accumulates: it goes on increasing rolling upon itself, as a snowball on the snow" (CE: 4). As Georges Mourélos makes clear, creative evolution proceeds "in the present" (Mourélos 1964: 229). While analysis distinguishes between a range of emotions, such as fear, envy, joy, and sorrow, these are in truth but artificial constructs that do not give back the entirety of the snowball's accumulative movement. Right now, the snowball is of a certain shape and diameter. But in a moment, it shall differ. Differentiation can only ever accumulate within the context of a certain actuality. Elsewhere, in an unrelated text, Gilles Deleuze writes, in an apparently Bergsonian vein, of how "concealed things unceasingly accumulate and grow larger like a black snowball" (Deleuze 2000 [1964]: 78). This description is telling, for a blackened snowball graphically displays the fundamental impurity of becoming. Snowballs become darkened by mixing snow with muddy, dirty land. No state is entirely separable from the other. When viewed analytically, each feeling differs from the other. Surely sorrow cannot be the same as joy. But how can we square this observation with the widely held neuroscientific view that, to quote a textbook, "all emotional elicitors activate the same neurophysiological structures (. . .) and all produce the same arousal condition"? (Lewis and Michaelson 1983: 104). Bergson's reasoning differs from contemporary neuroscience, the latter being founded upon the view that mental states can be brought into a correspondence with neurophysiological data, yet some evidence does point toward an agreement with neuroscience on this single point: close analysis of emotional states reveals their underlying indistinguishability. True, in Bergson's case, this is more of a premise than an empirically verified fact. Be it a material circumstance or a psychological datum, "the state itself is nothing but change" (CE: 4). What Bergsonism does is eliminate the distinction between constancy and inconstancy. No permanence is to be discovered, change is the sole substance of things, yet this impermanence displays an endurance. Insubstantiality entails a radical permeability between the entirety of pasts and futures. The actual moment is the sole effective temporal mode, but this must not be understood to mean that memory can be separated from the present. The virtual entwines with the actual. It is for this reason that, to quote Elizabeth Grosz, "becoming" is "open duration" (Grosz 2004: 156). Both past and future are impotentialities, as compared with the vibrancy of the present, yet no power can be imagined without also positing a void of powerlessness. By consequence of this original hybridity pertaining among the states and modalities of becoming, we cannot assent to Deleuze's characterization of the present as "only the most contracted level or degree of the past" (Deleuze 1997 [1966]: 91). Such a description entails a needlessly redundant "backwardation" of Bergsonism. If the present were but another form of memory, then duration would lack any means of transcending itself. Thrusting into the future would be an impossible feat without an admixture of durations. The past is a composition of various former presents, all struggling to permeate this actual present, while the future is an invasion of presents, fighting to burst forth from the stomach of serpentine duration. Were we to follow Deleuze's erroneous subordination of the present to the past, the snake of time would suffer from perpetual indigestion. Unable to consume previous novelties, it could never be ripped apart by the triumphant advent of an unprecedented emergence. The present is, in truth, always more than mere potential. Actuality pierces through the impotent scales of virtuality.

Relationality produces a range of spins, currents, and flows. The doctrine of insubstantiality would hold that there is movement without any mover. In the words of Bergson's 1904 text, "Introduction to Metaphysics," "movement does not imply a mobile" (CM: 173). The objection could be raised that it is nonetheless possible to differentiate a range of states, say in terms of their intensity. For example, the freezing point of a liquid would be a definite, measurable occurrence differing fundamentally from other temperatures. A liquid material subjected to a change in temperature can be said to endure through its modifications. For those who believe in substance, this would entail the necessity of there existing an objective essence or substance guaranteeing the persistence of the chemical in question. Just because it has solidified, the material is there, only in frozen form. Surely, it remains intact and its changes are merely accidental elements additive to its substance. The thing itself, for the objectivist, endures. A fear of those who adhere to individual substances persisting independently of interobjective relations is that if we reduce things to processes or actions, their individuality will be in some way hampered or dissolved. As Graham Harman sees things, the danger of a relational ontology is that substance can potentially be undermined by "granting preeminence to a massive system of relations in which all objects are stationed" (Harman 2002: 227). If we are uncharitable to Bergson, we could say that the individuality of living beings is degraded by their subordination to the larger process of life, organisms being but mere instances of the self-elaboration of the *élan vital*. Beneath relational ontology, we therefore have the bugbear of holism, the reduction of every single thing to an all-encompassing unity. According to Harman, the best guarantee against such a move is to maintain the status of entities as "both withdrawn from all relation," themselves being "composed of relations" at the same time (Harman 2002: 283). Hiddenness of individual execution is the guarantee of individuality, preventing processes from being reduced to one another. Does Bergson subsume individual durations within the immense temporality of evolution? On our part, we fail to see why process philosophy should automatically entail holism of any kind whatsoever. One can maintain that there is nothing but movement without thereby subsuming all movements to the self-elaboration of a single global mobility. We hope to prove that *Creative Evolution* contains an essentially irreductionist ontology, constituting a process philosophy that refrains from holism. Not unlike Harman, Scott holds that Bergson specifically affirms the individuality of the various tendencies of living beings, only to separate movement in its pure form from actual things. Real species are instances of stoppage, immobilities situated within the impetus of evolution. Movement alone is real, but if this is the case, then the individuation of species represents a halt and hence, an unreality. As a consequence, nothing in particular can be said to be authentically real. Scott alleges that Bergson's concept of evolution ends in complete pessimism: the separation of life forms makes them ever less real, separating them from movement in its pure state. The absence of any substance apart from change undermines the real status of individual changes, leading to the recognition that real, existing things are absent. Every individual is emptied of its being by the absolute affirmation of movement as the "ultimate reality" (Scott 1913: 356). In spite of his assurances to the contrary, Bergson actually represents the penultimate form of pessimism, as there is no way left for us to reconcile relative immobility with the circumstance of absolute, global immobility. There are many confusions surrounding process ontologies. In the two examples cited above,[3] we find common misconceptions which are worth addressing in the context of this chapter. The first is the idea that the privileging of processes and relations involves a slippery slope, leading inevitably to the negation of individual objects. Without individual substance, the very basis of individuation is supposedly endangered. The second, not entirely unconnected view, as expressed by Scott, holds that the privileging of movement as a general principle leads to a type of pessimism which, in the final instance, negates particular entities, inasmuch as the latter are relatively immobile. If "reality is a continuous undifferentiated movement," then a frozen, solid liquid has lost something of its status as a real entity (Scott 1913: 355). Similarly, a species which has, for some reason or other, stopped mutating, has also been interrupted, and must thereby be excluded from the class of really existing living things. If process is all there is, this leaves us with no manner of filling in the blanks between a mobility and an immobility. In Bergsonian cosmology, a differentiation would be, in Scott's interpretation, something of which "nothing can be made" (ibid.). The suggestion here is that the differentiation of species in *Creative Evolution* entails a "progressive sundering" of the life force "into greater and greater discord," implying that nature never was unitary. There never was anything like nature to begin with, if discord is the sole result of evolution (Scott 1913: 357). If this is the case, pessimism also presents itself in the form of the absolute absence of any direction. Not only can particularity be brought into adequation with the universality of modification, but no unity whatsoever can actually be postulated. Bergson would therefore, on Scott's reading, simultaneously undermine the reality of individual species while also managing to destroy the idea of teleology and progress in the universe. Quite an accusation indeed! As we hope to show, in Bergson's system we are not compelled to posit a grand duration of durations. Neither must the constancy of change and the inexistence of stoppage imply the negation of individual objects. It is the absence of permanence that vouches for the perpetuity of change. The discordance of reality need not automatically suggest anything like a pessimistic conclusion, although optimism too would have little place within the insubstantial framework.[4]

Certainly, in some relational forms of theoretical biology, it would appear that a holistic move is being enacted. Robin Durie purports to show a structural similarity between Bergson's project and that of complexity theorists Stuart Kauffman and Brian Goodwin (Durie 2002: 357–383). In Goodwin's case in particular, the process of evolution changes everything "except certain aspects of the organization of the materials, their dynamic relationships" (Goodwin 2001: 36). Impermanence is the law, but certain recurrent structures in both organic and inorganic contexts remain relatively unchanged, repeating during the course of evolutionary history. Certain chemical reactions appear to mimic forms of life, for spatial patterns exhibit similarities in relationships endogenous to the molecular connections pertaining to organic and inorganic structures. Matter crystallizes in analogous ways across the spectrum of life and nonlife. What matters therefore are these "similarities in the relationships," the individual manifestations themselves are secondary (Goodwin 2001: 47). Adherents of substance would be correct in pointing out that such a view does seem to subordinate individual differences to broader, holistic affinities between structures. Our goal here is not to decide whether such a possible critique does justice to Goodwin's view. Rather, what we seek to uncover is whether Bergson's idea of a vital impetus or life force can truly be subsumed under such broad and vague concepts as "holism" or "vitalism."[5] Under "vitalism," we understand any idea that "life is somehow to be understood as possessing a mysterious 'vital force' or 'vital principle,' *apart from the causal, experimental world studied by natural science*" (emphasis ours, Bognon-Küss et al. 2018: 115). The latter point is of particular importance. Most frequently, vitalist positions have been criticized by the mainstream scientific community for their lack of empirical vigor. Supposedly, vitalists ignore or bracket empirical facts, inventing all manner of speculative entities and fabulations lacking any basis in actual scientific research. Most importantly, vitalism is supposed by its critics to be superfluous. It can be maintained that by introducing ideas such as a vital force, we cloud our judgments with words that do not add anything of explanatory power. It must be emphasized that vitalism comes in many forms. As Bognon-Küss, Wolfe, and Chen point out, in this regard we must always speak in the plural: "Vitalisms (. . .) are various and do not necessarily entail metaphysical claims or the positing of immaterial life forces" (Bognon-Küss et al. 2018: 125). At the bare minimum, any position can be said to constitute a type of vitalism that commits itself to "the irreducible nature of living processes to a set of straightforward chemical reactions" (Bognon-Küss et al. 2018: 126). Far from constituting a reduction of individual organisms to a whole then, vitalism can be conceived of as a form of indeterminism and irreductionism. The organism need not be reduced to a set of chemical reactions or relationships. Individuation, as a form of stoppage, is irreducible to relationships pertaining among material properties or elements. In this day and age, definitions of life revolve around "operative criteria" and not essential properties. Science has surrendered any attempt to provide a definition of life that seizes the substance or essence of the vital. The authors of the aforementioned study quote the official NASA definition of life, which is striking for its epistemic and ontological minimalism: "Life is a self-sustained chemical system capable of undergoing a Darwinian evolution" (Bognon-Küss et al. 2018: 132). Goodwin and other complexity theorists would represent examples of what Bognon-Küss, Wolfe, and Chen call "functional vitalism." What matters when it comes to life is the affinity between functions or structures, and not the material composition of an entity as such. Life forms cannot be constructed simply by recreating the primordial soup supposed to have been prevalent upon Earth at the dawn of life. Rather, on the functionalist vitalist view, a certain patterning conducive to the production of further degrees of self-organization is required. Life would be an emergent irreversible pattern irreducible to preexisting material compositions. This implies a thoroughgoing indeterminism. Bergson's entire concept of creative evolution hinges upon the absence of deterministic laws. The living is that which is free and unconditioned. Not being determined in advance is the primary characteristic of all that can be called creative. Durie seizes upon this, showing that indeterminism plays a similarly important role in complexity theory. For example, Kauffman maintains that life can be viewed as an autocatalytic process that does not necessarily require the presence of a genome. Autocatalysis, in other words, as an emergent would predate genetically based reproduction (Kauffman 1993: 329–337). Similarly, Goodwin states in a Bergsonian vein that a "relational order (. . .) cannot be predicted from a knowledge of the properties of the component parts in isolation" (Goodwin 2001: 73). An atomistic decomposition of entities into fundamental properties is what Harman characterizes as "undermining." We can also call this a downward reduction. Such a position is no less threatening to the individuality of objects than holism, for "it cannot account for the relative independence of objects from their constituent pieces or histories, a phenomenon better known as emergence" (Harman 2016: 9). Holism would constitute the opposite extreme, namely, "overmining." Under the latter, we must understand any position which defines the being of objects as the sum of their effects. Overmining is also dangerous to individuality, for "it allows objects no surplus of reality beyond whatever they modify, transform, perturb, or create" (Harman 2016: 10). Again, Harman collapses relationalism into holism. The relational view cannot be prevented from unifying the world into an interconnected whole, resulting in the eliding of difference. If an object is not more than its present effects or relations, Harman contends that we are bound to lose our grasp upon its singularity and objectivity. Does Bergsonism and process philosophy result in this slippery slope? Is the entire world turned into an indifferent mush by processual thinking?

In our view, Bergson's ontology avoids this trap. As we have already shown at numerous points, the multiplicity of durations is a guarantee against untrammeled holism. There is no duration of durations in Bergson's system. Material entities too are endowed with a type of vibratory duration, as Elena Fell has pointed out (Fell 2012: 61). At this stage, we must return to the text itself, to better flesh out our understanding of impermanence, multiplicity, and their relation to the process of evolution. Specifically, we must figure out whether the self-elaboration of the *élan vital* does indeed represent a global, holistic process, or, alternately, if such a view can be reconsidered. But first, a methodological stricture is in order. Impermanence applies to descriptive terms as we all the existents which form the object of description. Jean Gayon warns that any general description of life "cannot be more than a stipulative definition," and it "will always be conventional, and for this reason, the wisest attitude" to follow is "to accept that it is open to change" (Gayon 2010: 236). It should therefore come as no surprise to us that some aspects of Bergson's doctrine appear outdated. This is a natural consequence of Bergson's commitment to integrating the most cutting-edge developments in early twentieth-century biology into metaphysics. What is duration? At the outset, Bergson gives the broadest possible definition: "Duration is the continuous progress of the past which gnaws into the future, and which swells as it advances" (CE: 7). The project of *Creative Evolution* is to give an operative description of duration as it occurs on level of life and matter. But in order to do so, we must first define what these two terms signify. What makes the job of definition somewhat difficult is the interpenetration of states. There is nothing outside of interpenetration. If only that which endures exists in any real sense, this also means that separated entities do not have any substantial being. Is Scott correct in stating that "nature never was a unity" in Bergsonism, for there is no real bridge between the separated (the static) and the interpenetrating continuum (the dynamic)? (Scott 1913: 358). Yes and no. That which does not change cannot be said to endure. Only those things which undergo changes are existent, or executant. In fact, the being of a process is this very undergoing of alteration (CE: 6). This appears to be a tautology. The bridge, arguably is provided by memory.[6] No tendency disappears entirely into evanescence. Starting from direct connection within singularity, the being of things-in-the-making can be extended along numerous lines. The following assertion must be taken in the most literal sense: "Our past (. . .) is made manifest to us in its impulse; it is felt in the form of tendency, although a small part of it only is known in the form of idea" (CE: 8). This sentence is of profound methodological relevance to the project of *Creative Evolution*. After all, Bergson's aim is none other than the revelation of a general tendency of life. Bergson is careful to avoid the charge of vitalism. The goal is not the positing of a nonempirical living energy or vital principle which would in any way substitute for empirical research. Biology as an empirical science revolves around the "small part" of life which is given "in the form of idea." Bergson is interested in the broader tendency that can be extrapolated through a speculative extension of empirically verifiable data. The evolutionary past in general is accessible only through "its impulse," tendencies can be reconstructed solely by a selective rendition of certain facts. An uncharitable commentator would of course characterize this method as a form of science fiction, being the speculative extension of scientific facts in fantastic directions. Why should we believe that the philosopher has any sort of privileged access to the general tendency of life? Before we label it as pseudo-science and reject Bergson's philosophy of life point-blank, we ought to remember that science in general works with "stipulative definitions," which consist in assigning "a meaning to a word, for the purpose of clarifying arguments. A stipulative definition may agree with the common use(s) of a word, but it may also contradict it (or them)" (Gayon 2010: 233). In scientific and philosophical work alike, we are under no compulsion to follow conventional forms of language. Innovation consists precisely in deviating from norms. The modification of existing concepts, or even the invention of altogether new ones, is not, in and of itself, unscientific, if it leads to a simplification or clarification of practice. What matters, from a scientific viewpoint, is whether Bergson's attribution of tendency to evolution does indeed add something to the fields of biology, philosophy, or both. Bergson is adamant that each change is an "original moment of a no less original history" (CE: 9). From this optimistic viewpoint, there can be no such thing as a superfluity. Every novelty is additive, furthering the accumulation of becomings. To exist means to create continuously. Foreshadowing late twentieth-century relational philosophies, Bergson observes notes that "we are (. . .) what we do" (ibid.).[7] A thing is the sum of its effects and behaviors, but that does not mean its individuality can be decomposed into the mathematical sum of its relations. Retrospective recomposition is impossible, because of the unpredictability of evolution. Everything is decided in the actual present.

EVOLUTION AS PLASTICITY

Intellect works by selecting and immobilizing certain aspects of its environment, so as to better utilize such elements. An idea is always a selection, sliced out from a chaotic underlying heterogeneity. According to this constructivist understanding of scientific knowledge, "all our operations on the systems that science isolates, rest in fact on the idea that time does not bite into them" (CE: 11). The scientifically validated extract (at least until its falsification) is treated as unmodifiable by change. Practically, it would be very difficult indeed for us to do things with objects that we cannot differentiate from their broader contexture. Action necessitates that we treat tools as solid, discrete entities. It simply will not do to confuse hammer with nail or to start driving a nail into a metal hammer using a piece of wood. On a more abstract level, scientific research would swiftly become impossible if the basic concepts of science were not treated as evident at a particular moment in time. A radical questioning of our ideas is certainly needed, but routine research generally gets along fine without such radical moves. Bergson does not deny the practical necessity of extracting discrete objects from the flow of change. What his doctrine does negate, however, is the final ontological status of objects as discrete entities. Things have individualities, but only in a temporal sense. To return to our prosaic example, the piece of wood and the hammer differ because they enact different trajectories in time, differing according to the intensities of their respective temporal vibrations. Similarly, scientific concepts and notions also vary in their durability. Some theories are invalidated swiftly, while others persist, surviving even seemingly devastating setbacks. It was thought for a time that life can be reduced to chemical processes and genetic codes. Bognon-Küss, Wolfe, and Chen have registered a revival of interest in such broad definitional issues relating to the ontological status of life. The question of what a living thing is has come to the fore once again, due in large part to newer developments in astrobiology, evolutionary biology, molecular biology, as well as research in artificial life and synthetic biology (Bognon-Küss et al. 2018: 129). The definition of life has become an issue of contention once more. In this regard, Bergson has gained a new actuality. Objects have durations of their own. We refer here to the famous example of sugar melting in water. As Bergson writes, "If I want to mix a glass of sugar and water, I must, willy-nilly, wait until the sugar melts" (CE: 12). This seemingly trivial fact of experience is meaningful, for it shows that material things have durations. The time of a thing is "no longer a relation, it is an absolute" (CE: 13). A thing is what it does, but this temporality itself can be viewed as a type of substance.

Bergson's ontology is revolutionary, for it consists in a reengineering of what substance means. *Impermanence is the substance of things*. This temporalized substantial element corresponds to what Harman characterizes as the "veiled performance or execution" of objects (Harman 2002: 19). That being said, it cannot be maintained in a Bergsonian framework that substance can have any stability whatsoever, if under "stability" we understand even a relative absence of change. Only fictional extracts, as molded by scientific or practical activity, have a relative immunity to the bite of time's fangs, and even these are affected by longer term historical transformations of knowledge and society.[8] Time is not a quantity but a quality. The amount of time it takes for the sugar to dissolve corresponds to its sugary nature: "The key Bergsonian argument in favour of the claim that time is a quality is his proof that processes, taking time to unroll, need exactly that much time, not more or less, for their authenticity" (Fell 2012: 117). What makes the idea of duration momentous is that it objectifies authenticity. A thing exists authentically in the way it *is* because of the specificity of its particular duration. In other words, an object is exactly the type of temporal flow it represents, and not just the amount of time it occupies. To say that sugar "takes" two minutes to melt in a cup of water is an erroneous use of words, for this would be to reduce the quality of sugary dissolution to a quantity. The authenticity of the sugar is its duration. An object exists to the extent that it endures, but this persistence is qualitative and not quantitative. Suppose the sugar-water mixture we have created boils. What happens then? The water has a boiling point, which has been raised by the addition of sugar. It is already qualitatively and quantitatively different from pure water. But from a Bergsonian perspective, the boiling of the water is irreducible to any temporal relation. Its duration will remain the same. Given the same temperature, the duration of the boiling water is an authentic fact, unique to this liquid. Duration is a temporal quality, in fact the sole substantial quality pertaining, albeit differently and individually, in all things. As Bergson declares famously, "Duration is immanent to the whole of the universe. The universe endures" (CE: 14). This must be meant to imply that all durations interpenetrate or affect one another in equal measure. On a faraway exoplanet, WASP-76b, it has been shown that rain takes the form of molten iron. Its surface temperature exceeds 2,400°C, and the planet's close proximity to its star causes almost unfathomable levels of magnetism. These two factors result in a queer condensation of metal in the atmosphere (European Southern Observatory 2020). Apart from the conversation value of the distant exoplanet's sheer astrophysical abnormality, the duration of molten iron rain cannot really be said to affect us, these lonely inhabitants of a relatively marginal planet located in a segment of a mediocre galaxy. Luckily, our duration does not meet with the temporal flow of molten iron rain, except through the mediation of scientific data. Various temporal qualities are indeed interconnected, but sometimes only very tenuously. As Fell reminds us, "Temporal relations with events that are irrelevant to a given temporal process cannot be equated with the qualities of that process" (Fell 2012: 118). This has important, indeed, vital ramifications for the concept of duration. A far-flung exoplanet has much importance to astronomers but is not of much use when dealing with the mundane example of sugar mixing with water. Nonetheless, endurance does—pardon the pun—boil down to the elaboration of a process within a world. Scientific constructs or extracts have a type of second-degree reality borrowed from their correspondences with real-world durations. Ruse, commenting on Bergson's epistemology, notes that existence can be differentiated along the lines of "primary" and "secondary" degrees. Meaning and other constructs are secondary realities, but "to be functional" they "must beat some relationship to primary reality" (Ruse 2002: 283). While this distinction is itself far from epistemically unproblematic, it does correspond in large part to Bergson's ambition. Scientific abstractions and artificial systematic descriptions borrow their reality from real, actualized durations, just like memory borrows its being from the present. The analytical extract is therefore yet another manifestation of the virtual.

At times, Bergson's use of language is far from clear. The commentator must be careful not to be led astray. For instance, it is all too simple to confuse the import of Bergsonian philosophy by characterizing it as a "philosophy of science." Gayon has been careful to emphasize that, in spite of his wide-ranging usage of scientific results, Bergson cannot be considered a "philosopher of science," for science as such does not compose the primary object of study in *Creative Evolution*. Rather, it is reality as such which Bergson is trying to account for (Gayon 2005: 43–59). In Gayon's view, the entire Bergsonian project is a work of spiritualist metaphysics which positions itself, if not against science, then outside or beside the constructs of the natural sciences, while seeking to remain broadly true to the scientific outlook. It is metaphysics, in the sense of constituting a different manner of approach to the topic of reality than the positive sciences. This does not mean that the use of scientific facts constitutes a mere foil, designed to disguise or camouflage Bergson's metaphysical intentions. Data are the ideas from which a tendency is to be elaborated, without taking the given task as already finished. This latter practice would correspond to Herbert Spencer's mistake, which consists in extrapolating the future from already evolved products and organisms (CE: 206). Nothing is closed or resolved, all is in the making. Bergson's project stands out from many speculative philosophical systems (Hegel's dialectic of spirit comes to mind as another example) for its agonistic self-reflexivity. The significance of time implies that any complete, universal definition of life is impossible from the outset: "A perfect definition applies only to a completed reality; now, vital properties are never entirely realized, though always on the way to become so; they are not so much states as tendencies. And a tendency achieves all that it aims at only if it is not thwarted by another tendency" (CE: 16). It would be a grievous mistake indeed to expect the life force to furnish us with a complete definition of what life is. Readers expecting a clear-cut definition of the organic from *Creative Evolution* will be sorely disappointed, this book containing no such thing. What we do obtain is a broad outline of the directions living things have taken during the course of their evolution.

There is a type of complexity at play in the history of living bodies which does seem to differentiate them from inert, material objects. For example, life contains "a diversity of functions," something which cannot be said about crystals or rocks (CE: 15). But, in and of itself, such a functional approach will never exhaust the totality of life nor definitely explain its tendencies from those of matter. Individuality is made possible by the necessity for the organism to separate itself from its environment. But "individuality (. . .) harbors its enemy at home," for each living being is also forced to open itself up so as to reproduce (CE: 16). The necessity of reproduction means that differentiation and accumulation make perfect closure impossible. In many species, the reproductive instinct surpasses even the survival instinct. The will to differentiation trumps the urge to persist. We have all seen in documentary films or real life the manner in which male spiders willingly sacrifice themselves, being consumed by the females they fertilize. Even more strikingly, virgin female velvet spiders (*Stegodyphus dumicola*) not only assist their kin by raising the progeny of other relatives but even allow themselves to be consumed by spiderlings (Yirka 2017). Such forms of extreme altruism point toward the impossibility of defining life through closure alone. Reproduction introduces a hole into life, allowing for the, at times, horrendous blending of organic durations. Accumulation demands sacrifice, in certain cases even on an intraspecies level. The tendency to individuation is therefore limited by the need for reproduction. *Differentiation gnaws upon individuation.* As Bergson holds, "Individuality is never perfect," for it is "sometimes impossible to tell what is an individual, and what is not," but it is far from erroneous to say that "life nevertheless manifests a search for individuality" (CE: 18). Reproduction is strange, for it encompasses a continuity of form, while also making the persistence of individuals impossible. Organic duration is at once continuous and split. Durie for one has provocatively argued for the immanence of discontinuity within the Bergsonian concept of duration. Time always already splits during the course of its self-elaboration (Durie 2000: 152–168). Every temporality is also a heterogeneous collectivity of durations that differ. Reproduction could be one such form of epochal breakage immanent to the broader continuity of the *élan vital*, but there are others as well. Specifically, *Creative Evolution* will be seen to revolve around the manner in which divergent tendencies within the broader current of life become ever more discordant, leading to the greater complexity of evolution and its products.

What is the relationship between individuality and creative evolution? The answer to this question depends in large part on how we interpret the metaphor of "creative evolution." Before inquiring into the nature of individuation in Bergson's philosophy of biology, a few remarks are in order regarding the status of metaphor in biology more broadly. We ourselves share

Ruse's position in this regard. It seems that the use of metaphors in biology is unavoidable. Despite the, at times misunderstandable connotations of certain words, there is a need for "function-talk" in the life sciences, for the disarmingly simple reason that "organisms, unlike planets and particles, really do look as if they were designed" (Ruse 1989: 52). Teleological language is a natural manner of describing living beings, constituting an extension of our own tendency toward artifice. Organs resemble the artificial tools humans have built to help them cope with an otherwise relatively inhospitable environment. Ruse's favorite example, one which recurs in several of his works, is that of the stegosaurus' fins. Paleontologists have speculated for a long while on the reason for the development of these bizarrely shaped elongated bony structures. What the introduction of a restricted teleology allows for is the asking of the right questions. By viewing the fins of the *Stegosaurus* metaphorically as artificial tools or appliances, we are able to unveil their function, thereby furnishing a speculative explanation that can then be put to the test. The fins "look remarkably like the blades one sees in hydro-electric plant turbos, designed for cooling. Could the fins be serving the same ends?" (Ruse 1989: 53). An unfulfilled function calls forth a morphic solution. This is not to suggest that teleology is global. Rather, there is a dynamic process of mutual adaptation in play, between the organism, other life forms it comes into contact with, and their shared environment. Teleology is purposeful, but always on a local level. For Ruse and other theoretical biologists, the universal application of any metaphor whatsoever is a great deal more problematic than localized, circumscribed use. In the case of the *Stegosaurus*, there are clearly defined limits to the goal-based functional explanation. We are dealing with a single type of organ, intent upon discovering the reason it evolved. It is not even an issue of explaining the whole organism itself, or why dinosaurs appeared on the face of the Earth, let alone the general tendency of life. The problem which we are intent upon unpacking here is how the individuation of species relates to Bergson's life force. Oliver Quick is accurate when stating that "the unity" Bergson "seeks is that of a trans-subjective activity," a form of effort which is not particular to any single individual (Quick 1913: 220). The issue is if life is a transindividual current, then how can the development of individual species be squared with the emphasis on becoming and transience? This problem is far from trivial. An adherent of substance like Quick is, for all his conservatism, correct in pointing out that the positing of change on a global level does appear to relegate the individuality of organisms to a type of second-class citizenship. It is difficult for us to imagine an activity lacking a specific identity, a difference without sameness. What would substance look like, if divorced from its attachment to a particular body? From a classical Aristotelian standpoint, such as that represented by Quick, "in the caterpillar and the moth, in the acorn and the oak, we find a real identity," for these organisms display a "subjective vital activity" attributable to living things (Quick 1913: 221). For a substantialist, life must differ in some real manner from nonlife. The organic does appear to display a higher degree of individuation than the inorganic in *Creative Evolution*. But if life and matter alike are nothing but modification, transformation, and mobility, then the whole dualism collapses into an undifferentiated flux. Life at its bare minimum is movement, but so is matter! The full implications of Bergson's radicalism cannot be grasped if we maintain a belief in the discrete substantiality of the vital realm. The error of substantialist critics such as Quick lies in their rigid adherence to a separation among essences. Retracing life to its origin within matter, we find that life is nothing more than "a bare principle of inexplicable spontaneity" (Quick 1913: 224). How then can individuality pertain if change is the stuff of things? The duality of living and nonliving dissolves into absolute becoming. For Quick, the concept of the *élan vital* is disturbing, because of its unusual plasticity. A concept ought not to display such fluctuation: "at once individual and universal, at once original and sustained, at once discontinuous and immanent, causing at once divergence and likeness, a characterless spontaneity itself yet determining the character of spontaneities," the *élan vital* displays a threatening plasticity, one that Quick dismisses hastily as "an abstraction," a queer product of mere "mental gymnastics" (Quick 1913: 225). Substantialism, at least the standard garden variety, cannot reach the depth of what the vital impetus means as long as substances are conceived of as mutually exclusive. Following Deleuze, we must never forget that in Bergson's philosophy dualism is "only a moment," a preliminary "to the re-formation of a monism" (Deleuze 1997 [1966]: 29). The difference between life and matter is a difference of degree. Bergson is leading us back to a certain reunification, although this new unity will be seen to be an open, chaotic, in a word, problematic heterogeneity. The reconstituted Bergsonian unity is an agonistic continuum. Individual bodies are temporary eddies, miniature whirlpools forming within the broader disharmonious current. Certain strands of contemporary biology speak of "morphogenetic fields" that generate certain bodily constitutions. Particular types of living things would then recur throughout the history of evolution, bodies being the products of a preexisting morphogenetic space or, at the very least, being influenced by the structure of such fields (Bizzari et al. 2019: 35–42). Any scission between field and organism is uncertain, making the separation of substances from one another impossible.

There is no differentiation that is not unproblematic, for difference itself is a product of life's problem-solving faculty. What Quick sees as a weakness of Bergson's philosophy of evolution is, in our view, a strength. Either we view the life force as the "universal activity" of a "superconsciousness," reducing the individual organism to an "epiphenomenon" or as a vague aggregate of individual species while keeping their autonomy intact, at the expense of turning the *élan vital* into an abstraction. Bergson "continually seems to halt and oscillate" between the notion of the vital impetus as a transindividual unity or a community of individuals (Quick 1913: 226). In our view, a commitment to absolute impermanence cannot mean a negligation of either collectivity or individuality. In a similar vein, Quick's contemporary Nann Clark Barr, also proved incapable of conceiving a transindividuality which preserves difference on the level of the relativity of durations. Surely, the *élan vital* must be progressing toward the unity of one subjectivity, Barr supposes. Substantialism, in its conventional form, weds itself to an anthropomorphic theism or absolutism: "There can be no mere change without continuity, no direction of movement without a permanent standard, no difference without identity. Life and matter, if they are not after all to be separate entities whose dualism is final, if they are genuinely to unite in function, must be processes constitutive of a Self" (Barr 1913: 651). What neither Quick nor Barr seem ready to accept is the necessity of discord in a system of indeterminism. The Bergsonian theory of evolution is an anarchistic philosophy, because it posits the ever-greater accentuation of discord in the universe. When it comes to transcending dualism, Deleuze's own solution is to introduce the idea of virtuality: "All the levels of expansion and contraction coexist in a single Time and form a totality; but this Whole, this One, are pure virtuality. This Whole has parts, this One has a number—but only potentially" (Deleuze 1997 [1966]: 93). Applied to *Creative Evolution*, does this equation of individuality with the virtual really do justice to differentiation? As Durie holds, with Bergson "time's movement (. . .) consists in differentiation" (Durie 2002: 377). If this is the case, then this movement must be real, effective change. Bergson's critique of retrospectivity would have been for nothing were we to place distinction within the realm of potential, as Deleuze seems to do. Earlier, we have seen that Bergson rejects the idea of potential altogether, for changes create themselves. Difference cannot be prior to its actualization. The difference is created during the course of the elaboration of becomings. True, Deleuze does not equate virtuality or potentiality with temporal priority. The organism, just like the psyche, is characterized by its endurance. If this is so, where does difference emerge? In which direction is the living thing heading? "Like the universe as a whole, like each conscious being taken separately, the organism which lives is a thing that endures. *Its past*, in its entirety, *is prolonged into its present*, and abides *there*, *actual* and *acting*" (emphasis ours, CE: 19). Every type of duration is alive, inasmuch as it persists within a present. The past is preserved, but solely if it proves capable of connecting to an actuality. There is nothing here that would suggest the equality of the virtual and the actual or, worse, the priority of the former over the latter. Everything endures in a present saturated with remnants of the past that have proven capable of surviving the bite of time. If "differentiation is always the actualization of a virtuality that persists across its actual divergent lines," as Deleuze himself admits, then it becomes difficult to maintain the necessity of having to attribute *any* kind of positive, affirmative existence whatsoever to virtuality (Deleuze 1997 [1966]: 95). It is only the presence of *actual* lines of evolution which allows for any persistence of biological time. Virtuality is fatally dependent upon actuality, there being no manner of overcoming this essential asymmetry. Without a trajectory of evolution, expanding in the direction of all hospitable morphological fields, memory cannot persist. The constitutive blind spot of Deleuzian philosophy is its reliance upon a virtuality that is nothing more than a redundant abstraction.

The horticultural example of artificially induced vegetative reproduction in seedless plants is illustrative of the point we are making. Although many plant species can be reproduced in such a manner, resulting in greater short-term yields, such methods are generally deleterious to genetic diversity and plant health (Scarcelli et al. 2006: 2421–2431). There is a lesson to be learned here. In this case, we can view the asexual reproduction of seedless fruit as an example of a virtuality which is ontologically dependent upon the more heterogeneous sexual methods of reproduction through the dissemination of seeds. Vegetative reproduction has no independent plausibility or sustainability of its own. In and of itself, the virtual is asexual, infertile, destined for disappearance. Without an actual, the virtual is dead in the water. Keith Ansell-Pearson's suspicion that "Deleuze has" in a very real sense "reified the virtual by positing it in terms of an independent, albeit simple, power" is entirely correct (Ansell-Pearson 2002: 105). The virtual has no power, obtaining all of its energy from an actuality. Why should one maintain that differences are given "virtually"? The ostensible reason for this would supposedly be the impossibility of real creation. Were individuality preexistent, no process of evolution could be discerned. There would be no actualization without the persistence of virtuality. Hence, Pearson claims Deleuze's equation of differentiation with the virtual is justified; this would be the ontological guarantee of the continued self-elaboration of life (Ansell-Pearson 2002: 110). In this, we beg to differ. The interpretation of Deleuzian philosophy would no doubt necessitate such an emphasis on the ontological equality of virtuality and actuality, but Bergson differs markedly from the Deleuzian conception by prioritizing the latter. There is nothing like a virtuality in Bergsonism: the snowball of evolution always proceeds in the present. When Bergson writes that "wherever anything lives, there is, open somewhere, a register in which time is being inscribed," we ought to take his words literally (CE: 20). The phrase "inscription," with its connotation of handwriting, is of course a metaphor. But the registration of time is nonetheless an endurance within the context of a now. Try as we might, actualism is an inescapable conclusion when it comes to the delineation of evolution's duration. The present forms the environment of morphological alteration. We could even call this productive mode of temporality the timespace of epigenetic transgenerational commerce.

Reality in general, including the development of organisms, demonstrates "a perpetual change of form" (CE: 22). The whole of life is an extension of change into a present. In contemporary biology, dominated by a mechanistic and deterministic outlook, morphogenesis is generally explained through causal mechanisms such as "cross-talk" and "feedback." Structures can be deduced from endogenous relations, mostly within the genome. The morphology of the living is held to be endogenous to a certain genetic structure and its inner relations, and environmental concerns play at best a secondary role. As one recent review of the relevant literature states, it has been found that "long-range pre-patternings," that is, the effects of morphological environments, are not necessarily a "requirement" for "organoid formation," leading to the conclusion that "some aspects of organoid formation may show self-organizing properties," although the whole process of morphogenesis is also influenced profoundly by "cell heterogeneity and patterned gene expression" (Gilmour et al. 2017: 318). From a philosophical perspective, such findings entail two consequences. Firstly, differentiation, even on the biological level, is endogenous to a duration given in a present. Secondly, heterogeneity is always present prior to a new elaboration. Differences among cells must already be present if a new speciation or morphology is to proceed. Deleuze is correct to emphasize that duration can be called living when it manifests as an actualization within the movement of differentiation, but commits a grave mistake when equating this mobility with virtuality (Deleuze 1997 [1966]: 94–95). There is no sense in which the emerging, mutating germ-line is virtual compared to an already present genome. The prolongation of what Bergson calls "organic memory," defined as the persistence of the living, is irrevocably bound to actuality.[9] Evolution as creation involves conceiving of individual organisms as extending backward into immemorial lengths of time. As Bergson writes, "The present moment of a living body" can only be fully explicated through encompassing "the whole of a very long history" (CE: 24). An interesting example of the persistence of traits is atavism. According to one prevailing theory, the existence of the hiccup is a remnant, or "vestigial reflex," that has persisted from the gradual amphibious evolution of fish. Survival on land necessitated an, at first problematic transition from breathing through gills to breathing with lungs, and the hiccup would represent an atavistic trace of the common origin of all complex terrestrial animals in the ocean, being an attempt of the organism to breathe with, by now, nonexistent gills on land (Madrigal 2008). Countless other facts, minute in and of themselves, but momentous when combined, point toward a general solidarity or even symbiosis of all creatures across the vast stretches of evolutionary time inside an atavistically implicated present. Without a doubt, this appears to smack of a type of holism that blurs the boundary between past and present. Our point of contention revolves not around any supposed impermeability of past and present states—no such mutual exteriority pertains—but rather around the relative status of these levels of temporality.

The privileging of the virtual is just not a plausible reading of Bergsonian philosophy. If the future is open, so is the present. As Paola Marrati emphasizes, from a Bergsonian perspective "our concepts" themselves "are framed in an open-ended process, and, as a consequence, they can be enlarged" (Marrati 2005: 1102). Crucially, we regard the Bergsonian idea of an efficient, actual present in the making as amenable to ontological enhancement. The present can be extended to include an endogeneous past subordinate to actuality.[10] On our part, we assert that the difference between concrete, that is, actual time, and abstract, or past time, is crucial to an adequate understanding of Bergson's process philosophy. Impermanence implies the absolute priority of a rolling present. Real time flows in the interval; it is the unquantifiable excess escaping reconnection to any plane. This does not contradict the mutual connectivity between different durations. Particularly because of phenomena such as atavism, the past can be viewed as being "bound up with the present"; in the case of "concrete duration," such an entwinement is very much the norm (CE: 26). But it does not follow from this mutuality or community of past with present that the concrete nature of duration forms anything other than a dynamic, flowing actuality. The challenge of Bergsonism is to imagine change without anything changing, to envision an evolution without any subject or purpose whatsoever. This is indeed a sundering, a laceration of Being into multiplicities. Bergsonism is the decomposition of Being into becomings. Duration right now cannot be anything other than actual, forming "a connecting link," indeed, the sole mode of communication between the flows (CE: 27). The actual is the key, the bridge between processes. Creation could not go forth without a connection between the mutually determining levels. Life is above all a chaotic system of energetic relations, an explosive impulse tending toward the accumulation of energy and the delay of entropy. Bergson's idea of life as impulse "admits of much discord," for each individual organism "retains only a certain impetus from the universal vital impulse and tends to use this energy in its own interest" (CE: 57–8). There is something undoubtedly thermodynamic at work in such a characterization of life as "impetus" or force. James DiFrisco describes Bergson's concept of creative evolution as a "thermodynamic ascent" which "describes the process of organization that subtends the duration of living things" (DiFrisco 2015: 66). Explosive divergence is the norm when it comes to the spontaneous elaboration of life. The interrelatedness of organic things is a retroactive consequence of their perpetual discordance. Life is that which ascends, while matter is the descending motion. The two movements are each other's inversions. This also implies in no uncertain terms that the organic and inorganic have no independent meaning outside of their interrelation. We can also describe Bergson's philosophy of life as an "organology." Many different life forms display anatomical and behavioral similarities. Bergson's own example is that of the eye. The basic structure of visual organs is similar in the case of both vertebrates and invertebrates. Does an organ express a function, or is a function the expression of an organ? This debate between determinism and finalism rests upon a false dichotomy, for in truth, the recurrence of the eye along various lines of evolution is a product of the common origin and unity of living things. Instead of comparing functions with organs, the real problem to be explained is how different life forms give rise to similar characteristics. An organ is the end product of the problem-solving faculty of life in general. The development of an organ is a response to a problem, namely, how to orientate oneself in an environment. This in turn cannot be separated from the agency of matter itself. Light also plays a role in the growth of eyes. Both mollusks and vertebrates "have remained exposed to the influence of light. And light is a physical cause bringing forth certain definite effects" (CE: 78). We find here a dynamic interrelation between life and matter. As Kebede emphasizes, resistance, be it fossilized social convention, the limitations of an overly solidified body, or matter in general, plays the role of a "stimulant" (Kebede 2019: 220). Just as virtuality would remain absent without an actuality it can feed upon, so life could never continue evolving without a matter that resists its ascent. There can be no explosion without the buildup of a previous tension, although this by no means implies that life always succeeds in its quest for uplift. As processes of solidification such as carcinization demonstrate, even relatively mobile and soft creatures can become hardened and, relatively speaking, more inflexible.

What is the place of matter in Bergson's ontology? Barr is entirely on point in her description of matter as "fossilized consciousness" (Barr 1913: 640). Matter is life solidified. Without a prior stoppage and concentration, life could never explode into a variety of directions. Life is activity, a force spreading itself as far as its ecological and bodily limitations allow. The body and its organs are ways of responding to problems posed by environments. Light provokes, so to speak, the emergence of the eye. "Life," writes Bergson, "is, more than anything else, a tendency to act on inert matter" (CE: 107). But this matter, as evidenced by the example of light acting on bodies in the process of growing eyes, is far from passive! Bergsonism most definitely represents a break with the masculinist ideological centering of activity in living bodies or minds, at the expense of making matter passive.[11]

For too long, metaphysics has operated by denying agency to the material realm. In this sense, Bergson can be said to constitute one of the early forerunners of contemporary new materialism, for in *Creative Evolution* matter is ontologically equal to life, providing the source of stimulus for the entire evolutionary process (for more on new materialism, see Coole and Frost 2010; Dolphijn and Van der Tuin 2011: 383–400). Each organ displays, in its own manner, the significance of the interconnection between transgenerational organic memory and materiality. As Pierre Montebello has highlighted, "Life, matter and consciousness are durations," which are always "intertwined" to a greater or lesser degree (Montebello 2007: 97). Unambiguously, Bergson holds that the source of variations among species, as well as the isomorphic similarities among the organs of different life forms, must not be sought in the endogenous structure of individual species. Broader explanations are required. The life force represents a metaphorical resolution to the question of the tension between similarity and individuality. By retrospectively projecting a unity without forgetting that all this fulfills a speculative function, real explanations can be uncovered. It is the "impetus, sustained right along the lines of evolution among which it gets divided," which forms "the fundamental cause of variations, at least of those that are regularly passed on, that culminate and create new species" (CE: 98). Later in his book, Bergson suggestively describes the *élan vital* as a "wave flowing over matter" (CE: 273). This view suggests that the movement of evolution implies an inseparability of its component tendencies, and bears a similarity to the quantum physical idea of wavefunctions as being heterogeneous in composition but unitary in their structure. As Roger Penrose highlights, in quantum physics "wavefunctions have a strongly non-local character; in this sense they are completely holistic entities" (Penrose 2004: 512). Accompanying the inseparability endogenous to the wave of life, we also find a relationship of mutual inseparability pertaining between the *élan vital* and matter. Light accelerates the endogeneous, already present spontaneity of life, defined as the process of differentiation. Montebello reminds us that *Creative Evolution* should be read as composing a part of "a larger movement of dematerialization of matter," Bergsonian philosophy being "a movement of going beyond the image and the object" (Montebello 2007: 97–98). Plasticity is at once cause and effect. Previous differences create spaces for new organs, but difference must also meet with a prior accumulation of tension. "The construction of an eye," writes Deleuze, "is primarily a solution to a problem posed in terms of light" (Deleuze 1997 [1966]: 103). More than that, the problem itself is posed at once by the dissemination of light and the proliferation of life. Without the provocation of matter, the living would never have the opportunity to break out. Life is that element of movement which, upon materializing into a body, is compelled to create forms for itself. The organic is this perpetual amplification or creation of novel forms. It is not so much a case of responding or adapting to an environment, although adaptation certainly plays a role in the process. Rather, living things "respond" to their environments by "building up a machine" made of organs, such goal-oriented adaptation being "not repeating, but replying" (CE: 66). During the course of its "fossilization" into matter, each organism is endowed with a certain intensity of duration, allowing for a dynamic responsivity to its surroundings. A body is not a heaping up of passively received affections or mechanically coordinated responses. Rather, it is an always unfinished machinery, already mutating into something else. For a while, we discern a certain species, but this is only the snapshot of a broader mobility. Ancient forms persist in very distant evolutionary eras, biology choosing to call these atavisms. A telling example is that of the "reptilian brain," a section of the brain, mostly the basal ganglia, which has persisted in relatively unmodified form over a very large period of time. Decision-making behavior depends in large measure on these areas of the brain (Lee et al. 2015: 66–74). There is something reptilian and cold about pragmatic intelligence, for it is a common characteristic of reptile and mammals alike. This persistence of the past within the present of radically different creatures entails a rejection of any linearity when it comes to describing evolution. Neither can a teleological concept such as progress so justice to change. Neither determinism nor finalism as biological ideas can grasp life in its becoming, for "reality," viewed intuitively, "appears as a ceaseless upspringing of something new" (CE: 53). Determinism reduces the living to a set of natural laws, while finalism reduces the organism to function. Both obscure the free, indeterminate, and discordant nature of life. It is because the present is *here* that living beings can engage in activities, striving to free themselves from any resistance whatsoever. Inasmuch as another life form poses a threat or an impediment to a creature, the former too can become a type of "matter." Even an organism's own body can function in the manner of a material obstacle. The material is, for Bergson, anything that presents itself in the form of a resistance or limitation to change, while the living is mutability. Informed as he was by contemporary theories of electromagnetism such as those of Thompson, Faraday, and Maxwell, Bergson's entire project conceives of "a universe of energy, which bursts forth" (Montebello 2007: 92). The organic is the ascending movement of energetic emancipation. Duration, on the vital level, is the time it takes for tension to explode into relaxation, being both "the foundation of our being" and "the very substance of the world" (CE: 45). Given absolute impermanence and mobility, there is no other substance to be found apart from this impermanence. The time it takes for a new morphology to burst forth, spilling out new organs, this is the manifold temporality of the organic.

DISCORD AND SYMPATHY

Each new species is the advent of an original situation. The story of vitality is the evolution of perturbation. A deterministic science will only ever be able to access end results, the solidified, fossilized remains of a former mobility. The plasticity of the individual, as well as the impermanence of the process, necessitates a mode of description which corresponds to a de-signing function. Concepts themselves are subordinated to the same evolutionary pressures as any other intermediary level object. Life is an explosive force unreducible to static categorizations. Bergson impels us to reverse our habits of thought. When it comes to conceptualizing becoming, "we must do violence to the mind, go counter to the natural bent of the intellect" (CE: 35). Intellect is naturally analytical, carving up the world into sets of objects suitable for manipulation. If anything, this has given occasion for confusing Bergson with pragmatism. In a limited sense, his doctrine *is* a pragmatism, for it consists in retracing intelligence back to its function as a survival tool. But at a certain point, intelligence can outgrow itself into intuition. The latter is much more than a merely epistemological instrument. Intuition is a way of seeing things, even more, being together with them while they are in the making. Heterogeneity truly is a real feature of Being; there is no denying that there are countless differences within the vital realm. But when viewed intuitively, we find the borders of supposedly discrete entities break down. Multiplicity and continuity can be made whole again through metaphor and analogy. What metaphors do is create new connections among objects. Following the philosopher of language Max Black, Harman has argued persuasively for this productive power of metaphor. The phrase "man is a wolf" results in the creation of an entire "wolf-system," unleashing new hybrids into the world (Harman 2005: 116–125). In a way, humans really do become wolves after the metaphor is created, provided of course that the entities brought into connection share a certain degree of similarity. There must be something wolflike in human beings if the metaphor is to prove effectual. Good metaphors come equipped with a suggestive power and are productive of hybrid actualizations. In Bergson's case, several metaphors describe the interconnection of tendencies in the history of evolution. Arguably the following section displays some of the most effective metaphors deployed in *Creative Evolution*: "The original impetus is a common impetus, and the higher we ascend the stream of life the more do diverse tendencies appear complementary to each other. Thus the wind at a street corner divides into diverging currents which are all one and the same gust" (CE: 58). For all its apparent simplicity of language, this passage is a great deal more intricate than it seems on first impressions. Life is described here not only as a flow, a "stream," but also a gust of wind. By introducing two metaphors at once, Bergson creates two sets of relations. Not only is biological duration a stream, but the flow itself also resembles a gust of wind. The stream is similar to the wind blowing, and wind bears a resemblance to the evolutionary process as a whole. It could be objected that all this is highly tenuous, even carrying the risk of abstracting or spatializing change. Streams are usually located somewhere, which seems to go against the intentions of the philosopher. The image of life as a gust of wind works better, for it is a great deal more difficult to locate. There is no method of mapping which can entirely reconstruct the pathway of a gust of wind with any certitude, whereas streams and other bodies of water are more easily amenable to a cartographic treatment. After the wind has moved on, it is almost impossible to subsequently retrace its movements. Mutability is an evanescence that cannot be spatialized. Philosophy should not consist in the diagrammatization of reality, because the real escapes any rendition.

The female *Ammophila Hirsuta* wasp paralyzing larvae with its sting will be Bergson's prime example of how instinct operates. The automatic behavior of the wasp is taken to display the mechanically repetitive nature of instinct as compared with intelligence. Indeed, the Bergsonian preoccupation with the uncertain boundary between intelligence and instinct is a salient feature of *Creative Evolution*, but we are interested here with another aspect. The essential plasticity of the living thing is displayed in Robert Frost's remarkable poem, "The White-Tailed Hornet." It is known that Frost read and was influenced by *Creative Evolution*, hence the poem is far from irrelevant to our discussion. B. J. Sokol advises us to "be alert for" Frost's "ulteriority." On the surface, the reader is treated to a detailed entomological description of a real wasp. For instance, we read, "he's after the domesticated fly To feed his thumping grubs as big as he is" (Frost 1964: 360–361). This animal is capable of committing errors, stinging first a "nailhead" and then a "huckleberry." For all the minute observations regarding the behavior of this species, there is actually no such thing as a white-tailed hornet! Among wasps that feed their young with paralyzed insects, it is the female which does this particular activity. There are also no known instances of wasps mistakenly stinging huckleberry fruit in frenzy. Furthermore, a house fly would make a poor choice of nutrition indeed! As Sokol observes, "I suspect he [Frost] may have been aware that there is indeed a white-tailed hawk native to North America and also a white-headed hornet, but no species of wasp or hornet called white-tailed" (Sokol 1990: 49). Frost's wasp is a poetic hybrid composed of two very different animals. What is the poet's purpose in inventing a fictional animal, and then proceeding to undertake a detailed, almost scientific process of observation? Everything revolves around the topic of doubt. Since the advent of Darwinism, human dignity and anthropocentrism have become discredited. In a nostalgically humanist vein, Frost exclaims that "once comparisons were yielded downward,/(. . .).We were lost piecemeal to the animals/Like people thrown out to delay the wolves" (emphasis ours, Frost 1964: 362). Must we view Frost's "disillusion," felt over the dethronement of the human, in exclusively negative terms? Is "The White-Tailed Hornet" intended exclusively to leave a foul taste in our mouths, or is the work more constructive than a mere piece of nostalgia for the good old days of anthropocentric ontotheology, when the world was created for Man, in turn created in the likeness of God? For Sokol, the entire poem is symptomatic of "a feared utter lack of divine caring or moral meaning in life" (Sokol 1990: 54). This interpretation is certainly functional. It manages to furnish an efficient and economic explanation. Instead of reading the poem in purely nostalgic terms, however, it could be equally productive to capture what is at stake here in terms of epistemology. The fallibility alluded to among the final lines permeates the entire poem. In a performative sense, Frost pollutes his text with epistemic uncertainty. For all its concreteness, it refers to nothing in particular, but it could very well be the case, given the fallibility of all human knowledge, that natural science has gotten things wrong, and the poetic description does indeed capture something genuine about some organism out there in the world. More speculatively, a mad scientist, inspired by Frost's poetry, might even undertake to deliberately engineer such a life form, constructing an aggressive white-tailed hornet, naming it *Chrysididae Frostus*, and introducing the artificial species illegally into the wild. Such an eventuality surely cannot be excluded. But neither can the mutability of life itself be predicted with any real certainty. Adaptation is not a mere passive adjustment to a given environment. A living thing or tendency is not an imprint of matter, for it "reacts positively, it solves a problem" (CE: 79). What Frost's poem suggests is that the use of the word "thing" is problematic when it comes to contingently entangled existents such as living organisms. Any and all categorizations are fallible. We find ourselves questioning our sense of reality. How can a poem tell us so much about something that does not exist? The empirical apprehension of a simultaneously real yet fictionalized object constitutes the core of a science-fictional outlook.

As a metaphor, the life force is at once psychological, spiritual, and physical. Bergson is clear that beneath the elaboration of biological forms, there is a type of "effort" involved. But this term should not be used in the everyday sense. Part and parcel of Bergson's method of reversing thinking is the extension of concepts beyond their usual confines. If the process of evolution is to be described as an "effort" of overcoming obstacles, then this phrase must be given "a very unusual extension" (CE: 87). The philosopher, so as to better explicate reality, should not shy away from introducing new usages of prior semantics. In itself, there is nothing pretentious in such an exercise, provided that the new definitions are sufficiently clear for an effectual utilization. The concept is a sign allowing the traversal of a foreign, difficult terrain. Duration itself is also an effort, for "the very act of holding in the present a condensed mode of existence" requires an expenditure of energy (Kebede 2019: 70). In the case of the broad current of life of course, recognition of this "effort" or "act" does not necessarily entail the presence of any underlying subjectivity driving the process of evolution.[12] The prolongation of the past into the present is a spontaneous process, which nonetheless requires activity on the part of individuals participating in this temporal interpenetration. What is to be made of the usage of "virtuality" on Bergson's part in *Creative Evolution*? In our interpretation, virtuality cannot have any effectiveness. It is almost as if matter itself were pure virtuality and life pure actuality. The phrase "virtual" occurs but rarely on the pages of *Creative Evolution*. Later on, in the second chapter, Bergson refers to the similarity between evolutionary tendencies and "psychic states," the latter "virtually" including in themselves "the whole personality" to which they belong. A tendency located in the present contains "a recollection" of "what developed along other lines," but in a "rudimentary or latent state" (CE: 131). From atavisms such as the so-called reptilian brain mentioned above, we can reconstruct an affinity or solidarity between life forms. Reptiles and mammals can be said to constitute elements of the same broad current of vertebrate evolution, but this definitely does not mean that every single mammal ought to be characterized as primordially "reptilian." Instead, an element of the reptilian survives in all mammalian brains. And some human individuals certainly resemble reptiles. Pieter Willem Botha, notorious Apartheid-era prime minister of South Africa, was known as the "Great Crocodile," for his obstinacy and unresponsiveness to demands for change. The imperviousness of this political leader translated into a metaphor that proved effective as an epithet. Returning to the issue of virtuality, for us the difficulty of placing virtuality on a par with actuality is insurmountable, as long as we remain true to Bergson's philosophy. This problem is recognized by Rudolf Bernet, who, regarding *Creative Evolution*, comments upon "the difficulty of understanding how the affirmation of a pre-existing virtuality is compatible with the affirmation of a dynamism of life where nothing is 'given' in advance" (Bernet 2010: 50). Life in Bergsonism is a ceaseless process of spontaneous invention and adaptation, stimulated by both the resistance and suggestive influence of matter. It cannot be a case of life realizing any plan or unrolling a preexisting potentiality. Were this the case, Bergson argues, then we could see ever-greater harmony in the realm of living things, yet quite the reverse is the case. "In communicating itself, the impetus splits up more and more. Life, in proportion to its progress, is scattered in manifestations" which, in spite of their complementary aspects, "are none the less mutually incompatible and antagonistic. So the discord between species will go on increasing" (CE: 115). Instead of any harmonic finality, an unproblematic reunification of beings on a single plane, evolution tends toward the maximization of chaos. The ceaseless becoming of difference is a lawless formability. In Bergson's definition, "life is tendency, and the essence of a tendency is to develop in the form of a sheaf, creatng, by its very growth, divergent directions among which its impetus is divided" (CE: 110). Life itself is a problematic totality of dynamisms. Even an immobile organism such as a plant forms part of a broader mobility. If there is an absolute, it is an agonistic universal, defined as the impermanence pertaining to all things. Nothing lasts forever, each process takes a certain quality of time as its own.

By expressing its duration, an organism strives. *Each act a fruition, each move a-making: a ripening preparing for the taking.* Intuition lends exposure to the absolute reality of movement. The dilemma of just how virtuality is to be actualized is a false problem, for the actual is an absolute release of power, an expenditure of energy. Radhakrishnan cites mobility as a concept cutting across both matter and mind and/or life, making dualism impossible to maintain. The differentiation of life from matter is abandoned by Bergson, because while "mind is change," this does not genuinely distinguish it from the material dimension, because the latter "is also movement" (Radhakrishnan 1917: 331). Bergsonism is a monism that reunites the organic with the inorganic through the affirmation of impermanence. Real potentiality requires a corresponding false impotentiality. What proves irritating in Bernet's discussion is the ceaseless emphasis on virtuality, without noticing the full import of this concept. For instance, Bernet characterizes the *élan vital* as "a network of virtualities," without really explaining why the self-actualizing force of life should be considered virtual (Bernet 2010: 53). Bernet does provide an answer of sorts to the dilemma raised by Radhakrishnan and other like-minded commentators, by recognizing in a fully adequate manner that matter already contains duration, even classifying it as "a fixed duration, a stretched living tension" (Bernet 2010: 54). The problem is that, without doubt driven by the influence of Deleuze's view of Bergsonism, Bernet proceeds to always introduce a loose usage of virtuality at almost every subsequent turn. By the end, our head is swimming in virtualities. They are literally everywhere, contaminating the interpretation of Bergson's philosophy. In the end, it becomes impossible to recover the dynamism of the Bergsonian conception from this commentary, convoluted as it is with an overgrowth of virtualities and "drives." Were he to realize the real meaning of virtuality, which is lack of power, Bernet would have discovered that impotentiality does indeed have a place in Bergson's philosophy of evolution. What Bernet resents is the supposed absence of endogenous limitation in the current of life. "The *élan vital*," he writes, "is a *dynamis* that doesn't know about the fate of the *adynamia*, that is, impotence. It doesn't even know about inhibition. Each of the forces in which the *élan vital* explosively develops is a disinhibited force that is void of inner restraints" (Bernet 2010: 61). This description stands in stark contrast to Bernet's own recognition of the status of matter in *Creative Evolution*! When speaking of life, we must see that the virtual is the impotentiality endogenous to the organic. The virtual is powerless, unable to exert any further effects on its environs. At its core, the life force is a method of liberating energy from circular enclosure. The virtual material is a restraint upon this transformative explosion, and definitely not that which is realized through actualization. The revolutionary promise of Bergsonism lies in the slim possibility that, finally, the dynamic of life can someday completely destroy the mutual implication of the actual and the virtual. At the very least, there is hope for sundering the actual from any virtual encumberments.

Commentators such as Bernet who place an undue stress on virtuality miss the point of what makes Bergson's concept of evolution "creative." A genuinely creative evolution cannot consist in the realization or unfolding of a preexistent or even a simultaneous (Deleuzian) virtuality. It is not enough to say that the virtual is contemporaneous with actualization, as Deleuze and Bernet hold. We are obliged to state that from the continuity of change it follows that the virtual can never be a given. Biological time does not occur as the realization of a project or the unfolding of a potential. Rather, the life force proceeds "like a shell, which *suddenly* bursts into fragments, which fragments being themselves shells, burst in their turn into fragments destined to burst again, and so on for a time incommensurably long" (emphasis ours, CE: 109). From a relative perspective, life can be viewed as the gradual emergence of multiplicity over time. A molecule came into existence somewhere somehow in the primordial soup of the primitive Earth, a thing which proved capable of moving itself.[13] From this simple, inauspicious beginning, all the more complex living systems evolved. This is the usual story, as told by most standard reconstructions of terrestrial life. But Bergson tells another tale. The process, when viewed as a whole through a durational lens, comes to paradoxically resemble something which is usually opposed to Bergsonism. When compared to the exploding cannon shell, the *élan vital* becomes strangely instant-like. Is the tyranny of the moment not what the very idea of duration is resolutely opposed to? That which takes an enormous amount of time when viewed quantitatively is, when perceived from a qualitative standpoint, actually a single duration, one vast moment of creative detonation. Both prospects are correct in their own ways. Just as "effort" can be extended, so the idea of a "moment" can also be expanded to infinity, without compromising its meaning. Doing violence to our own minds does not necessarily result in a defective philosophy. Bernet sees a danger in the ostensibly "uninhibited" nature of life, as conceived in the Bergsonian doctrine: "could not one suppose that at a certain moment it [life] would put an end to its fertile dialogue with matter, and that nothing could hold back its force of self-affirmation any longer, allowing it to go mad?" (Bernet 2010:

60). Self-inhibition is, on this reading, absent in the idea of the *élan vital*. But as we have seen, limitation is in fact already endogenous to the life force: the virtual is a constraint indigenous to the actual. What Bernet describes pejoratively as "madness" we define as relaxation and ascent. Each body is a possibility, a channeling which refers to a brief interval between dissolution back into matter and the peculiarity of self-assembling negentropic rebellion. Matter, as Bernet himself admits, is also life, defined in terms of elementary tension. A body is nothing more than a passageway for the "breath of life," blowing through materially embedded corridors (CE: 111). At this point in time, it seems the breath of life requires an effort. But life is also tending toward detension, a continually recurrent birth of selves into materiality. Put bluntly, life is forceful only inasmuch as it is limited. Force is a direct consequence of the partial riveting of terrestrial beings to their material and structural limitations. Later, Bergson mentions the possibility of a life form not dependent upon such an attachment. If life were "pure consciousness, *a fortiori* if it were supraconsciousness, it would be pure creative activity" (CE: 268). The material dimension is an internal, virtual interruption of this flow. A situation has no objective truth. Neither can any evolutionary tendency be brought into adequation with anything resembling an idea. Bergson is not intent upon manufacturing intellectual constructs. The life force is intended as a performative, inventive use of language, not unlike Frost's construal of a fictive wasp. Almost hypnotically, the positing of a life force actually contributes to a modification of the reader's consciousness. When Deleuze writes that the human, through the practice of the intuitive method, is "capable of scrambling the planes," of transcending its condition, it is this reversal of consciousness is in play (Deleuze 1997 [1966]: 107). Oddly enough, the future is rarely mentioned by Bergson. Most of the talk of duration centers around the relationship between present and past, perception, and memory. In one of these few instances, Bergson remarks in passing that the "future overflows its present, and cannot be sketched out therein in an idea" (CE: 114). If anything, the future bears a greater resemblance to the present than the past, for the latter is ineffectual, lacking in strength, unable to lay a claim to the truth in a now which has become foreign and inhospitable to it. The past depends on the present in a fundamental sense. On the other hand, the future is relatively more open to modification. If nothing further can happen to the past because of its absence, then in the future anything can happen, because it is not yet here—so we suppose it may constitute a positive absence, full of potential. This latter conclusion would, however, be an error on our part. The future overflow is also relative to a present. Without an actuality, the future degrades into yet another virtuality. Tomorrow is a superior power to yesteryear, but try as one might, the addition of an infinity of absences will never yield an effective presence. Everything which acts is a becoming, a coalescence, while intellect "cuts out cross sections of this flow" (Radhakrishnan 1917: 332). In truth, the present is what cuts across both past and future planes. The actual is the coalescence wherein ascending and descending motions meet. The future is infinite openness, but this production gains its contents nature from the openness of *this* present in particular. A moment is a nucleus around which a fringe of parasitical pasts and futures orbit. "Before the evolution of life," notes Bergson, "the portals of the future remain wide open. It is a creation that goes on forever" (CE: 116). Without the present, this creative process could not even begin, let alone persist. Endurance depends on the presence of a passageway through which flows canalize. Flexibility results from the deserving succession of the four fundamental motions, more or less voluminous, viscous, and active. The living is "animated," the creature is animation. Animal life is "mobility in space" (CE: 120). Some organisms appear to be rigidly held in place by certain characteristics, but to live generally means to be the source of one's own motions. Greater control of the body allows for a greater intensity of actuality. Heightened mobility leads to an accentuation of intelligence. The following passage illustrates the actual nature of the animate: "Fixity, in the animal, generally seems like a torpor into which the species has fallen, a refusal to evolve further in a certain direction; it is closely *akin to parasitism* and is accompanied by features that recall those of vegetable life" (emphasis mine, CE: 121). If we seek to know how living something is, we must ask how far it can move or, more specifically, what intensity of movement its endogenous efforts can produce. Immobility is nothing but a mobility that has fallen into a state of virtuality, the absence of movement having no independent reality. That which appears to be at rest is an activity waiting for unleashment.

As DiFrisco has argued, the status of matter as "flux" allows for a reconciliation of the organic and inorganic dimensions of being in the Bergsonian cosmology (DiFrisco 2015: 62). At the most fundamental ontological level, change is all there is. Impermanence applies even to the overtly immobile and unchanging aspects of becoming. Just because all is in motion though, we cannot infer any information relating to the future pathways the elaboration of evolution will trace out. This is a direct consequence of the binding of duration to its own present. An organism exists insofar as it is endowed with a "power to move freely" (CE: 123). This freedom cannot prove effective outside of its now. Marrati's characterization of the actual power of time in the Bergsonian system is entirely precise: "The power of time, its own agency, *is* the open-ended possibility of the emergence of radical novelty. The whole of the universe *is* the open" (emphasis mine, Marrati 2005: 1104). The phrase "possibility" here should not be taken to imply a stable emplacement within a predictable future. Neither does it denote a potentiality. Instead, radical novelty is forever *emerging*. The entirety of the universe is open, for the very reason that it is authentically *here*. Growth is occurrent, synonymous with a regained adaptive autopoietic durational power. *A fruition is in progress. Fertility is law reinstated, through discord.* The power of free movement also corresponds to the degree of enlightenment. As long as an organism is not rigidly attached to its place, it is capable of operating in the manner of a transition system, traversing and scrambling the four directions. Ascending along the continuity of becomings, organisms discover an ever-greater ability to create new movements. To use a simile of our own, life is the florescence tending toward an unlocking of the inner intensities contained by matter. Both are movements, heading in opposite directions. DiFrisco emphasizes that it would be erroneous to describe Bergson's philosophy of life as vitalist, for "in it neither life nor matter is a substance" (DiFrisco 2015: 63). If under substance we understand any type of inert essence, any inner content resistant to impermanence, then Bergson is an insubstantialist. On the other hand, if we define duration as an empty substantial temporality, then Bergson is a strange kind of substantialist who grades objects based on their temporal characteristics. What something is will be found to depend upon the qualitative timespace it occupies. "Life," writes Bergson, "is a movement, materiality is the inverse movement, and each of these two movements is simple, the matter which forms a world being an undivided flux, and undivided also the life that runs through it, cutting out in it living beings all along its track" (CE: 272). Evolution, as emergence, is the continual elaboration of independence. The discord observed in nature is a direct consequence of its contingency. According to Bergson's view, two preconditions are necessary for life to emerge and proliferate: an "accumulation of energy" and "an elastic canalization of this energy in variable and indeterminable directions" (CE: 278). To be classed as living, an object must be capable of producing surprises. Recent developments in the empirical life sciences, such as the paradigm of synthetic biology, point to the renewed relevance of the idea that between life and matter there is a continuity. In Bognon-Küss, Wolfe, and Chen's view, the field of synthetic biology in particular, through the invention of artificial organisms, has contributed greatly to the reinvigoration of an epistemological principle which is par excellence present-bound, namely, that of "maker's knowledge" (*verum et factum convertuntur*) (Bognon-Küss et al. 2018: 132). The life forms presently emerging can only be known once they are already here. Maker's knowledge is a demonstration of the restiveness and irreducibility of the now to any givenness of a past or future. It is not unimaginable that life be free of its concentration in a body. Bergson writes elsewhere, in a manner befitting the best authors of the science fiction tradition, that "it is not even necessary that life should be concentrated and determined in organisms properly called" (CE: 279). As long as it proves capable of connecting to a flow of energy, even a non-corporeal life form could proliferate, introducing novel trajectories into becoming, pulsating its indeterminate actions into its world. The concept of life, if extended beyond what we presently know to be "possible," possibly surviving detachment from the predominantly descending (entropic) motion of matter. At the very least, a purified ascent cannot be ruled out. A pure life would be activity untainted by any obstacles. It is important to see that effort, when viewed intellectually, is at first a struggle, but intuitively speaking, striving "in itself is" also a "letting go" (CE: 232). The resistiveness of matter comes naturally to it, as does mobility to most living things. An organism strives to gain a hold in its environment, emplacing itself in a situation suitable for reproduction. "Vitality" is cosmically expanded by Bergson to include all indeterminacies, incorporating any emissions of contingency whatsoever: "At the root of life there is an effort to engraft on to the necessity of physical forces the largest possible amount of indetermination" (CE: 127). Within any world, or level of duration, if there is indeterminacy at work it can be suspected that a certain vitality is emerging from the impotence of slumber.

From whence does the plasticity pulling the totality of life forward originate? When all is said and done, what *is* the *élan vital*? How can we account for its operations in an empirical manner? Madeleine Barthélemy-Madaule provides a clue when she writes that "the *élan vital* does not mean anything other" than a "dynamic vision imposed by the pace of evolution of the living" (Barthélemy-Madaule 1970: 122). What this choice of words suggests is that the life force as concept is forced upon our thinking from the outside, by the real momentum of changes under way in the organic realm. Reality does violence to our thinking, liquifying it, mashing up our concepts until they are suitable for accepting a process-orientation. While the translation between philosophical concepts and scientific metaphors is never an endeavor without its risks, the authors of a highly interesting 2016 study propose to empirically account for the presence of plasticity in the process of creative evolution by positing the symbiosis of viruses with host organisms. According to this view, "viruses are the essential agents of life," their ubiquity allowing for a role as mediators of communication among the various organic strata (Artuso et al. 2016: 79). Artuso, Roldán, Scolaro, and Carlucci argue that the resemblance of the virus to the Bergsonian *élan vital*, with its ability to traverse through various organisms, insinuating itself along various lines of evolution simultaneously, is more than a mere accident. "Massive viral colonization" would be the empirical equivalent of the life force, the secret driver of plasticity and adaptation among organisms. Indeed, research suggests that many of the most significant functions of living things, including photosynthesis and gene translation, have in large measure been influenced by the infiltration of host bodies by virally introduced genes (ibid.). Bergson himself uses the language of permeation frequently. For example, individuated complex organisms are described in one place as "depositaries" of the vital impetus (CE: 252). The viral relates to the body in much the same manner as life to matter. Viruses and retroviruses alike have been implicated in the development of the DNA replication in general. The resistive power of foreign agents seems to be an important requirement for the quickening of evolution. This insertion results in a creative effort of the organism, to either rid itself of the invader or, failing that, integrate it effectively into its operations. A key phenomenon is that of "attenuation," a process wherein gene coding is halted, resulting in the production of mutations advantageous to either the organism or the virus. If intelligence lies in hesitation, this could be perceived as a subconscious effort of self-preservation on the part of bodies, resulting in a cooperative relationship conducive to the further acceleration of Darwinian competition. As the authors note, "Attenuation is linked to the immunological response elicited" by viruses (Artuso et al. 2016: 81). Another research piece suggests that the introduction of "novel viral genes" into hosts constitutes the basis of genetic material transference mechanisms among eukaryotic genomes (Villarreal 2009: 165–253). Coming up against the immune system, the virus must preserve its integrity, ensuring the conditions of its multiplication without killing the host too quickly. Recent advances in biology show that beneath the apparent distinction pertaining between individuated beings, there is a more fundamental synthetic level of interconnectivity. Infecting mice with herpes simplex virus (HSV), Artuso, Roldán, Scolaro, and Carlucci found that those viral strains proved hardier which had been combined with carrageenans, chemical structures originating in the cell walls of red algae (Artuso et al. 2016: 80). Viral adaptability was enhanced through the complex interplay of calcified algae cell wall substances, viruses, and the reactions of host immune systems. In this regard, evolutionary success lies in the ability of the virus to become integrated in the female lab mouse, without killing the host in a brief period of time, nor being in turn killed off too quickly by the immune system. Beneath their differences, there exists an affinity or community between red algae, HSV, and complex animals. The evolution of complex organisms is inseparable from the evolution of virality. Rapid variability is basically a by-product of infections which have come to be domesticated by organic memory. HSV is, after all, an "ancient DNA virus," the structure of which has modified but little over the course of its evolution, and neither have the heparan sulfate glycosaminoglycans to which they attach themselves when infecting host cells (ibid.). The memory of the virus resonates with the structure of its host, the two then persisting in a state of equilibrium. Two series of repetitions coincide in the context of an impurely coalescent moment. A more recent study suggests that the gene responsible for the self-stabilization of mitochondrial cells in plants originated in a virus (Wu et al. 2020: 1–8). Some have gone so far as to suggest, rather anthropomorphically, that viruses are the "genetic engineers" of the natural world, being the underlying entities responsible for changes in speciation (Witzany 2012). The use of metaphors such as "engineering" is a little troubling, especially considering Bergson's original intention, which was to posit an all-encompassing, yet impersonal explanation for plasticity. But the virus as "natural genetic engineer" must not be taken to imply the presence of intentionality or planning. The activity of the virus is creative precisely because it cannot be foreseen or derived from its prior structure. Such complex results as the emergence of new species or attenuation result from the interplay between virus and host, together forming a community brought into harmony through mutual resonance. The virus must deceive the immune system of its host, duping the latter into accepting it as a functional component of the body, in the final instance transforming into an authentically useful function in its own right. Viruses that have come to be integrated or domesticated by the host play a role in forming immunity against similar viruses. In their summary, Artuso, Roldán, Scolaro, and Carlucci make an astounding claim: "Viruses may be considered as the *élan vital* of creative evolution" (Artuso et al. 2016: 82). Bergson's *élan vital* could very well be the *élan virale* underlying life's evolution. We simply do not know enough about the microbial realm at the moment to validate such a speculative view. Whether a metaphysical concept can ever be verified in an empirical manner is a question open to debate. At the very least, we must as philosophers keep an open mind in relation to the directions research projects could take. The dependence of plasticity upon sub-individual causalities such as those transmitted by viral infection does seem to point toward the presence of a broad, collective tendency mediating evolutionary processes. The translation between a general metaphysical concept such as the *élan vital* and particular scientific empirical results ought to be handled with a degree of care. Nevertheless, it is heartening to see that Bergsonism, far from constituting yet another dated philosophy of yesteryear, can still serve as a speculative point of departure for those working in the life sciences, quickening the minds of scientists long after *Creative Evolution* was written.

We must return here to the issue of consciousness, and how it relates to the broader topic of evolution. Resisting the downward pull of materiality, the nervous system exhibits a tendency of extending its trajectories, creating ever more complexity around itself. Freedom is resistance to immobility. Because of its proximity to general consciousness, each particular nervous system is "a veritable reservoir of indetermination" (CE: 140). Just as the body is a canalization of the life force, so the brain is a channel through which consciousness flows. Matter itself is dormant consciousness; insofar as it becomes the source of its own movements, the material realm folds onto itself, creating vitality. The entire process, the sum of ascents and descents, is creative evolution. In positing a commonality between matter and life,

Bergson represents a manifestly Heraclitian position. In Fragment LXIX of Heraclitus, we read the following: "The way upward and downward are one and the same" (Heraclitus 1889: 101). Both are movements. From an absolute perspective, ascent and descent coalesce into the absolute instantaneity of change. Life is just matter waiting for subsumption in the material flow, but matter is also life waiting for awakening. The degree of awakening is the intensity of actualization. The organic is closer to what pure, unadulterated movement could become, if left to itself. Does this mean that matter and life necessarily correlate with one another? It would be an error to view the *élan vital* as an absolute, universally present force. Just because matter can become life, it is not predestined to be so. Creation can fail, for "the impetus is finite" (CE: 277). We have no basis to accuse Bergson of correlationism, as the coexistence of life and materiality is not inevitable, for two very important reasons. Firstly, Bergson does not rule out the possibility that life can fall back into the inorganic. Extinction is a failure of the creative drive and is a very real possibility. But the obverse scenario is also an entirely conceivable development, namely, the triumph of life over all its constraints. In the latter case also, the coexistence of life with matter would be undone. Because of the open nature of evolution as a time-bound process in the Bergsonian system, neither of these scenarios can be excluded. An actualization completely free of inhibitions is just as imaginable as actuality remaining forever locked within powerless virtuality. What must be recognized is that the vital impetus consists, above all else, in the effort of actualization. In a world without material constrains, the impetus of life could go on increasing the level of complexity forever, "the complication may (. . .) go on to infinity in all directions" (CE: 274). As we have seen, Deleuze misguidedly privileges the virtual, making it into a reified object of contemplation ontologically superior in stature to the actual. Pete A. Y. Gunter argues that the virtual is attributed a greater, more general level of being than the actual in Deleuzian philosophy, functioning as the primordial source of differentiation. The elaboration of actualities would in fact be a diminution of the virtual, conceived of as final source, not unlike a Platonic idea. Bergson never tired of critiquing such a view of differentiation. Just because the impetus was harmonious and unitary in the past, we cannot conclude that harmony has the final word when it comes to the description of the evolutionary process. Discord is the law, and harmony can only be a retroactive restoration. The virtual is a by-product of chaotic, purposeless actualization. As Gunter explains, "The *élan vital*" is "entirely immanent in the world (. . .) needing no virtual reality outside of actuality to sustain the creative advance of nature" (Gunter 2009: 175). The Deleuzian virtuality is, in other words, a transcendental fiction, a reincarnation of Plato's fabrication of ideas situated outside the actual world. There is almost something authoritarian in subordinating real movement to the unraveling of something alien to the plane of immanence. Nothing is added to our explanations by the introduction of such a fallacious abstraction. To quote Gunter, "*élan vital* would be first actual in God and then actual in the actual world. It would nowhere have the pure 'virtuality' with which Deleuze wishes to endow it" (Gunter 2009: 177). This interpretation accords remarkably with one of the earliest British commentators of Bergson. We refer here to Herbert Wildon Carr's 1911 book, *Henri Bergson: The Philosophy of Change*. In many respects flawed and overly simplistic, written as it was for a wider public and not specifically for the philosophical community, this work is nevertheless on point in its emphasis on the *actual* nature of the life force. The *élan vital* is, in Carr's interpretation, an "*actual present now* in which all existence is gathered up (. . .) The past is *gathered into it, exists in it*, is carried along with it, as it presses forward into the future, which is continually and *without intermission becoming actual*" (emphasis ours, Carr 1911: 15). Both past and future are gathered into the now, having no independent existence apart from this present duration. Becoming *is* actualization, without any gaps or intermissions. This perfect continuity makes the positing of any separate, sovereign virtuality all but impossible. *Virtuality is immanent to actuality*. It is precisely for this reason that Bergson speaks of the dependence of the former upon the latter: despite everything, "the virtual has to become actual" (CE: 175). Without a manner of canalization, the impetus would have remained a mere possibility. But as we have seen from Bergson's critique of the notion of potential, the possible can have no being. Being is becoming, and becoming is actualization.

Underlying the broad duality of matter and life, other transitory antagonisms can also be discerned at work in *Creative Evolution*. Arguably the most important of these transitory dualisms is the opposition between intelligence and instinct. Both represent two lines of the evolution of consciousness, which have extended along divergent canalizations. At the one extreme, we find human beings, capable of abstraction and inventing tools. We stand out from other organisms because of our ability to invent completely artificial instruments. Properly speaking, "we should say not *Homo sapiens*, but *Homo faber*," for "intelligence" at its root "is the faculty of manufacturing artificial objects" (CE: 153). The degree of intelligence is a direct corollary of an organism's inventive abilities. To the extent that a species creates its tools, it can be said to be intelligent. Instinct, on the other hand, would be a built-in type of consciousness, furnishing the animal with ready-made capabilities that must not be learned through experience. Already tool use by certain wasp species was described by the pioneering entomologists George and Elizabeth Peckham in their book, *Wasps, Social and Solitary* (1905), cited by Bergson himself in *Creative Evolution* (CE: 190). Bergson was aware that, in and of itself, the use of tools is not enough to delineate human intelligence from animal instinct. Therefore his careful choice of words is far from accidental. To the best of our knowledge, *Homo sapiens* is the sole living organism which has the ability to *manufacture* completely artificial implements. How then can the remarkably complex behavior of certain hunter wasp species like the *A. Hirsuta* be explained? This particular species "gives nine successive strokes of its sting upon nine nerve-centers of its caterpillar, and then seizes its head and squeezes it in its mandibles, enough to cause paralysis without death," proceeding to then bury its victim and lay its eggs upon the juicy, helplessly paralyzed caterpillar (ibid.). With its cool, calculating cruelty, the wasp appears to display a keen pragmatic intelligence. Its actions are finely adjusted to the structure of its prey, as if it knew the location of the nerve-centers to be targeted. In Bergson's view, it is absurd to suppose that the wasp has any practical knowledge relating to the position of the nerve structure of the caterpillar. Rather, there pertains a queer type of solidarity between predator and prey. We must "suppose" the presence of "a sympathy (. . .) between the *Ammophila Hirsuta* and its victim, which teaches it from within, so to say, concerning the vulnerability of the caterpillar. This feeling of vulnerability" is the product of "the mere presence together of the *Ammophila* and the caterpillar, considered no longer as two organisms, but as two activities" (CE: 191). In his copy of *Creative Evolution*, Frost wrote a cryptic note, which appears to contradict Bergson's optimistic description of vitality as a process of cumulative ascent: "Life dives under" (Sokol 1990: 53). To propagate themselves, hunter wasps are forced to commit acts of cruelty. Thrusting *downward* into its prey, the female wasp paralyzes its victim, which is then transported *below* ground. The full significance of this example lies in the rigidity of instinct, as well as its allusion to descending motion. Another word for instinct is habit. Hunter wasps have grown accustomed, through various hereditary processes, to their actions. For them, these complex acts come automatically, with relatively little mental effort. A wasp does not, at least in Bergson's view and to the best of our present-day scientific knowledge, reflect on what it happens to be doing. It just does those vital activities that must be done. It has no qualms about killing another life form; murder comes effortlessly to the female wasp because its own larvae must feed. Were we, through some accident or technological invention, to be capable of asking a female of the *A. Hirsuta* why it commits these acts, it probably could not understand the terms of our query. On the one hand, the very translation of a question relating to human morality, the latter being primarily (though not exclusively) a product of intelligent reflection, into "wasp-talk" or any other animal communication, is extremely problematic. Ludwig Wittgenstein remarks famously that "if a lion could talk, we could not understand him" (Wittgenstein 1986 [1953]: 223). How much more so in the case of a hunter wasp! Moral reflection is an evolutionary luxury of complex organisms, apex predators that have been able to achieve a state of relaxation, their domination over their own lifeworlds unthreatened by any other rival species (apart from each other and the usual microbial dangers faced by any monstrously over-populated animal species). On the other hand, we can imagine that questions relating to the wasp's own lifeworld, such as how to dig a burrow and lay your eggs on a caterpillar, could be posed to the wasp with somewhat greater ease, were there some manner of transmitting our communications to the specimen. Intelligent knowledge would be the awareness of external relations as they relate to our vital concerns, based upon a connection and mapping of causalities, resulting in an analytical carving up of the world into discrete objects along the lines of the said concerns. Instinct is like "an intuition," inasmuch as it resembles a "divining sympathy," an interconnection with other beings felt from within (CE: 193). Occupying its position within its environment, the organism exhibits instinct when it behaves in a habitual manner, automatically adapting to circumstances, while hesitation shows the presence of a reflection, intelligence exhibiting some degree of detachment from the environment. "Instinct," writes Bergson, "is sympathy," being the web connecting, in a solidary way, seemingly discrete animals (CE: 194). Another example highlights the Bergsonian theme of sympathy, while also illustrating its pitfalls. In a video, a decapitated wasp can be observed (the species is unclear) as it proceeds to pick up its own head and fly away, apparently unbothered by the severing of its head. What is the secret of this damaged, yet sickeningly functional wasp? On a Bergsonian view, we could posit the presence of some type of sympathy pertaining among the various sections of its body, but in the case of insects we know that the nervous system is distributed throughout the body in the form of ganglions. Hence, there is an anatomical explanation for the decapitated wasp's improbable feat of surviving its own decapitation. Yet science has still to fully account for the workings of insect nervous systems. Jewel wasps (*Ampulex compressa*) prey upon cockroaches in much the same way as the parasitical wasp species described in *Creative Evolution*, injecting victims with nerve toxins, preventing the conducting of voluntary movements. The cockroaches are thereby transformed into passive, zombified living prey. Researchers have found that this phenomenon raises interesting questions relating to the nervous systems of insects more broadly (Gal and Libersat 2010: 1–10). It would appear that jewel wasp nervous systems have adapted to cockroach nervous systems, but how this could have happened is still a mystery. Sympathy again seems pertinent as a working hypothesis for the extremely prey-specific adaptations of such wasps. Without the hunter wasps, as well as other parasites or predators to regulate their populations, various insect populations would become overgrown, leading to resource depletion. And lacking fat caterpillars or, alternatively, cockroaches stripped of their free will, the wasps could not feed their larvae. Bioregulation functions effectively, almost without any effort, hence its mechanical, repetitive, automatic appearance. But if instinct is sympathy, and so is intuition, does this mean also that intuition is a form of instinct? Bergson is careful to qualify his statement, adding that "by intuition I mean instinct that has become disinterested, self-conscious, capable of reflecting upon its object and of enlarging it indefinitely" (CE: 194). Quite frankly, it is one of the most boorish and stupid of mistakes to equate intuition with some type of return to "animality," as Bertrand Russell did in his atrociously poorly argued article (Russell 1912: 323). What differentiates wasp from philosopher is the fact that intuition extends beyond instinct. Extracted from its environment and placed among completely different organisms, the *A. Hirsuta* would not be able to continue acting in the manner its instinct has intended. Without a supply of familiar caterpillars, the wasp simply could not act. Its sympathetic capabilities, in the absence of organisms it has adapted to, would freeze up. Bergson's intuition, defined as a philosophical or even spiritual method, differs from instinctual sympathy, needing no ecology. It can be applied to any and all things, and brooks no limitation. Intuition is a "sympathetic communication" which "transcends intelligence," breaking through any and all pragmatic limitations (CE: 195).

Intellect has evolved to help action. Intelligence consists in the preparation of actions, the degree of spirit corresponding to the ability of an organism to enact movements. In and of themselves, potentialities do not have a being of their own. "Consciousness," states Bergson, is "proportionate to the living being's power of choice. It lights up the zone of potentialities that surrounds the act" (CE: 197). Without consciousness, potentials would remain in darkness. Life is the light of the world. Instinct, like matter, is consciousness that revolves in a circle, unable to break out of its entrancement. Life forms entrapped within the circle of instinct act automatically. Habit is a threat lying in wait for all life forms, the risk of becoming locked in mechanical repetition. The living can, when left to its own devices, transform into a self-referential nontrivial autonomous machine, but inertial obstacles prevent its efforts from ascending to perfect, limitless relaxation. In most organisms, intuition is restricted to a certain lifeworld, remaining mere sympathy, it proves unable to break free of limitation. The human, defined as *Homo faber*, exemplifies the possibility of life exploding limitlessly, populating the remotest areas of the universe. The task of intelligence is the conversion of matter into a device for the unleashing of energy. Intuition can become a light shining through matter, sending rays to the beyond. As Frederic Worms writes majestically, intuition is "the immanent light of intense drive," a striving which rushes to rid itself of materiality even while enwrapped with corporeality (Worms 2010: 249). From a jail, matter can be reengineered into a tool for enhancing freedom. Intelligence allows for the possibility of transforming matter "into an immense organ by the industry of the living being" (CE: 178). The material dimension is the organ of life striving to ascend. Such a perspective introduces the organological register into Bergsonism. Coined by Georges Canguilhem, a follower of Bergson, "general organology" holds that "technology is a production of organisms as living beings, by which they deal with their environment" (Hoquet 2018: 259). Canguilhem's innovation builds on the Bergsonian insight that technology should be viewed as a vital strategy. Intelligence is not opposed to life; rather, it is one of the products of life's broader evolution. Commenting on the place of organs in *Creative Evolution*, Hisashi Fujita explains that "our organs are natural tools, and tools themselves are artificial organs. Thus the Bergsonian 'organ' transcends the boundary between organic and inorganic" (Fujita 2007: 119). *Ascendance resides by striving. Something is struggling for birth, trending*. An organ is not a local, endogeneous structure. The Bergsonian method of conceptual extension should by now be familiar to us. As with "effort," so it is with "organ": the concept is extracted from its original semantic environment, and, loaded up with new contents as well as fuel, is blasted up into space, sending it farther than could ever have been formerly imagined. Instead of a definite structure, endogeneous to a certain individuated body, the "Bergsonian organ," to use Fujita's expression, can be anything that assists living activity. Returning to our metaphor, life is the fluorescence that enlightens the dimensions, making them into props for its qualitative enlargement. "Life *organizes* matter," not only in the banal sense of creating unprecedented forms or bodies but also in the more radical sense of ontologically transforming matter into a set of tools for the enhancement and proliferation of its own tendencies (emphasis ours, CE: 182). The implications of an organological interpretation are momentous. In a footnote to his essay, "Machine and Organism," Canguilhem characterizes *Creative Evolution* as "a treatise of general organology" for the reason that Bergson extends the scope of instrumentality far beyond the human realm, "mechanical invention" being "a biological function, an aspect of the organization of matter by life" (Canguilhem 1992 [1952]: 69). Instinct is an *effort of organization*. The organ is the extension of the organism's own channeling of energy, and is, therefore, a machine, inasmuch as it displays a purposiveness.[14] It is as if the organic were striving to escape from entrapment within the inorganic. Invention, conceived of as intelligent enactment, proceeds "as though the grip of intelligence on matter were, in its main intention, to let something pass that matter is holding back" (CE: 201). Left to its own devices, the current of life would ascend, in the manner of a hot-air balloon, floating beyond any enchainment, swimming in the sky, denying any similarity or solidarity with a particular environment. The correlation between matter and life is a momentary relationship that has been forced upon organisms by external circumstances and the endogeneous limitations of their bodies. Becoming upright, *Homo faber* has unleashed unprecedented geological forces. Overproduction threatens life on this planet. Two possible resolutions emerge, either the inversion of this production of surplus in a spiritual direction, or the release of this pent up effervescence in an expansive, cosmic direction. In the words of President Johnny Gentle of David Foster Wallace's *Infinite Jest*, "Let's Shoot Our Wastes into Space" (Wallace 2018 [1996]: 382). A combination of space exploration with spiritual enlargement is also a far from impossible eventuality. If life is to continue in its striving, it must expand its power of organization beyond the bounds of this planet, transforming everything into an organ.[15] Although the organic seems limited by certain material constraints, the organological is infinite. It is a question of inscribing "the mechanical into the organic" (Canguilhem 1992 [1952]: 64). On the basis of the Bergsonian ontology, nature does not stand opposed to technology. The lesson to be learned is that the natural is always already technological. Technology would then be viewed not as the enemy of life, alien to nature, situated outside of the organic dimension. Instead, it is an artificial product of nature's own inventive power. Bergson could not be more explicit: "Consciousness is synonymous with invention" (CE: 287). The living, especially (though not necessarily exclusively) when it attains a degree of consciousness, is a power of responding in often innovative ways to challenges and resistances. To think means to introduce new patterns into becoming, through the exercise of a certain intensity of reflection.

Is everything then a cyborg? Thierry Hoquet notes that Canguilhem's and, by extension, Bergson's "view of cyborgs" is "highly deflationist," because all organisms are, in the final instance, "organisms extended with tools" (Hoquet 2018: 269). Hoquet recommends a neologism of his own, "organorg," denoting all tool-using organisms, while ignoring the historical circumstance that neither Canguilhem nor Bergson actually use the phrase "cyborg." This is a deliberate move, designed to favor a retrospective reading which registers the latent presence of the concept of a "machinic" organism in Bergsonism. In all but name, cyborgs populate the pages of both *Creative Evolution* and Canguilhem's "Machine and Organism." Such a speculative thought experiment is to a certain degree justified, but it seems excessive to take two authors to task for not clearly defining a concept that did not exist at their time of writing! Hoquet himself notes that the phrase "cyborg" was coined in a 1960 article by Manfred E. Clynes and Nathan S. Kline, decades after Bergson's death and some years after Canguilhem's article had been published (Clynes and Kline 1960: 26–27, 74–76; Hoquet 2018: 258). Without question, Bergson would have found the idea of the "cybernetic organism" of profound interest, but Hoquet seems too strict in limiting the concept of "cyborg" to autonomous artificial systems. On his view, the stage of cyborg is only reached once technology becomes autonomous from its dependence upon organic bodies (Hoquet 2018: 268). This move ignores the unavoidable hybridity of the cyborg concept. Ian Hacking's more liberal definition is conducive to our own goals, allowing for a greatly diverse application of the concept, without needlessly restricting its scope to a certain quality. A cyborg is "sort of mechanical but also sort of organic," and may be described as "alive, living," being an artificial entity "patterned upon living beings" (Hacking 1998: 212). As we have seen, in Bergson's case even living things, whose status as biological entities none of us would seriously question, contain something repetitive or machinelike. The wasp's movements are reiterations of genetically ingrained habits, constituting behaviors bearing a resemblance to preprogrammed automata. Any feedback mechanism capable of displaying an inventive capacity beyond mere automatism can be termed a cyborg. Hacking argues for a liberal interpretation in line with Canguilhem's view that cyborgs, or for that matter, machines in general, ought not to be viewed as "separate from life" (Hacking 1998: 214). Anything displaying the presence of a feedback servomechanism can be characterized as a cyborg, as can "any coupling of humans and machines" (ibid.; see also Pickering 1995: 1–48). Fascinatingly, Hacking does not segregate the reality of the cyborg from its status as a productive metaphor. Just as Bergson connects the underlying fluidity of reality with the need for liquid concepts, so Hacking views the sheer productivity of the cyborg metaphor as constituting a real characteristic worthy of note. If the cyborg has proven so fruitful for both science fiction and real science, then it contains an excess that cannot be derived from any exogenous source. The sheer productivity emancipated by this word points toward an underlying truth that cannot be readily dismissed. Inflationism paradoxically results in the narrowing of a metaphor's breadth, while deflationism allows a much broader use of words. By deflating the idea of machine, removing its singularity and antagonism vis-à-vis organism, Canguilhem allows for a broad use which allows for a reintegration of technology into nature. The organological perspective, "rather than explaining organism by machine," strives to "explain machine by organism" (Fiant 2018: 158). The technological cannot really overwhelm the natural, because the machine itself is a product of life's broader inventiveness. The machinic is a downward tendency of life. When a living thing becomes enmeshed in habit, it degrades into something less than a cyborg. As Canguilhem outlines, "Finality or purposiveness would be more applicable to a machine than to an organism" (Canguilhem 1992 [1952]: 57). On the level of creative evolution, indeterminism precludes any plan or project. Bergson's philosophy of life excludes teleology. The tendencies are heading toward maximal discord. Leaning back, observing the entire process of evolution from an intuitive, that is, absolute standpoint, we seek it as a pure act of charity, transcending any finality or machinic purposiveness: "It is as if a vague and formless being, whom we may call (. . .) man or superman, had sought to realize himself, and had succeeded only by abandoning a part of himself on the way" (CE: 290). The world which has undergone "organization" forms a wreckage composed of remnants strewn everywhere by an immense love, which gives itself in, through and to all things freely. A light, upright mass of tendencies, the "formless being" alluded to can be called many things. Without seeking to reduce it to any particular register, be it ontological, mythological, or theological, we view it as the primordial originality of mobility in general, the pureness of activity prior to any canalization, cabalization, or cannibalization.[16]

THE PHILOSOPHER AND THE CINEMA

Many further elements remain which could be productively unpacked, but such a broad undertaking would take us well beyond the scope of our book, requiring a general discussion of the present condition of theoretical biology and how subsequent developments impact the idea of creative evolution. Such a project, however worthy an undertaking it would be, is outside the purview of this book. At this point, we must proceed to the final, fourth chapter of *Creative Evolution*. After the above considerations, Bergson's relation to cinema and what he calls the "cinematographical view" of reality must be addressed. Although Deleuze's *Cinema*-books have proven useful to our own delineation of the durational moment, providing us with the concept of "time-image," regarding the relationship of Bergson and cinema it must be said that Deleuze proves less than helpful. On the pages of *Cinema 1*, Deleuze argues that Bergson's critique of the cinematic view of reality, as expounded in the fourth chapter of *Creative Evolution*, only applies to the early stages of cinema. Supposedly, the fixity of the camera resulted in the localization of the shot. Fixity was the norm at this stage of cinematic technology: the immobility of the camera resulted in moving filmic bodies "passing from one spatial shot/plane to another, from one parallel slice to another, each having its independence or its focus: (. . .) We can, therefore, define a primitive state of the cinema where the image is in movement rather than being movement-image. It was at this primitive state that the Bergsonian critique was directed" (Deleuze 1997 [1983]: 24). The idea here is that Bergson simply did not have access to later cinematic techniques. Later forms of cinema, as exemplified by legendary directors such as Alfred Hitchcock and Orson Welles, proved more capable of deterritorializing the camera, so to speak, resulting in more effective representations of movements, changes, modifications, and temporal shifts. Bergson would have written differently, had he been able to view technically more sophisticated films of a later era, only having access early, imperfect cinematic illusions. *Creative Evolution* was published in 1907.

Considering this date, it is almost miraculous that Bergson even mentions cinema at all. Had Bergson watched say, *Citizen Kane*, with its mixing of temporal layers, he would surely have had a more generous view of what cinema is capable of. The original Deleuzian insinuation is repeated by Paul Douglass, who writes that during the 1930s, a "period of dynamic change for cinema," Bergson was "an ailing man preoccupied with finishing his last book, *The Two Sources of Morality and Religion* (1932). Perhaps he simply never encountered the cinematic innovations that might have changed his mind" (Douglass 1998: 27). Deleuze goes so far as to claim that Bergson's ontology actually itself composes a cinematic view of reality, corresponding to later forms of cinematic representation such as the montage, which proved far more capable of capturing the mobility of reality. As Deleuze writes, "The material universe, the plane of immanence, is the *machine assemblage of movement-images.* Here Bergson is startlingly ahead of his time: it is the universe as cinema in itself, a metacinema. This implies a view of the cinema itself which is totally different from that which Bergson proposed in his explicit critique" (Deleuze 1997 [1983]: 59). To say the least, all this seems implausible. Ontologically speaking, a cinematic representation of a real object, defined as a movement-image, will never be the same as the process itself. A simulated movement remains a virtuality. The whole of Deleuzian philosophy is based on a conflation of the virtual with the actual. For Deleuze, a cinematic image will be just as real and authentic as a "movement-image." Deleuze is quite explicit in his rejection of any duality between appearance and reality. As he writes in *Cinema 2*, "truth is not to be achieved, formed, or reproduced; it has to be created. There is no other truth than the creation of the New" (Deleuze 1997 [1989]: 146–147). Again it must be reiterated that this differs markedly from Bergsonian philosophy. On his part, Bergson consistently maintains that metaphysics ought to give us access to an absolute, a reality independent of any fabrication, or rather, a becoming which is creating itself. Evolution is truly more than a forgery, being a real, objective, actual achievement. Neither Deleuze nor Douglass, nor any other commentators influenced by the Deleuzian interpretation of Bergson's anti-cinematic stance, adequately address what is really at stake. Tom Gunning has argued that while Bergson could very well have been unaware of later cinematic developments, and may even have "confused" Marey and Muybridge's "chronophotography" with "Lumière's cinématographe," it is more plausible that Bergson "was referring to the cinematic apparatus and its components, its mechanical operation, and not its final effect" (Gunning 2014: 5). In the 1931 introduction to his collection of essays, *The Creative Mind*, Bergson explicitly references cinema as a mode of representation which is not adequate to the becoming of reality: a deterministic and linear perspective views change as "the unfurling of a fan, or (. . .) the unrolling of a cinematographic film.

Calculable ahead of time, they existed prior to their realization in the form of possibles" (CM: 20). Bergson's point of contention has nothing whatsoever to do with the *aesthetic effectiveness* of cinema as a media of representation, and everything to do with the uniterability of the real. Authentically objective becoming contains nothing of the possible. The tendencies at work in evolution are contingent explosions. We never know what is going to occur next. That which appears distinctive can resolve into a heterogeneity of elements, while continuity too can break down into transitory hesitations, preparations, and differential resolutions.

An illustration is in order, one which can contribute immensely to clarifying our train of thought. Gunning mentions an anecdote dating from the classic era of cinema: a man "went to see a certain Greta Garbo film more than a hundred times. When asked about the nature of his obsession, he explained that he returned because, in one scene, Garbo undresses in front of a window, and right at the crucial moment, a train rushes by, obscuring the desired body from the viewer. 'I figure,' the man explained, 'that one time that train just has to be late'" (Gunning 2014: 6). Can cinematic re-presentation ever fulfill our desires? Gunning argues that the story is open to several interpretations, and points to the ambiguity of cinema. From a Bergsonian perspective, "cinematic images may move," but the movements they reproduce are mere simulations, remaining "fixed and predetermined" (ibid.). One can go to the movies an infinite number of times, never shall the goddess of the silver screen ever reveal the entirety of her bodily charms. But Gunning also takes Bergson to task for appearing to ignore the possibilities of a transformed human relation to technology. Cinematic motion also holds in store an "animating" of "the mechanical," a "blurring of the dichotomy between animate and inanimate" (Gunning 2014: 7). Are the various critiques of Bergson we have summarized above fully justified? Is it a case of an old philosopher having been left behind by the advance of media technologies, or is there something more fundamental to Bergson's criticism of cinematographically transformed perception? We must discover exactly what the Bergsonian critique is direct at, before passing any judgments. From the outset, in Bergson's usage the "cinematograph" merely represents a comparison, a symbol for "conceptual thought" (CE: 296). To think "cinematographically" means, above all, to slice the world into discrete images, with fixed frames. In itself, reality is composed of an infinite heterogeneity of becomings, which interpenetrate. Different functions, characteristics, and colors are separated by the operations of analytically oriented intelligence, but these are adaptive, instrumental products, and not the real thing. Becoming is opposed to any technique of representation. Intuition is, therefore, a nonrepresentative philosophical method. In its desire to go beyond representation, it is in accordance with the broader ambitions of Deleuzian philosophy. As have labored to point out, where the two diverge is the status accorded to the virtual and the actual, respectively. In Bergson's view, the simulation of a movement cannot ever be brought on the same plane as the movement itself. The latter's indivisible continuity eludes analytical thinking. The operations of intelligence are, in a pragmatic sense, effective, helping the organism navigate a challenging environment filled with dangers. It matters a great deal whether we can separate inedible plants from edible sources of nutrition. Our chances of survival are greatly enhanced if we can tell prey from predator, friend from foe, and so on. But such a partial view, nor the sum of all such partial, interested perspectives, will ever add up to an intuitive grasp of real becoming. Writing of a marching regiment, Bergson describes the analytical way of perception as a cinematographically oriented decomposition of movements into static images. Conceptual thought is akin to taking

> A series of snapshots of the passing regiment and to throw these instantaneous views on the screen, so that they replace each other very rapidly. This is what the cinematograph does. With photographs, each of which represents the regiment in a fixed attitude, it reconstitutes the mobility of the regiment marching. It is true that if we had to do with photographs alone, however much we might look at them, we should never see them animated: with immobility set beside immobility, even endlessly, we could never make movement. In order that the pictures may be animated, there must be movement somewhere. The movement does indeed exist here; it is in the apparatus. (CE: 331–2)

It almost beggars belief that a line of thought phrased in such starkly clear terms could have been misinterpreted so gravely by Deleuze and others. The Bergsonian critique of the cinematographic view of reality is not the product of a frail old mind, unable to keep up with the times. Neither does it relate to the technological method of cinema. Media can register movements, even creating new, mechanical ones in the process. Bergson's contention relates to the manner in which we approach these technologies of mediation, as well as the territorialized nature of film technologies in general. Put bluntly, the subsequent evolution of cinema does not in the least impact the import of Bergson's critique. As long as it is localized in any manner, be it restriction to a screen or dependence upon the presence of any type of camera, film technology remains a partial entrapment of recorded movements. Think of the scene of horror in the found-footage film, *The Blair Witch Project* (1999, dir. Myrick and Sánchez). In this work, we never actually observe the thing or entity responsible for the disappearances. The magnificent achievement—and horror—of this cult classic lies in its paradoxical representation of the unrepresentable, so successful precisely because it makes no pretence to an exhaustive revelation. Reality similarly eludes any and all contact in the

Bergsonian philosophy: becoming is too mobile for any definitive representation. The filmic apparatus is a way of deterministically locking movements in place, and not even a Hitchcock or Welles could modify this fundamental circumstance. In *Citizen Kane* (1941, dir. Welles), there is a fading shot where a memory momentarily coincides with the present. For Deleuze, Welles represents a revolution in cinema, for the latter was the first director who proved capable of not only delocalizing the camera, introducing moving shots and dissolves, but also effectively destabilizing temporal progression itself. Cinema since Welles has, reputedly, transcended its originally static nature. The gist of Deleuze's argument is that "like philosophy, cinema has evolved towards a greater awareness of reality-as-mobility" (Douglass 1998: 31). But even today, what does the viewer do in a cinema or in front of a laptop? She is mostly riveted in one place, sitting or lounging around, staring at a screen replete with the illusions of movement. In fortunate cases, the film will elicit certain spontaneous affective responses such as crying, laughter, or screams. But the integrity of the body will remain mostly untouched. Film usually fails to penetrate our reality in quite the same way as an actual becoming. One can survive a death undergone in the first-person view of the main protagonist, but that life never was a real duration to begin with. To qualify as real, a duration must constitute a singular novelty. Is there then such a thing as a simulated duration? For Bergson, the reproduction of a movement would compose a virtuality. In this regard, film shots are not unlike memories. A memory cannot ever be returned to unmediated immediacy. Stasis is still the norm when it comes to film. After all is said and done, the screen has also remained intact. In the early decades of the twenty-first century, media is only just beginning to emancipate itself from its original territorialization within the confines of a frame.

The solitary male viewer shall never see Greta Garbo's full beauty. Her actual duration is forever situated somewhere else. In today's world, internet pornography has become a fundamental part of sexual life, informing desires, triggering transgressions, and raising concern among medical professionals and conservatives of various stripes. Are the worries justified? What Bergson seems to suggest is that we must be on our guard in relation to the operations of media technologies. We ought not to confuse the reality endogenous to the screen with the real becoming indigenous to our own durations. Cinema as contrivance consists in "extracting from all the movements peculiar to all the figures an impersonal movement abstract and simple, movement in general, so to speak: we put this into the apparatus, and we reconstitute the individuality of each particular movement by combining this nameless movement with the personal attitudes" (CE: 332). Deluded consciousness would consist in a thoroughgoing, almost hopeless confusion of the planes. A frequent subcultural practice has to do with giving gifts to webcamera models. Actual amounts of money are exchanged for virtual promises of love, enacted upon screens or "tubes" functioning as channels for hopeless, pointlessly unfulfillable desires. It is difficult to imagine a more degrading self-impoverishment. The adherent of Deleuzian philosophy is bound to commit a similar mistake to the love-sick gift-giver, exchanging the actual for the virtual, being cognitively and conceptually ill-equipped to differentiate between movement and its imitation. The import of the Bergsonian critique of the cinematographic mode of perception boils down to an almost disarmingly simple point: becoming can be neither thought, nor conceptualized (ibid.). *Thoughts are growing, but outside of this increase, clouds abound. The abundant abides in the swelling sprout.* The cinematograph is effective at capturing movement, but by its very nature it can only integrate duration as a static abstraction. Does this mean that we must abstain from the consumption of visual technologies? Should the philosopher refrain from visiting the movies or watching streaming video services? The ambiguity of Bergson's position is brought in a 1914 interview. In this piece, Bergson briefly explains his opinion of cinema, casually relating a series of methodological observations. The philosopher must, above all else, study the object of contemplation as closely as possible. This applies to anything whatsoever, including cinema: "Several years ago, I went to the cinema. I saw it at its origins" (Georges-Michel 2011 [1914]: 81). New technologies have just as important a role to play in philosophical practice as scientific disciplines. Out of the interplay of the philosopher's mind and the screen, there can emerge new analogies, metaphors which have a productive value of their own. Mentioning contemporary directions in painting, Bergson states that cinema has contributed to a revolution in art. Impressionism specifically is, for Bergson, a manifestly cinematic genre, "the cinematograph having taught the painter that photography was wrong. By reproducing movement on the basis of personal impressions, the artist had recomposed, fused into one, several successive attitudes, giving the illusion of life and therefore of movement. They found these attitudes again on the screen" (Georges-Michel 2011 [1914]: 82). If it manages to connect to a contemporaneity, the virtual gains a secondary energy, borrowing its potency from the world of movements it connects with. Yet for all that, the movements and modifications depicted on a painting or a screen are still illusory. Art, while originating from the same creative impulse as life, is nevertheless an arrested movement, a form of nonliving memory. As Bergson states explicitly, "Memory is like cinema, a series of images" (Georges-Michel 2011 [1914]: 81). The image, once neutralized by incorporation inside the confines of a screen, is depotentialized, for it has been extracted from the dynamic interplay of its causal ecology. Let us be attentive to what Bergson is communicating. Cinema and other media technologies are just as worthy of philosophical treatment as other elements of reality. They too are becomings. But never should we confuse a movement *located* within memory or media with a real, creative, *unsituated* movement. By filming transitions and trajectories, the director is recomposing mobilities, functioning in the manner of a *bricoleur*. Many genres have formed which deliberately exploit this characteristic of cinematic technology, slicing linear narratives into a chaos of loosely coherent, or even completely incoherent scenes. The postmodern films of Leos Carax and Peter Greenaway are particularly paradigmatic examples of cinema-as-bricolage. Slicing is a way of approaching reality, and on its own terms, is as legitimate as any other method. But Bergson's metaphysics seeks a direct, immediate access to real change, and cannot, therefore, remain satisfied with simulations. In a playful tone, Bergson recounts having been asked to pose for a film, only to reject this offer for the following reason: "The philosopher cannot really push his recognition to the extent of becoming a subject himself for the cinematograph, no matter how amiably he is invited" (Georges-Michel 2011 [1914]: 82). Is this the result of a phobic, even Luddite approach to technologies of representation? Or is it merely a manner of remaining mystically inaccessible, as the idolized master of a whole school of philosophy? Something more seems to be in play beneath this gesture of rebelliously punkish rejection. If we are to remain true to the priority of real impermanence and contingency, we cannot be satisfied with mechanical repetitions of movement. Loyalty to the actual means a ceaseless affirmation of the explosive spontaneity of life. On an ethical level, we are forced to preserve indetermination, preventing it from incorporation within the confines of repetitive mechanisms. It could even be said that, far from acting as an enemy of technology, Bergson is looking forward to an age in which technology has proven capable of reproducing the spontaneity of movement. A Bergsonian cinema would be an unpredictable film, in which anything can happen. Greta Garbo would unveil her naked breasts, and we could feed upon then, even consuming her entire body. Or the reverse eventuality could also materialize: cannibalization by the idolized goddess of the screen, transformation into black excrement, preliminary to being flushed out of a virtual toilet into a perpetual silver sea. A Bergsonian cinema would give us unimaginable durations, without any guarantee of survival and in the absence of a restrictive screen. Perhaps instead of "Virtual Reality" (VR) technologies, we require Actual Reality Technologies (ARTs). In a way, the Augmented Reality movement already goes further than VR, in its emphasis on the holistic affective engagement of the whole body. Viewers exchange mere visual fixation for participation and become inhabitants of technology. In the early years of the twenty-first century, we could be in the early stages of an ART revolution.

In the final instance, what is reality made of? It is a gradation of energies, composing degrees of intensity awaiting a final liberation. Life, defined as spontaneous movement, escapes mechanical reproduction; like shrapnel, "bursting before it falls to the ground," real mobility "covers the explosive zone with an indivisible danger" (CE: 335). Intensive intuitive perception shows us the intertwining of our own duration with the broader, cosmic heterogeneity of durations, limitlessly expanding upward, tracing a pathway to infinity. For all our criticism, Deleuze is correct in pointing toward the presence of a "creative emotion" or "cosmic Memory" in Bergson's ontology which "actualizes all the levels at the same time," mobilizing every kind of duration in an upward rush (Deleuze 1997 [1966]: 111). Alongside cosmic memory, we may posit a cosmic explosion, striving to energize everything it reaches. Each organism is an explosive expenditure of energy, its body emplaced so as to "arrest its dissipation." The time it takes to spend energy is the time of living creativity (CE: 269). How then do death and negativity figure in this system? Is there any place for the negative in such an apparently overwhelmingly positive worldview? In his famous criticism of Bergson's denial of the negative, Gaston Bachelard states that there pertains "a perfect correlation between emptiness and fullness" (Bachelard 2000 [1950]: 30). To claim that nothingness is a mere derivative of Being or a deficiency of a certain plenitude, is a begging of the question. In a Buddhist vein, Bachelard notes that "emptiness from one point of view is automatically fullness from another" (Bachelard 2000 [1950]: 31). Therefore, it is not legitimate to deny the intuition of nothingness, as Bergson does, by simply claiming that the negative has no independent being apart from the positivity which it negates, for the reverse could also be claimed with just as much basis. Bergson's criticism of the idea of nothingness boils down to a very simple idea, originating from Parmenides and others. Even if we imagine the elimination of all things, annihilating them in our consciousness, the absolute zero or "nought" so defined is still "the annihilation of everything" (CE: 305). Nothingness has no positive being of its own, constituting a parasitical concept not unlike virtuality or memory. The illusion of nothing having a power of its own stems from the experience of obstacles and absences that are waiting to be overcome. Things truly can disappear, and our activity can at times be prevented. For Bergson, this possibility does not mean that we are obliged to attribute any being to absence. The idea of nothingness originates from the substitution of one thing for another, as well as the real frustration of an effort (CE: 306–7). When speaking of absence, we are not actually negating anything, nor even describing any object. Nothingness is a virtual, and therefore unnecessary addition to reality. For Bergson, ontology could get along fine without any such supplements. A "nonexistent" constitutes an addition to reality, a semantic excess lacking stature, causal power, or plausibility (CE: 310). Is Bergson's proposed rehabilitation of Parmenidean metaphysical plenitude convincing? This attack on the idea of nothingness forms, arguably, the weakest point of *Creative Evolution*. From a metaphilosophical perspective alone, the negation of negation is less than convincing. John Mullarkey for one has argued that the absolute, all-encompassing denial of the power of negation on Bergson's part makes it very hard for us to actually take his claims seriously: "totalising philosophies lack explanatory power, for in claiming that everything is (directly or indirectly) x, y, or z (. . .) they lose the ability to account for their opposite, even as a derivative illusion or error" (Mullarkey 2000: 257). By absolutely denying the validity of negation, Bergson undercuts his own ontological claim, making it all but impossible—at least within his own framework—to account for how on Earth nihilistic philosophies affirmative of negation could have ever come into being in the first place. Similarly, Bergson's own negation of the nihilism and other opposing views also becomes untenable. There is no normative basis upon which we may be justified in reducing the intuition of emptiness alluded to by Bachelard to the status of mere illusion. It is more convincing to deny substance altogether. The triumph of truth must take place against the backdrop of destroyed illusions. Creation necessitates rupture and discontinuity. Nina Power, paraphrasing Bachelard, notes that "Bergson's 'psychology of plenitude' is (. . .) too full, too rich, too entranced by the trace of life in biological entities to truly account for specific change" (Power 2006: 121). Knowledge itself is, Bachelard states, "the annihilation of appearance" (Bachelard 2000 [1950]: 35). Rather than denying negativity altogether, on our part we would propose its integration within the ontology of creative evolution. Is Power's characterization of Bergson's philosophy as being entranced by vitality fully justified? We hold that the explosivity of the energetic constitutes a way of transcending the hopelessly false dualism of something versus nothing, affirmation versus negation. Bergson compares the current of life to a "rocket," which, "passing through the fragments" of materiality, "lighting them up into organisms" (CE: 285). Could the entire cosmos be conceived of as a process of enlightenment? In this day and age, such a view smacks of excessive optimism. But words such as "positivity" and "negativity" do not come close to visualizing the sheer vitality of becoming. In Jeanette Winterson's novel, *Gut Symmetries*, we read the story of a love triangle between three lovers, Jove, Alice, and Stella. The complications of their entwinements coalesce with the immensity of the world, vividly displayed by the title. "Gut" refers to entrails, but also to the Grand Unified Theory long sought after by theoretical physics. Enacting a correspondence between bodies results in a cosmic intuition of pure energy and activity, the lovers engage in a panerotic act of transgression, melting into a single duration: "His body. Her body. My body. Unseparated, twisting, dark (. . .) The silent gravity gone somersault of she on he on she (. . .) He was me I was him are we her?" (Winterson 1997: 36–37). In the final instances of the novel, the vision of creative evolution is the intuitive recognition of the power of the now, the centrality of the continuous instant, extending to infinity: "History and futurity are now. The universe curving in your gut" (Winterson 1997: 219). As Mihaela Cristina Oancea notes, commenting Winterson's novel, this melting of durations into one another brings the cosmic back into intimate correspondence with the personal, "the public" being "fused" with the personal "to the point of suture: just like the Grand Unified Theory seeks to unite the three major forces of the universe (weak, strong and electromagnetic force), the three lovers, Jove, Stella and Alice form an 'Eternal Triangle'" (Oancea 2018: 38). Creation is the actual movement of change, coupling the unimaginable with balance, growing spores of spontaneity upon deterministic surfaces, contracting galaxies into innards, only to finally break the mold, shattering all frames, letting energy flow above and beyond any body.

NOTES

1. Even in one of his later books, Stephen Jay Gould, while rejecting a wholesale teleology immanent to the whole of the living world, still admitted that it was possible to posit a general increase in the degree of complexity of intelligence in living things (Gould 1996). More recently, biologist Daniel McShea and philosopher Robert Brandon have argued in their 2010 book, *Biology's First Law*, for the existence of a "Zero-Force Evolutionary Law" predestining life to the increase of complexity. When left to its own devices, life tends to increase complexity, although in real cases this is limited by selection pressures: "In any evolutionary system in which there is variation and heredity, in the absence of natural selection, other forces, and constraints acting on diversity or complexity, diversity and complexity will increase on average" (McShea and Brandon 2010, quoted in: Ruse 2016: 105).

2. The battleships so much admired by Huxley and used by the biologist as a metaphor for the supposed "warfare" between animal species were actually hardly even used during World War I, with the exception of the indecisive Battle of Jutland (1916). The naval dimension played barely any role in the mostly terrestrial war. The relegation of battleships to a kind of second-class status was accentuated by the advent of aircraft carriers in World War II, which lent a new actuality to naval warfare, at considerable expense to the former class of ship. The armored battleship as technology proved improductive from a tactical standpoint, and was surpassed by new, more mobile technologies. In the second major global military conflagration of the twentieth century, battleships played a subordinate role at best, being hopelessly outmaneuvered by aircraft and, even more importantly, greatly improved submarines. The victory of the U.S. Navy against the Japanese Imperial Fleet at Midway (1942) was due to a large extent upon the former's greater utilization of aircraft carriers, while the latter had remained stuck in a pre–World War I emphasis on building battleships. The Japanese had fewer aircraft carriers, placing them at a tactical disadvantage, with catastrophic ramifications. The armored battleships had adapted to one another too successfully, only to become outdated by changes in the broader

technological landscape. Today's tactical field has no place for these slow, lumbering machines. The 1991 Gulf War was the last occasion the U.S. Navy employed battleships in combat, and all such vessels were eliminated from American service during the first decade of the twenty-first century.

3. We refer here to Harman and Scott's fear regarding the supposed holism of relationalist ontologies.

4. As Keith Ansell-Pearson has pointed out to us in a private communication, Bergson only mentions Arthur Schopenhauer but once in his writings, and fairly disdainfully to boot. Therefore the French philosopher's idea of the *élan vital* (life force or vital impetus) can hardly correspond to Schopenhauer's life-as-will. Whereas the *élan* is supposed to function as an empirically grounded metaphysical speculation, Bergson characterizes Schopenhauer's concept as redundant and incapable of adding to our understanding of life. The significant divergence between the two thinkers was evidently lost upon German philosophers H. Bönke and Baron Cay von Brockdorff, who, no doubt motivated by anti-French nationalism and perhaps also a degree of anti-Semitism, accused Bergson of having "plagiarised" Schopenhauer's ideas down to several fine details (Bönke 1915; Brockdorff 1916; for more on this issue see François 2005: 469–490).

5. The phrase "holism" was invented by South African statesman-philosopher Jan Smuts, and is outlined in his 1926 book, *Holism and Evolution* (Smuts 1926). Although Smuts's idea of "creative Holism" does bear a resemblance to Bergson's conception of creative evolution, we may nonetheless accept the process philosopher C. Lloyd Morgan's view: "He [Smuts] seems again and again to accept Bergsonian conclusions, though he rejects the Bergsonian foundations on which they are based" (Morgan 1927: 96). The most that can be said is that Smuts was a contemporary influenced by the Bergsonian view and the problems Bergson focused on, but cannot be said to have been a disciple of the latter on the whole.

6. Of late in the biological community there is a renewed emphasis on the importance of epigenetic processes in the reproduction of traits, as opposed to the hardcore genetic reductionism of yesteryear. An influential study published in 2020 emphasizes the importance of epigenetic processes in the evolution of yeast subjected to externally imposed environmental pressures (Luo et al. 2020: 1–13). Rather than vindicating a mechanistically Lamarckian view though, could one not say that the study's conclusions are open to a different interpretation? Could we not say that life, instead of merely adapting to external effects, is actually an active process of creation? Bergson would find some vindication in a denial of genetic determinism, but would most definitely reject environmental determinism as well.

7. We think here of certain ontological strands of contemporary philosophy, such as those which utilize Judith Butler's concept of "performativity."

8. Although it is a stretch, one could speculate that society in general also has a "time," possibly the *longue durée* of sociologist Fernand Braudel. We are not sure to what extent, if any, Bergson explicitly influenced Braudel, but one study points to an affinity between these two registers. As Ulysses Santamaria and Anne M. Bailey explain, in Braudel's model "temporality (. . .) is defined by duration, the measurement of which is both 'mathematical' and 'intersubjective.' The second foundation

of Braudel's notion of time is that of simultaneity, that is, the combined presence of the past, presence, and future in an object of historical examination" (Santamaria and Bailey 1984: 79). In a footnote the authors suggest that Braudel may have been influenced by Bergson in the positing of a distinction between measurable (homogeneous) and simultaneous (heterogeneous) temporalregimes. We could also point to Maurice Halbwach's ideas about social memory.

9. The predominance of actualization by no means excludes the seemingly repetitious evolution of certain structures. For instance, biologists have identified at least five distinct periods in which carcinization, the evolution of crab species from noncrablike crustaceans, can be identified. Even in a relatively short timespan, carcinization can be observed (Keiler et al. 2017: 200–222). A similar morphospace, along with the presence of organisms with suitable plasticity, results in the solidification or, alternately, softening, of certain durations, leading to repetitive results. But all this takes place upon the surface of a becoming, and always nonetheless differs somewhat from previous examples.

10. A plausible case can be made for instance that Alfred North Whitehead proved more capable of reconciling the notion of "instant" with the perpetuity of change than Bergson. Whether this evaluation, originating from Maurice Merleau-Ponty's lectures on the topic of "Nature" is adequate or not would require a separate study that goes beyond the remit of this book. cf. Khandker 2013: 64; see also Merleau-Ponty 2003: 118–119.

11. Iris Van der Tuin has done a fine job of defending Bergson against one particularly orthodox feminist critique. In an excellent and perceptive piece, Van der Tuin convincingly demonstrates that Bergson's model does not privilege spirit at the expense of matter (Van der Tuin 2011: 22–42). Just how such a gross misunderstanding could ever have occurred in the first place is beyond us.

12. In this sense then we cannot really assent to Catholic philosopher Teilhard de Chardin's divinization and personalization of Bergson's creative evolution. The elaboration of life does not lend itself to a creationist treatment, because in Bergson's conception there is no central organizing agency involved, neither is any final state or destination ever posited. Nothing like Teilhard's Omega Point can be discerned in the pages of *Creative Evolution* either. If anything, this spurious construct better resembles the outmoded ideas of Herbert Spencer. If Teilhard posits evolution as convergence, then Bergson would be the philosopher who posits "life as divergence" (Barthélemy-Madaule 1963: 276–280).

13. Provided we discount the Panspermia hypothesis, according to which life is not exclusive to Earth. Bergson himself adhered to such a view.

14. The commonality with Gilles Deleuze and Félix Guattari's idea of objects as self-organizing ("desiring") machines is striking. Bryant summarizes the Deleuzo-Guattarian concept of "machine" in the following, minimalist definition: "A machine is a system of operations that perform transformations on inputs thereby producing outputs" (Bryant 2014: 38). A machine need be neither rigid nor automatic.

15. Of course, meteors and spacecraft alike can play a similar role in the interplanetary, perhaps even intergalactic, proliferation of life. There is no need to tie the fate of life on Earth to any anthropological or civilizational register. If anything, the

rapid success of human beings on our planet has been deleterious to life on Earth in general, but it is an open question whether this necessarily need be the case or whether technological civilization can harmonize with a "nature" that has never been anything except "artificial."

16. To dismember and consume another is, in the imaginations of cultures which practice cannibalism, a method of reenacting the primordial unity of tendency, of getting back into contact with the formlessness of pure movement and actualization. Properly speaking, underneath the appearance of cannibalism, a more profound "cosmic consumption" may be discerned, an ambition to digest the entirety of becoming by partaking of the body of a being that symbolizes the whole (Kaliff and Oestigaard 2017: 105). Eating the sacrificial victim in this case becomes a way of becoming one with becoming in general. The consumption of a personage transports ritual participants into a sympathy with the original, primordial charity, generative of all things, the giver of life, and source of all movements.

Chapter 4

Leaping into Materiality

How can we expand our definitions of reality while accounting for all modes of being? If the image of reality is a translucent clarity, and the mind shall be taken to constitute a mere conductor of ecological information, as will be shown, then how may we account for the ontological circumstance of transmission? Supposedly intractable barriers segregating "matter" and mind, "memory" and perception, "virtuality" and actuality, will be proven illusory from a metaphysical standpoint. Instead of a dialectical reason, Bergsonism represents an accomplishment which reorients comprehension toward the affirmation of temporality purified of substance. The bad mixtures, the impurities, must all be sliced apart, preliminary to replacement by categories more in sync with reality. As Valentine Moulard-Leonard notes, the "Bergsonian 'unmixings' are always only the result of a violent effort aimed at working out the exact nature of the mixes" (Moulard-Leonard 2008: 125). In the case of *Matter and Memory*, the key question will be how to transcend the false dichotomy of matter and mind. As we have seen regarding the relationship of matter and life, the Bergsonian survey reveals the unity underlying supposed oppositions. But this union is never an unproblematic one. The impetus of life is animated by discord. A rejection of one dualism need not necessitate its replacement by a new pair. Instead of implementing a new duality, we ought to bring our philosophical intuition into as exact an accordance with the image as possible. A whole is not to be discerned, for no duration contains all durations, yet the accumulation of endurances will give the appearance of a cone-shaped set of worlds. As Deleuze emphasizes correctly, an impure mixture, inadequate to the task of revealing the contours of reality, can always be traced back to a badly stated problem (Deleuze 1997 [1966]: 19). The object of this chapter is to outline how, in *Matter and Memory*, Bergson transcends the mind-body problem, or rather, in what manner the entire problematic is displaced and deterritorialized in his philosophy of mind. At issue here is the ontological status of the "image" and "perception," as well as the impossibility of restricting these concepts to any idea of representation. Temenuga Trifonova's characterization of *Matter and Memory* as a philosophical project in terms of an "image ontology" is an apt description (Trifonova 2007: 25). The idea of an image which is *more* than a representation shall prove the key to unmixing the false dualities of "mind/body" and "mind/matter." At issue is the transformation of opposition to discordant heterogeneous continua. In the final instant, the image, in itself, will be the most pervasive element of reality. The image is the final issue of ontology, the sole category connecting all durations; omnipresent, images are the stuff of the world. Change is the eternal, modification is the persistent. However paradoxical it may appear, Jacques Chevalier's description of duration will be discovered to be sound: "The substance of duration is eternity; it is non-temporal, but rich in all that duration develops in time, an eternity characterized by movement and life, not by repose and death" (Chevalier 1928 [1926]: 77). The goal of unmixing is the reconstruction of a problematically agonistic unity. The engendering of representations is marked by the eternity of duration's permanent revolutions. An image is not the deceased remnant of a distant reality, the fading echo of a memory, but rather a component of becoming's continuity. There *is* no such thing as impotence. The privileging of ideas leads to rejection of immanence, a lazy ensconcement akin to the dream state. We have critiqued Deleuze's idea of the virtual extensively, but we have not yet explicated the final consequence of such a subordination of duration to virtuality. While Deleuze gives us the time-crystal, Deleuzian philosophy fails to draw out the full consequences of this discovery. In Deleuzianism we are left with a neutered impotentiality. The condition of virtuality is a passivity, a reactive state: the cinematic subject, the spectator immobilized or hypnotized by the pure (i.e., virtual) artificial image of time "records rather than reacts. He is prey to a vision, pursued by it or pursuing it, rather than engaged in action" (Deleuze 1997 [1989]: 3). Those enthralled by the idea of the virtual are enchained by the simulation, slaves of the matrix, degrading themselves to the status of neutralized, infertile representations.[1] The virtual never allows us anything like *unmediated* access to the eternity which is duration. We ourselves are images, unleashed into the world by contingency! *You are an image, but no screen can limit you. Releasing the fruit, enfolding the moment, holding it, time shoots out, spinning, into the infinite resolution of unfathomable distance*. The pure optical situation would, in Moulard-Leonard's view, correspond to immersion within duration, for in this "motor impotence" we would supposedly find a point of indiscernibility between "the real and the imaginary" (Moulard-Leonard 2008: 109). There is a substantial problem with such an emphasis on impotence and the imaginary, namely, that it has absolutely nothing to do with the ipseity of the image. Thinking in duration differs radically from impotentiality. Generally speaking, "potential" is absent from the Bergsonian doctrine. Neither can its negative mirror image, the posited "absence" of potential, have any substantial role to play in the framework we envision. Actuality is more than the unfurling of a preexistent potentiality. Neither can the virtual be conceived independently of the actual. As Gunter makes clear, the privileging of the virtual is entirely "a Deleuzean invention," and differs markedly from the Bergsonian view (Gunter 2009: 173). The virtual introduces an element of transcendence, whereas Bergsonism is entirely immanent. The eternal is immanent to the ephemeral. We take issue with any privileging of the virtual in the case of Bergson commentaries, because the false dichotomy of the virtual and the actual obscure of the purity of Bergson's image.

REALISM BEYOND MATERIALISM AND IDEALISM

The object of *Matter and Memory* is to affirm both "the reality of spirit and the reality of matter" through the analysis of memory (MM: 9). It cannot be emphasized enough that memory does not represent anything like a symmetric opposite to matter. Memory is the example through which false dualism of matter and mind will be transcended. The explication of memory from the outset is thereby subordinated to a greater metaphysical project. Consciousness is the internality of all, the supersensuality within superficiality. Spirit is that element which is ceaselessly "lighting up matter" (Chevalier 1928 [1926]: 34). Indeed, we may go further, and observe that matter itself is a process of accumulating enlightenment. To unmix a previous false dualism, a new concept is needed: in this case, this function is served by the image. "Matter," states Bergson, "is an aggregate of 'images,'" the latter being something "more than" a "representation, but less than" a "thing" (MM: 9). Before the process of unmixing the false problem can be achieved, we are encouraged to forget everything we have learned hitherto about the relation of mind and matter, representation and world, perception and reality. If we are to get anywhere, we ought to forget "the dissociation which idealism and realism have brought about between" existence on the one hand and appearance on the other (MM: 10). This reduction is designed to loosen our thinking, dilating the mental faculties, preventing superstitions and prejudices from obscuring perception. Far from implementing elimination or subtraction, Bergson's reduction slices through strata of deposited, fixed ideas, separating the mind from its habitual patterns of thought. Leonard Lawlor characterizes this type of reduction as "an act of liberation," an act of creation through which "Bergson himself is inventing the terms of the problem" (Lawlor 2003:

2). The process of enlargement will take us beyond the notion of image-as-representation toward the self-existing image. Above all, as Frédéric Worms points out in his study, Bergson's unconventional concept of an *objectified image* is what allows for "the suspension of theories of consciousness," which we have described in terms of an emancipatory bracketing of our existing knowledge (Worms 1997: 20). Registering the multivalency of durational rhythms, the intuitive view transcends dualism, grounding us within integral perception of endurance. As Worms emphasizes, it is not a question of setting up an artificial distinction between ourselves and the realm of images. In general, there is no difference between us and the images of the world. The recognition "this is my body" is the revelation of an image embedded within a heterogeneity of other images (Worms 1997: 21). There can be no question of dismissing the outside as illusory, of rejecting objects in favor of subjectivism, or, conversely, reducing the subject to a mode of brute materiality. In-betweenness characterizes, concretes, so to speak, the life of imaging. As Bergson writes later on, "our body" is "one image among others" (MM: 181). Information is contextual, and the body itself is a property, a structure for enfolding that projects itself into the yonder. Yet embodiment is also paradoxical because we can also view our body and its world from a position of externality. Such a cognitive specification will then result in the appearance of a fundamental disembodiment. What the purity of the image suggests is the presence of multiple pathways leading toward an, as yet, inconceivable goal. At this stage, life is locked within a body, a solidity obscuring its will to distension. At the other extreme, the living would explode out into emptiness, resulting in the actualization of a disembodiment already latent within the vital impetus. *Matter and Memory* is a project of singularization in Worms's view, an attempt to prove the singularity of the living body. Matter, defined as the multiplicity of images would, on this reading, reflect the singularity of lived, felt corporeality (Worms 1997: 25). Preliminary to any justification, there is the circumstance of projection. Having reduced our prior knowledge of theories of consciousness to nothing whatsoever, we can permit the philosopher to rework our mind, reconfiguring our thought to accept permeability, impermanence, and interpenetration. As distinct from Worms, Trifonova suggests that the authentic goal of *Matter and Memory* consists in a work of dissolution. Instead of returning us to a particularity, defined as embodiment, the idea of pure memory will function as a channel "dissolving us back into a primordial, impersonal state that ontologically precedes (and exceeds in 'value') our particular embodied existence" (Trifonova 2007: 26). On this view, the concept of "pure memory" in particular serves as a synonym for impersonality. To be fair, Worms also highlights the dimension of impersonality, on a strictly atheological basis. There is no divine perspective uniting all the durations within itself. As distinct from idealists such as Berkeley, in

Bergson "there is no one who may play the role of the universal witness, of the immanent gaze of matter reflected upon itself" (Worms 1997: 54). Never do the durations unite into a single whole entirely lacking some heterogeneity. Neither can the accumulation of durations be said to be heading toward the creation of a person, or a type of personage bearing any resemblance to the anthropomorphic. In this sense, the Bergsonian conception differs markedly from Teilhard de Chardin's Christological philosophy, for the latter identifies the cosmic process with some kind of personalization. Bergson's goal is not a reconnection of thinking to its body, nor the strengthening of our imprisonment within corporeality, nor the restriction of reflection to any kind of personality or individuality. Quite the reverse! This is made abundantly clear in the following simile, which describes the relation of consciousness to the brain: "That there is a close connection between a state of consciousness and the brain we do not dispute. But there is also a close connection between a coat and the nail on which it hangs, for, if the nail is pulled out, the coat falls to the ground. Shall we say, then, that the shape of the nail gives us the shape of the coat, or in any way corresponds to it?" (MM: 12). Consciousness, as primordial formlessness, is absolutely and unequivocally prior to the brain.

The various scientific examples gathered selectively by Bergson are all intended to underpin the emancipation of consciousness from the brain as organ. As the poet T. S. Eliot's mother, Charlotte Eliot (neé: Charlotte Champe Stearns) recounts, "in Bergson's emphasis on life, its power and indestructibility, I think some persons found an intimation of immortality, which excited their interest" (quoted in: Childs 1991: 476). If pure memory is indeed ever-present, then it can be surmised that its activity is perpetual. We may cite a recent neurological example, incidentally an otherwise healthy French man, who lives with the vast majority (90 percent) of his brain matter having been destroyed by an unusually severe case of hydrocephalus. What makes the case extraordinary is the relative absence of any abnormality on the part of the patient, who lives an ordinary life as a civil servant with an IQ of 75. Since a 2007 *Lancet* article, scientists have been mystified by the reason for such an apparently improbable persistence of brain activity in such a damaged organ (Feuillet et al. 2007: 262). In 2011, neuroscientist Axel Cleeremans proposed the "radical plasticity thesis," arguing that consciousness is not an anatomically fixed datum, but rather a constantly self-renewing learning process (Cleeremans 2011: 1–12). Were we to follow Bergson, we could posit an even more radical thesis, namely, that consciousness cannot be localized in the brain at all, not in spite of, but rather, precisely *because* of its status as *materiality*. Consciousness will be seen to consist in a completely actual pure perception. The stress placed in some commentaries on the positive being, even independence of memory from matter is intriguing, because Bergson is adamant that "memory (. . .) is just the intersection of mind and matter," having no positive being of its own (MM: 13). The relative cannot be transcended by an effort sinking us back into any of the terms of the preexistent duality. It is in *Matter and Memory* that the idea of "attention to life" is first introduced. By "attention," Worms notes, "Bergson means the process which regularly intervenes to supplement perceptions with memories," relating it to forgetting (Worms 1997: 116). Perception implies a process of selection that relegates irrelevant information to the periphery of our individuated consciousness. Through intuition, we can traverse the planes of consciousness, reaching ever more complex dimensions, achieving "a greater dilation of the whole personality" (MM: 14). At the extreme, the person as such dissolves into pure memory, the absence of selection between the three temporal modalities.

The methodological reduction abducts us, conveying our perception back to a pretheoretical state. We are "in the presence of images" (MM: 17). Our entire ecology is composed of secret vibrations, whose pacts, conspiracies, and alliances are constantly shifting, like the sands of an inscrutable geopolitics. Relations are distances, quite simply given, near and far, not produced by our awareness but already, originally there, in the form of real becomings. As distinct from the phenomenologists, Bergson includes the extra-perceptive dimension in the scope of his philosophy. In an article published in 1960, Georges Gurvitch even called Bergsonian philosophy a "new and comprehensive realism," eerily foreshadowing the new realist turn in early twenty-first-century Continental philosophy (Gurvitch 1960: 308). If the Bergsonian idea of an image endowed with objectifying power is indeed the key to breaking the stranglehold of mind/matter dualism, then we must examine this concept in particular, before proceeding to the other elements of the Bergsonian synthesis. The world cannot be a product of our mind, because if the world were to be subtracted, consciousness would cease to exist. Differently put, "an image may be without being perceived" (MM: 35). It cannot by any means be maintained that Bergson is a correlationist. Invisible elements too can persist as images. Manifestation, in other words, does not depend upon the presence of an individuated form of sentience. Performing a reverse reduction of sorts, Bergson imagines a subtraction of the world from consciousness: "Eliminate the image which bears the name material world, and you destroy at the same time the brain and the cerebral disturbance which are parts of it" (MM: 19). The reverse maneuver, the subtraction of consciousness from the world, does not and cannot undo the latter. Suppose all sentient bodies were exterminated, and no intelligent creatures were left in the cosmos. We will then find that "the picture in its totality, that is to say the whole universe, remains" (ibid.). It can be objected that such a move is impossible for *us*, some kind of observer being always implied in the example. Bergson's thought here is not terribly distant from Quentin Meillassoux's idea that the world is "ancestral," designating an "event" or existent "anterior to terrestrial life and hence anterior to givenness itself" (Meillassoux 2008 [2006]: 20). This material givenness composes an exteriority prior to the possibility of any representation or recomposition. Bergson's philosophy "cannot be seamlessly assimilated to phenomenology," not only because of its positive interpretation of indeterminacy—in Alia Al-Saji's view this affirmation of contingency goes against the phenomenological primacy of lived experience—but also because of the mind-independent objectivity accorded to the image (Al-Saji 2010: 160–161). Differently put, Bergson cannot be a correlationist (somebody who posits a necessary linkage between individuated consciousness and world), because the cosmic image is independent of the brain-image. Neither is the body anything like a central category, for it too is "an image which acts like other images, receiving and giving back movement" (MM: 19). *All images are equal.* That being said, for us, our own body certainly is of key importance, forming the center of our world. Situated as we are inside of our corporeality, this latter element causes us to view all manifestations in the environment as projections of this fleshy mass we inhabit. Perception is mostly the preparation of activities that shall subsequently flow out into our ecology. The body is destined to extract certain forms of information from its world, only to surrender its energies, giving back the excess to the outside. Movement can only be held in storage for so long: activity demands release. The body is nothing more than "a center of action" which "cannot give birth to a representation" (MM: 20). In other words, the images cannot exclusively be the products of a deluded, muddied mind. Moulard-Leonard is keen to remind us that many words in Bergson's philosophy should not be considered as adjectives but as substantives. For example, "number is a multiple" is an adjective, while "number is a multiplicity" is a substantive (Moulard-Leonard 2008: 142). The same can be said of duration. "Time is a duration" means something very different from "time *is* duration." And the same applies, of course, to the idea that "matter is imaging." Although not expressed in such terms, this is the point Bergson is attempting to make. The process of imaging cannot be separated from that of materialization. An even more drastic reduction awaits us. Not only can we imagine a sundering of consciousness from the world, we can even experiment with a removal of our own brains from our bodies! Although such an act comes highly unrecommended, we can examine its meager end result. Removing our brain will not accomplish a great deal. This is not much of a reduction: "What will happen? A few cuts with the scalpel have severed a few bundles of fibres: the rest of the universe, and even the rest of my body, remain what they were before. The change effected is therefore insignificant. As a matter of fact, my perception has entirely vanished" (MM: 21). However cynical this description of the situation may appear—losing one's brain must count as a significant event indeed in one's life!—the fact of the matter is that the images outside our body have survived this subtraction.

Movement is primary, while individual perception is but a sampling of images. What Bergsonism allows for is the introduction of pure, undomesticated mobility into philosophy. Nothing exists apart from images or movements. These two words are all but synonymous with each other. The images persist outside the body, but there may be no corporeal constitution outside of change. *Reality is mobile imaging.* The implication of the "Bergsonian revolution" is outlined by Brian Massumi in the following terms: "Position no longer comes first, with movement a problematic second. It is secondary to movement and derives from it. It is retro movement, movement residue" (Massumi 2002: 7). Lose images, and we have lost everything. There can be no imageless thought, as an image need not be conceived of as a definite construct which has congealed into anything like a representation or, worse, likeness. Images exist even after individuated consciousness has been removed. Therefore the brain cannot be the storehouse of perceptions. What then would perception be, if not a passive recording device for images? Bergson gives us a clue: "I call matter the aggregate of all images, and perception of matter these same images referred to the eventual action of one particular image, my body" (MM: 22). Perception occurs whenever an image comes to be folded over or, conversely, over itself. It is not a case of attributing something extramaterial to consciousness. Rather, what makes it impossible for us to restrict consciousness to the brain the omnipresence of perception everywhere in the universe! To better illustrate his point, Bergson introduces two artificial categories which will be of immense interest. "Pure perception" would be a state of perception free of time, a perception unattached to any pragmatic interest whatsoever, a collective becoming not yet sliced into heterogeneous, discontinuous elements, "an uninterrupted series of instantaneous visions, which would be a part of things rather than of ourselves" (MM: 65). Pure perception is perception wholly without memory. As Suzanne Guerlac explains, it would amount to "an immediate and instantaneous vision of matter" but such a vision is radically inaccessible to any human consciousness, for it "implies a radical impersonality, a total transparency, and a total interactivity" (Guerlac 2006: 109). Conversely, pure memory would be a brute, raw inexistence lacking activity, completely bereft of effectiveness. The virtual is the closest we can get to something approximating nothingness. In Bergson's words, "virtual, this memory can only become actual through the perception which attracts it. Powerless, it borrows life and strength from the present sensation in which it is materialized" (MM: 127). The virtual is parasitical to the actual, the former gaining all of its being from the fruits of the latter. It is erroneous to claim that "in Bergson's philosophy the virtual is always privileged over the actual" (Trifonova 2007: 35). If this were truly the case, we could never attribute a positive existence to the multiplicity of images in *Matter and Memory*. Images have a presence in excess of any access. Had Bergson attributed such a priority to the virtual, he would probably have entitled his work *The Memory of Matter*. But the goal of Bergsonism is something more than the reiteration of a tired, antiquated dualism. Rather, this project is an attempt to explode duality by subverting it from within. The images must necessarily be something radically distinct from both individuated subjectivity *and* memory.

A WORLD OF NONCORRELATIVE IMAGES

The separation of images from representation creates a home for imaging outside the confines of any skull. All differences are figured within the new, rarified, artificial thought of pure perception and pure memory. Ideally, imaging has no need for a body. It could very well do without such a channeling. But the limitations of matter entail a coupling between the corporeal and the mental. The material is a process of enlightenment, tending toward absolute impersonality. At some point, we may speculate that life will have succeeded in sloughing off its embodiment. Resampling its surroundings, each act of perception on the part of the organism is a preparation for openness, a nascent birthing of exteriority. Bergson is unambiguous when he identifies the image with movement. "The movements of matter," we are informed, are to be regarded "as images" (MM: 23). Instead of mutually exclusive atoms, the Bergsonian cosmology, anticipating many insights of quantum physics, views matter as an assortment of waves. We encounter here another reduction, the dissolution of particle into vortex! "Condense atoms into centers of force, dissolve them into vortices revolving in a continuous fluid: this fluid, these movements, these centers, can themselves be determined only in relation to an impotent touch, an ineffectual impulsion, a colorless light; they are still images. It is true that an image may be without being perceived—it may be present without being represented" (MM: 35). Matter is nothing like solidity. No body is wholly impermeable. The exploration of matter leads to a recognition of the transcience of all states. Contemplation of materiality engenders the mind, filling it with vortices that occasionally condense here and there into aggregates, only to split up once more, rejoining the flow. The fourth reduction, the *vortex-reduction*, is used by Bergson to further underline the mind-independent status of images. These movements are present, even if the individual brain is absent from the scene, or merely potentially present, hence the characterization of scientific analysis as "an ineffectual impulsion" or "colorless light." It cannot be the case that these changes have anything to do with our subjective states. We cannot always directly participate in the vortices, being constrained by the limitations of our constitution to a mode of mediated scientific observation. There is no method of shrinking one's brain to the size of a molecule and inserting it into a stream of molecules, in the manner of Greg Bear's sci-fi novel *Blood Music*. Human perception can still elongate itself though, penetrating into pure immediate perception of matter through an act of intuition. Solidity is an appearance, underneath which the Bergsonian finds that "a body, even a corpuscule, is nothing more than a set of qualities" (Dolbeault 2012: 30). As we shall see, Bergson's view of matter as pure vibration has important cosmological implications. Vibration will come to symbolize the entirety of transcience and impermanence. Outside the confines of this body, the world is an assemblage of images. Matter is imaging. Contemplatively, Bergson invites the reader to exercise a submergence within this medial system. The image of the vortex functions as a spiral, a mode of immersing brain-matter into a larger organic form. "Intuition," the enlargement of consciousness beyond the human condition, "is the genuine experience of matter" (Lawlor 2003: 7).[2] A vicariously correlative structure does pertain occasionally pertain, but only between images in our vicinity and our body-image. If the microscopic vortex is inaccessible to our perception, being the object of pure impersonal perception, then the intermediate level is the realm of correlation *for us*. "Images themselves cannot create images," but when emplaced near an intelligent, individuated body, "they indicate at each moment (. . .) the position of a certain given images, my body, in relation to the surrounding images" (MM: 23). Pure perception represents the upper bound of intensity, ascending upward to a level far beyond our intermediate human state. Here, in the intermediate level dimension of individuated living organisms, perception has been canalized. The body pollutes or obscures pure perception, restricting its gambit, narrowing it down for the sake of practical utility. Not all things have relevance to a limited body. A life that observes all objects would, *per definitionem*, be of infinite extension. Such a being would have nothing to do with the world, but then what use would the world be for this infinite creature? Perception is a preparation for action, an extension of the organism's acts in virtual, that is, not yet existent, form. The "zone of indetermination" specifies the powers of the body, as encapsulated in the life of an individual, denoting the "degree of independence of which a living being is master" (MM: 32). Bodies are media writing freedom into materiality. It is the body, moving freely, which "introduces a 'zone of indeterminacy' into the system," unbundling complexity (Guerlac 2006: 110). The living is the unpredictable. Where there is a surprise, there is also life and freedom. With the introduction of a pragmatically oriented, selectively perceptive body, the Bergsonian philosophy itself starts to become more complex. The complication here is that among images, we are now forced to differentiate those which exist for themselves and those which are relative to the "central image" of the body. It is the body, defined as the interrelation of the body-image with external images, which Bergson somewhat confusingly calls "consciousness" (Mullarkey 2000: 53). The image can thus operate in the mode of an image-subject. A body, as distinct from an object, is capable of acting voluntarily. An important metaphysical goal of Bergson's philosophy, including *Matter and Memory*, is the replacement of the mind/matter dualism with a new difference, namely, the distinction between two "types of action, determined and undetermined" (Mullarkey 2000: 54). To avoid misunderstandings, we must speak of this "centrality" of the body in a qualified manner. One could get the erroneous impression that this amounts to an ontological privileging of the living body. But why then would Bergson claim that our body and brain alike are merely "images among images"? The centrality of "my body" is not a product of Bergsonian philosophy, being instead a given which is already relevant to our own experience. There is no question then of the body-image or what we term image-subject being in any ontological sense privileged. This is an appearance, stemming from our own practical involvements in a world. When divorced from the subject—remember, Bergson is no phenomenologist!—images do not depend upon the presence of corporeality centrality nor subjective experience. "Of the aggregate of images," writes Bergson, "we cannot say that it is within us or without us, since interiority and exteriority are only relations among images" (MM: 25). Memory, it will be shown, is a contraction, a folding of images upon themselves or one another. There is more than a merely quantitative difference between an image and a representation. The paradox of Bergson's use of "image" lies in the fact that, while the word "literally means 'copy,'" the "Bergsonian image is not a copy of a hidden thing" (Lawlor 2003: 6). This image stands on its own, actualizing itself independently of individual consciousness. Relative to the body, the image becomes the repository of a living body's virtual actions, while the noumenal or objective aspect of the image reserves its power all to itself.[3]

What Bergson takes issue with is the notion that perception must be thought of as an epistemological category. As Guerlac makes clear, in *Matter and Memory* Bergson advocates for the view that "perception serves action, not knowledge" (Guerlac 2006: 111). As a consequence, perception corresponds to actuality. In its pure form, perception gives the intuition of the instant as a momentum of imaging. The image-subject is the location where activity is entranced by sensual multiplicity, rendered separate from its otherwise pure actualization. Our body is the image in us, serving a public translation of latent image data into consciousness. Perception is a dissevering of the perceived state, being constantly reunited with a corresponding material continuum. Every act begins from scratch, as it were, writing itself into a graphically conceived unity of composites. At the most basal dimension, pure perception cannot be differentiated from matter. It is life lived within the impersonality of alternation. As Moulard-Leonard underlines, "there can only be a difference in degree, and not a difference in kind, between matter and perception" (Moulard-Leonard 2008: 24). The images cannot originate from the brain. They are already pervasively in the world, which is a scenery replete with graphics. Prior to the individuation of our consciousness, the images already display divergences among themselves. A perfected image would be wholly exterior to any body, but there are interior images too. Bergson calls the image which has been folded onto itself an "affection," an image subordinated, virtualized by a corporeal constitution (MM: 52). Therefore, we have here a new dualism, one pertaining between images for themselves and images serving as reflections for the virtual acts of bodies. The virtual is the degradation of the image, the diminishment of its power. In Guerlac's summation, "not all of the action received by external things is reflected or deflected. Some is absorbed—this is affection" (Guerlac 2006: 115). Virtualization entails a restriction of the image's scope of being. Neutralized, it is held in place by a corporeality that acts in a manner akin to a net. A virtualized state is a sorry condition indeed, a state of entrapment. When locked into an affection, the image functions as a signification of the body's range of decisions. "Consciousness," after all, "is choice," but once it has been localized, this indetermination is always necessarily inferior to the perfectly unbounded free image (Moulard-Leonard 2008: 28). The organism in general is a demoted, posterior manifestation of absolute freedom, being selectively closed with regard to its environment. The brain is a selection mechanism, a net sifting through elements, choosing among the true differences, reducing the authenticity of ontologically pregiven difference to something approximating a lie or caricature. In itself, the nervous system has no power of its own: "The brain is no more than a kind of central telephonic exchange: its office is to allow communication or to delay it. It adds nothing to what it receives" (MM: 30). Imagine a telephone conversation in which the technological object itself were to participate! This would be the stuff of comedy, but in the case of a brain that interrupts a communication, we lean more toward considering such a case in pathological terms. Images flow either way. In themselves, the becomings of a material universe reduced to ceaseless motion are independent of any mind. We need not stuff a brain into every nook and cranny of the cosmos in order for processes to endure, nor does perception necessitate such a move. All perception signifies is a choice, a partial indetermination, situated within a certain durational slice. In David Lapoujade's characterization, "a matter reduced to movement" necessarily implies its promotion "to the status of virtual consciousness" (Lapoujade 2018 [2010]: 43). Once again, we find in such a description the pernicious influence of the Deleuzian misreading of Bergson. The virtual pertains to the body and its diminishment of pure perception. Only after it has been specified and particularized, that is, imprisoned within an organic, finite body, does perception come to pertain to virtual actions. As distinct from any interpretation privileging the virtual, we must point out that pure perception is the immensity of activity, free of any localization whatsoever. Bergson's own words give a far more concise explanation than we ourselves can:

> That which distinguishes [the] present image, as an objective reality, from a represented image is the necessity which obliges it to act through every one of its points upon all the points of all other images, to transmit the whole of what it receives, to oppose to every action an equal and contrary reaction, to be, in short, merely a road by which pass, in every direction, the modifications propagated throughout the immensity of the universe. I should convert it into representation if I could isolate it, especially if I could isolate its shell. Representation is there, but always virtual—being neutralized, at the very moment when it might become actual, by the obligation to continue itself and to lose itself in something else. (MM: 36)

We propose a radical move, namely, to absolutize this view of representation as mere virtuality. In our view, Bergson is trying to say that representation in general is necessarily virtual, corresponding to the domain of powerlessness. The image comes to be stripped of actuality once it transforms into a representation. There can be no question of an actual present image inside of representation, for at the very moment a brain exchanges real sensations for memories, the image morphs into affection. Despite appearances, Lapoujade avoids the all too frequent mistake of Deleuzian commentators of Bergson by correctly identifying duration with a qualitative moment of actualization. We read further on that "duration exists once an instant, however brief it may be, conserves what it receives from the previous instant, even if only to transmit it immediately" (Lapoujade 2018 [2010]: 43). The admission that duration can indeed be instantaneous is of immense importance for our own investigation, for we maintain that *duration cannot be anything other than actual*. Nothing like a ripening can be discovered either yesterday or tomorrow. Fructification is happening right now, we hold the softening peach in the clasp of our hand at this moment. Once the future duration will have arrived, actuality shall reign, spreading its light, enlivening our taste buds, filling our stomachs with sweet juices. For this occasion of plenitude to happen, an actuality must be present. Never have the hungry been fed with virtual loaves of bread. In the pure, objective, real image, action brooks no delay. It is what it is, embedded in the simple moment corresponding to the entire range of positions it occupies along the spiral of becoming. Resonances are all that remain if we subtract representation from the scene. Individual consciousness is at once the injection and reduction of indetermination. Through the zone of indetermination it exudes into its environment, the living body produces surprises, but the outside will always remain infinitely more complex than this body. The ateleological world of images is central to nothing apart from itself. It is therefore only the individuated, impure perception which constitutes "a variable relation between the living being and the more-or-less distant influence of the objects which interest it" (MM: 33). Without affection, the image can be everything and anything. This is not to say that we are striving to demonize affect. Rather, we are emphasizing that incorporeality and impersonality are original features of actualization. Ironically, we take Massumi's description of virtual as being pertinent to the actual! "The body is as immediately abstract as it is concrete; its activity and expressivity extend, as on their underside, into an incorporeal, yet perfectly real, dimension" (Massumi 2002: 31). We advocate for a reversal of this description to bring it into accordance with the Bergsonian position. On the basis of *Matter and Memory*, we must say that the activity and expressivity of enlightening actualization extend into the inner, dark cavity of the individualized body. Incorporation represents a contraction of actuality, in terms of deactivation. Once deposited back into a body, activity exchanges a loss of overall energy in exchange for a concentration of attention. Attentiveness, in other words, is gained at the expense of an overall reduction in intensity. Against the commonsense view, Bergson argues that *intensity negatively correlates with concentration*. The more contracted the activity of the images becomes, the more their intensities are diminished. Unconscious pure material intuition is unfathomably more complex than any individualized form of perception: "The perception of any unconscious material point whatever, in its instantaneousness, is infinitely greater and more complete than ours since this point gathers and transmits the influences of all the points of the material universe, whereas our consciousness only attains to certain parts and to certain aspects of those parts. Consciousness—in regard to external perception—lies in just this choice" (MM: 38). The outside displays a greater intensity of energy and complexity than any localized, organic, pragmatically oriented manifestation. We would not be overstating our point if we were to call the organic body a simplified *caricature* of the infinite flow of transient imaging. The aggregate of images alluded to by Bergson is consciousness *prior* to its canalization. Absolute freedom can pertain exclusively to something absolutely inhuman. This point is of key importance and recurs in contemporary treatments of Bergsonian philosophy on a more or less constant basis. Trifonova refers to the intuition of matter as an "aggregate of images," as "the inhuman state of the world from which the human perspective has been excised, the mode of existence of things before the birth of consciousness" (Trifonova 2007: 36). In a similar vein, Lapoujade outlines the thought of the nonhuman and its relation to Bergsonism as a recognition of the circumstance that "there is basically nothing human about man. It is because intuition reaches the nonhuman tendencies in man that it can give the reciprocal impression of humanizing the nonhuman" (Lapoujade 2018 [2010]: 47). Beyond representation, there is the extro-correlative, the intuition of the outside absolutely foreign to thought. The image is absolutely different from a representation. The goal of *Matter and Memory* is to transcend what Keith Ansell-Pearson, following Edward Casey, calls "the language of containment" (Ansell-Pearson 2018: 77). On this view, consciousness can be localized, mapped in terms of brain states. Bergson's aim is to extend the concept of consciousness, exploding it out into the world. In the space of actuality, there can be no difference between memory and perception, for the intuition of materiality unites both in a moment of conservation (Ansell-Pearson 2018: 78–79). What Bergson will call memory will be something which differs from the actual, a lack of power and absence of being which nonetheless has a positive role of sorts to play in the cosmology of images, if only as a parasitical, vampiric element living off the graphic excesses it isolates.

There can be imaging without affection, but affection is unimaginable without a prior subtending imaging. By comparing the brain to a "central telephonic exchange," Bergson mobilizes a late nineteenth-century semantics of the "body electric," as exemplified by Walt Whitman's poem of the same title, to cite one example among many (Menke 2005: 638–664). The metaphor of life as electricity was a salient theme of contemporary scientific and pseudoscientific representations alike in the final decades of the nineteenth century. At the time the new idea of electromagnetism was used by many as a productive metaphor, linking various levels of reality together in a new cosmology. Even toward the close of the twentieth century, there have been recurrent attempts to articulate a view of life in electromagnetic terms (cf. Becker and Selden 1998). At a later point, Bergson himself references the latest advances in the then emerging field of electromagnetic physics, citing Kelvin and Faraday's discoveries and their ontological implications. For Bergson, these scientific results prove that materiality is a heterogeneous continuity, and not a collection of spatially discrete particles (MM: 200–1). Electromagnetic phenomena proved that "matter is a set of elements whose lines of force are more or less interpenetrated," while matter in general "can be understood as a continuum in which vortices occur, producing a kind of division in the continuum" (Dolbeault 2012: 28). By consequence, this means that matter cannot be conceived of as a set of "contiguous points or parts," a view that presages the concept of matter as wave in quantum physics (ibid.). If consciousness is choice, then complexity is the infinite heterogeneity prior to the decision. The zone of indetermination is but a screen, reducing complexity so the organism can make a choice among alternatives it has constructed for itself. It is not an issue of selecting among finished alternatives or pre-existing molds presented by the ecology to the individual. Certainly, these too exist. But in general, in order to become registered as an alternative, an external image must be recorded, that is, diminished into a representation. Only after this cognitive enclosure has been achieved does choice become possible. Eclipsed, rendered into a background feature, complexity does not for all this disappear into nothing. Quite the reverse: without a hypercomplex simulacral world, the organic body itself could not even manifest, let alone persist. We must remain true to the Bergsonian intuition of impermanence: "Everything is motion" (Fell 2012: 34). The individual mind is no exception to this rule. Brains are the products of the organization of the body. If impure individuated perception corresponds to the organized aspect of reality, the realm interiorized by corporeality, then pure perception must be that dimension which is the imaging aspect of the real. This latter is defined as a relative exteriority, relative to *this* particular center of action, *our* body. We can define this duality Bergson obtains as a set of mutually exclusive relations that bisect reality along the lines of a new differentiation of pairings (Figure 4.1.).

Corporeality can also be called a *limitation* of imaging. The difference between matter and the perception of matter will be akin to a difference between "the duration of the perceiver and that of the perceived object" (Trifonova 2007: 37). Materiality itself, defined as the aggregate of all images, has a duration of its own, an inhuman or prehuman temporality which differs incommensurably from the human experience of personal time. Images arrive, at the speed of light, from exteriority. Impure perception slows them down to a speed suitable for translation into enclosed and embodied information. Attention divides the real along the lines of a vitally determined pragmatic interest. Those elements we term "sensory" are the stuff of consciousness, as incorporated within internality. Of course, every duration has an inner aspect, an excess beyond contact. Therefore, the images in themselves have another interiority, one which differs from the corporeal. Matter has a foreign life, an inorganic, unincorporable mobility. One critique leveled against the Bergsonian philosophy of life is that it overly privileges mobility

Perception Mode	**Function**	**Product**	**Topographic Position**	**Mode of Embodiment**	**Liminality**
Impure perception	Organization	Organ	Internality	Corporeality	Finitude
Pure perception	Imaging	Image	Exteriority	Incorporeality	Infinity

Figure 4.1 The Relationship of Ontological Modalities in *Matter and Memory*. *Source:* Author created. Not required.

at the expense of immobile forms of life, connecting vitality to the sphere of intellect while defining the latter as "one among many instances of active evolutionary inventiveness" (Marder 2013: 165). The problem with such an activist view for Michael Marder is its downgrading of relatively passive life forms to a kind of second-class status. What of the plants and other, relatively immobile creatures? Marder points toward the necessity for reconceiving the relationship of life and movement. From specificity, we must shift in the direction of infinity and generality, without thereby positing any false universality that can be localized. The subversive import of Bergsonism lies in its positing of an "x" without restricting this agonistic universal indetermination to any regulative specificity: "Bergson encourages the kind of thinking that thinks with life, not against it. Whether it has to do with the plant or the human, it thinks points toward the thinking of life itself, a de-formalizing activity that, when inserted into the categories of conceptual thought, implodes them from within" (Marder 2013: 166). But what of the movements and vortices of matter itself? Extending ourselves beyond the *élan vital* toward the mobility of all images, we ought to dilate thinking, making it stretch until it expands beyond any center, into a thinking-together with materiality. Mysteriously, we shall then find our own durations coincide with instantaneity. Images cannot be intracerebral, for they are already unleashing their qualities and indeterminations in continua foreign to our limited modes of perception. It will be found that "the image is material life," insofar as it displays a continuity occupying a certain inorganic duration (Lawlor 2003: 8). What the nervous system does is utilize a certain share of material vibrations, integrating them into its closed self-programming. If the organism is a closed system, then the material life of the images is an infinity of openness. On the one hand, outside a body there can be no representation, only an excess of external vibrations, reflecting endlessly upon the surfaces of a spiraling heterogeneous continuity. The image, on the other hand, cannot be given in its immediacy to any body in particular; these two modes of being are radically distinct. To quote Bergson, "There is no image without an object," because the image itself is the real object, in all its infinite, irreducible complexity and inexhaustibility (MM: 44). While a nascent future is always crawling upon the body of the present, it shall only ever succeed in making manifest a novelty if coupled with an act. And the activity is only born if movement extends unopposed into a present moment. Duration is the processual fusion of elements in an energetically charged intensive continuum. "The present," writes Moulard-Leonard, "is essentially that which acts; it is power. The past, on the other hand, is that which does not act any longer; it is nonpower" (Moulard-Leonard 2008: 30). If pure perception is therefore immediate intuitive deliverance to the activity of materiality, then pure memory will be the absolute absence of power, the impotential nonpower living on sedimentations of dead activities. There is no place for potential in the Bergsonian system, and the virtual, in the form of memory, will itself be a passive, deactivated re-presentation of enfolded images.

PURE PERCEPTION AND MATERIAL LIFE

So far we have described the status of reality and matter as outlined in *Matter and Memory*, as well as the place of the body in this cosmology. To reiterate, "organized" bodies are considered as "centers of real action" (MM: 31). In their material aspect, living systems belong to actuality, like anything else which has achieved manifestation. Actuality is attained within pure perception. That being said, there is no individuated perception which is not already shot through with memories. Confusingly, Bergson uses the word "consciousness" in two very different ways. As Joël Dolbeault indicates, consciousness can mean a "reflective activity," a narrowed act of representation on the part of an individuated living thing, while Bergson also attributes a broader meaning to consciousness conceived of as the generality of "perceptual activity" (Dolbeault 2018: 557). The latter corresponds to what Bergson names "pure perception," the overflowing perception situated nowhere in particular. The elements psychology terms "sensory" are nothing if not localized, domesticated perceptions that have come to be subordinated to the vital interests of the organized body. Far from denoting an infant or dormant perception, it appears that the narrowing of consciousness into a body is a diminution of sorts. Centers of action take activity to a new level, but there is no escaping the feeling that the process of life is also a fall of sorts, a contraction. After all, Bergson will use the latter term as a byword for memory. As compared with the overwhelming luminosity of pure perception, defined in terms of an immediate intuition of materiality (i.e., the totality of objects or "images"), the body is something obscure, dark, foreboding, even menacing. There is a hint of the Gnostic at work here. Bergson's entire project revolves around the recapture of immanence and the elimination of dualism, but paradoxically it must utilize an asymmetric dualism to achieve this ultimate reintegration of reality. The body, as the site of production of representations, will be necessarily inferior to the aggregate of images, in terms of its overall complexity. The increase of intelligence, the accumulation of memory, is achieved at the expense of a narrowing of consciousness. Once organized, a body becomes that through which perception does *not* pass:

> The luminous point gives rise to a virtual image which symbolizes, so to speak, the fact that the luminous rays cannot pursue their way. Perception is just a phenomenon of the same kind. That which is given is the totality of the images of the material world, with the totality of their internal elements. But, if we

> suppose centers of real, that is to say of spontaneous, activity, the rays which reach it, and which interest that activity, instead of passing through those centers, will appear to be reflected and thus to indicate the outlines of the object which emits them. There is nothing positive here, nothing added to the image, nothing new. (MM: 37)

We are the internal darkness we cannot touch. To survive, the organism must maintain its closure vis-à-vis the outside world of exterior images. In the absence of closure, the rays of the sun would permeate our bodies, but nothing like an individual life could remain intact. *Obscurity* is the price for *having* a life. Centers of real action are those areas of reality that have succeeded in shutting themselves off from pure perception. They have built borders around themselves, enabling the implementation of intelligence. The brain is the organ allowing the living body to delay its reactions to outside effects (MM: 30). It is no exaggeration to state that in Bergson's philosophy of mind, the brain, far from acting as an enabler, is a retardant, an obstacle to perception. As a mere instrument of selection, the brain adds nothing to perceptions. No new image is created by individuated forms of consciousness. This model represents an example of a "selectionist" theory of memory and cognition. What happens when a living thing senses or perceives its environment? On the selectionist view, abundance is primary and representation secondary to this initial burst of information: "There is an initial proliferation of neurons and synaptic connections and a subsequent elimination or pruning of the initial diversity" (McNamara 1996: 219). In Bergson's case, diversity is an original fact of reality, a characteristic of the world conceived as pure perception. There is already an infinite wealth of images prior to the narrowing of consciousness in life forms. The Bergsonian "image" is very much non-correlational, for it is an objective process, extending itself in its world. We are therefore in complete agreement on this point with Trifonova, who writes that in *Matter and Memory* "representation is not conceived in terms of projection (i.e., addition) but in terms of immanence or virtuality (i.e. detraction)" (Trifonova 2007: 40). Representation very much belongs to the virtual dimension, which we define as the aggregate of domains of subtractive dimensions of being. Anything attesting to the presence of a subtraction corresponds to the virtual, the ineffectual, the less than real. Gathering independent lines of facts together, individuated consciousness forces them to correspond to its pragmatic interests. Every act of perception, when localized, becomes parasitic, a shallow recreation of complexity. For all this, we are not seeking to diminish the role of recollection. Such a view would be just as misguided as a privileging of the virtual at the expense of the actual. Perception, when applied by living bodies to their needs, is manifestly virtual and simultaneously existent. Our contention does not consist in a rejection of the very being of the virtual; rather, we merely strive for a recognition of the objective priority accorded by Bergson to the *real* qualitative multiplicity of images. We must remember that "the representation is less than the image, which is connected continuously to other images in the whole. A representation is a part cut out of the whole; representation is a decomposition of the whole" (Lawlor 2003: 10). For this reason also, it cannot be maintained that virtuality is ontologically equal in stature to actuality. Were Bergson to uphold such a doctrine, his philosophy would regress to the very idealism he seeks to transcend. The new realism Bergson advocates for is predicated on the recognition of the ontological universality of both impermanence and indeterminacy. Images too are endowed with a type of freedom, and while it may smack of anachronism, it is not entirely implausible to suggest that this ambitious speculative metaphysics anticipated many later developments in chaos theory as well. During the late twentieth century, the prevailing cosmology has substantially shifted in favor of a general recognition of indeterminacy on the level of material time (Antoniaou and Christidis 2010: 185–202). In their floating state, unattached to particularized forms of consciousness, images are even freer than the corporealized zones of indetermination. Abundance is the predominant characteristic of pure perception: individuated awareness appears to be nothing, if not an impoverishment of the initial pure heterogeneity. The brain "localises only one part of mind, which Bergson describes as the 'actual' phase of a continuous process," writes Mullarkey (Mullarkey 2000: 55). But this actuality is embedded, enmeshed within a broader actual world of indeterminate images. As a local zone of indetermination, the body is but one crest of a far vaster wave of chaos.

We are shown, for perhaps the first time in modern philosophy, raw brute being. Far from being separate, let alone sovereign, corporeality is from the outset subordinated to the presubjective dimension of becoming. As Maurice Merleau-Ponty says in a 1959 speech given in commemoration of Bergson's 100th birthday, what distinguishes the latter's philosophy from other approaches is its resolute commitment to the presubjective. By describing subjectivity in terms of "a circuit between being and myself," it is shown that "the spectator exists 'for being,' and Bergson thereby "regains at the heart of man a pre-Socratic and 'prehuman' meaning of the world" (Merleau-Ponty 1977 [1959]: 185). Each "discernment" on the part of individual consciousness is also a "dissociation" or disengagement "from the image" (MM: 45). In all cases, the abundance is primary and selection only comes after the initial wealth. We cannot say that representations are the products of coding either. In a selectionist theory of perception, there simply is no need for us to define perception in terms of an encoding or retrieval, for the images are already there prior to consciousness (McNamara 1996: 220). To describe the activity of the brain in terms of decoding information would be a misstatement on the Bergsonian view because this would attribute productivity and novelty to a space where there is none. Rather, the nervous system inhibits and impedes consciousness. It testifies to the speculative force of the Bergsonian system that it is capable of organizing itself around such a counterintuitive perspective on consciousness. Ordinarily, we are prone to thinking of consciousness as being a process localized in brains. But nervous systems are just half of the story. Pure consciousness is the second, broader form of consciousness, a full awareness that has not yet undergone organization and canalization. To be organized means to be restricted to an organ. Representations, certainly, are products of the brain, but these are diminutions and not products or even supplements. A representation adds nothing to an exterior image, Bergson is quite clear on this point. The centrality of corporeality is but one stage on the path to the recognition and metaphysical articulation of pure perception and pure memory. Instead of a fundamental ontological basis, corpocentrality is an artifice. The body is a center for us, and not for being in general (MM: 46). Therefore it also cannot be maintained that affect would have anything like a central ontological role to play in *Matter and Memory*. The affective is just as subordinate to ontologically active and prior imaging as memory is to activity. We must at this point refer back to the Bergsonian subtraction of the world of images outlined above. As will be remembered, this consists in the elimination of the world, defined as the totality of images. It is shown thereby that neither body nor brain can survive the elimination of the objective dimension. Nothing persists of consciousness without objects of consciousness (MM: 19). Lacking a material foundation, bodies lose any possibility to channel their surroundings. They cannot even exist, but the images nonetheless continue to actualize in the absence of any body. Lawlor contrasts the Bergsonian reduction with Edmund Husserl's move. On the surface, the two positions appear somewhat similar. In paragraph 49 of *Ideas I.*, Husserl also imagines the destruction of the world but draws a very different conclusion: while consciousness "would indeed be necessarily modified by an annihilation of the world of physical things its own existence would not be touched," as "no real being (. . .) is necessary to the being of consciousness itself" (Husserl 1983 [1913]: 110). The dividing line between phenomenology and Bergsonism can be discerned, in Lawlor's view, from the diametrically opposed conclusion the two philosophies draw from the annihilation of the world. Simply put, "for Bergson, after the annihilation of the world, there is no residuum of consciousness" (Lawlor 2003: 13). On Bergson's view, nothing can survive without a world, for even if we accept an extended, expanded idea of consciousness as ecologically pervasive, pure perception still depends, in the final instance, upon the presence of objective images. Therefore, no common ground can be found between the Husserlian and the Bergsonian positions, their answers to the question of "what happens in the absence of a world?" being radically incommensurable.

Perception, once restricted to a body, accumulating in the shape of interiorized, domesticated images, becomes an affection. Far from denoting an original aspect of the world, affection is a late by-product of evolutionary selection. Whether we are describing the feeling of a simple or complex organism, in affection, "perception and movement" are always "blended in a single property—contractility" (MM: 55). The negativity of this language is telling. As distinct from the, at times rather excessive contemporary emphasis on "affect," exemplified by such discourses as the so-called affect theory, Bergsonism does not attribute a central role to the affective dimension. Sensation is the contraction of images, as localized within a finite, striving body. Were we to manifest in the form of limitless beings, allowing our bodies to extend into the infinite reaches of outer space, unchallenged by any material obstacle, amalgamating with continuity itself, there could be no question of a definite feeling or any concrete sensation. However uncomfortable such an affirmation would appear, the status of affection as reduced, impoverished image implies that each feeling is a primordial suffering, an adversity resulting from its very status as limitation. Every enjoyment is but the blunt side of a blade of anguish. "There is in pain something positive and active"—what does this mean? (MM: 54). On the surface, it would appear that Bergson is expressing excessive optimism, naively attempting to project something meaningful and positive into a negative affect. Excavating deeper, however, we find a greatly more important meaning in this cryptic sentence. We must ask the following question: *from where does an affection borrow its actuality and activity?* Already, our query contains the hint of an answer. The affect, in itself, can contain nothing positive in itself. The positivity ascribed to the feeling of pain is, in truth, an element arriving from outside the subject, properly speaking. "Affection must," Bergson continues, "at a given moment, arise out of the image" (MM: 55). Differently put, the objective sphere, the totality of images, is the final origin of affection too. In itself, affect is nothing if not a representation, but its content is dependent upon an objective scaffolding. Verily, the true source of power lies in the objectivity of an image. Always, we must excavate further, until we have intuited the authentic wellspring, the originality underlying appearances. Without going beyond the given state, our intuition would remain a pale representation. The aim is to break the stranglehold of representation until our perception becomes one with the vibrations of the material images. Only in this way shall we go beyond the superficiality of constructive and destructive, wholesome and poisonous feelings. We experience pain because of the local nature of all emotional phenomena. The sense of isolation and impotence is what produces pain in the case of the organized, living body (MM: 56). The experience of suffering points toward an underlying pessimism regarding the status of affection, for the latter itself is, by consequence of its very structure, a tormented finitude. Emotion should not have the final say when it comes to the registration of becoming. There cannot be anything like unlimited, infinite, unbounded affect. Beneath each affection, there is the prehuman, even prevital image, itself inseparable from the process of its own imaging. The affective is a limitation of the graphic.

Individuals are becomings that function in the manner of restriction and contraction. If the function of the nervous system is to restrict the gambit of perception, then the organized body must be viewed as an instrument for the absorption of images. Affection is a sufferance, a passive absorption of exterior images by the organic body (ibid.). It is not a genuine approval, but rather a toleration of invasiveness. We are perpetually being violated by the outside. Corporeal integrity is a quaint illusion, considering the real permeability of corporeal boundaries. Nowhere can we find a body that is entirely closed. It must be emphasized that when speaking of closure and self-restriction on the part of the organism, we have been describing something more akin to a nascent ambition or tendency rather than a statement of empirical fact. There is much proof that the development of species is influenced in large part by cues furnished by other life forms. Speciation is at least in part a symbiotic process (Gilbert 2016: 415–433). Evolution can be viewed as the product of continuous violation, transgression, and insinuation. As we have seen, Bergson defines the *élan vital* in *Creative Evolution* as a subversive, invasive assortment of waves, smuggling life into matter. A similar relationship can be outlined in relation to the connection between organic bodies and images. To feel means to be already invaded by images: an affection is an undergoing. The circumstance of owning a finite body limits the range of encounters we are capable of surviving. Not all endurances were made for human access. In this sense, a tardigrade is far superior to a human being, for it is capable of accessing a wider variety of environments and images. The crux of the pessimistic Bergsonian idea of affection-as-limitation is that the individualization of images, as effectuated by the perception of a particular body, also results, of necessity, in a virtualization. Once rendered particular, that is, once the material life of an image has been decomposed into a representation, it begins to refer to a "virtual action" (MM: 57). Bergson describes the relation of action to corporealized and individuated perception (remember, we are speaking here of consciousness in its individuated modality!) as an increase or decrease in distance. At first, such a spatialization may appear puzzling, considering that the Bergsonian doctrine in large part rejects any equation of the durational with space. What is at stake here is the relation of corporeality to exteriority. Bergson hypothesizes that those organisms displaying a greater variety of abilities are also capable of extending their perception toward greater spatial dimensions. When situated far from the place of action, the organism is virtual. Drawing closer to the place where the act shall proceed, perception becomes more actual. The paradoxical implication of this description is that in order to actualize itself, the body must distance itself from its given state. To actualize ourselves, we are compelled to leave our present condition. Perception is the mode we actualize ourselves into an assortment of exterior images. As distinct from affection, perception is always an exit from a prior state, an ecstatic movement away from fixation. For this reason we agree with Lawlor's assertion that "Bergson" is "reattaching perception to the real. Since the image is one image in the whole of the material world, it acts like other images" (Lawlor 2003: 14). *When it acts, the body is exterior to itself.* Bergson could not be clearer: "My perception is outside my body and my affection within it" (MM: 57). Even in its nascent, individuated, narrowed condition, perception is tending toward infinity, dragging the body along in a circular swirl of accretion. The revolutionary import of Bergsonism lies in the recognition that perception does not originate inside a body. In Renaud Barbaras's words, "perception" here "does not go toward exteriority, it proceeds from it" (quoted in: Moulard-Leonard 2008: 27). Only once it has become enfolded within the darkness of a body can perception be said to constitute an inner aspect. This latter state corresponds to a diminishment of perception, hence the following description: "The truth is that affection is not the primary matter of which perception is made; it is rather the impurity with which perception is alloyed" (MM: 58). Resulting from the frustration of the aspirations of the body, affection is inherently negative, a subtractive emergence. This does not entail that inner states are ever simple. If anything lies further from Bergson's goal, it would be a simple description of mental states or memories in terms of a quantitative accumulation of homogeneous data. Pure perception is already operative, prior to the development of any localization in a nervous system. As Bergson states, "perception, in its pure state, isolated from memory, does not go from my body to other bodies; it is, to begin with, in the aggregate of bodies, then gradually limits itself and adopts my body as a center" (MM: 61). This latter limitation is a fall of sorts, a degradation, and diminishment of perceptive intensity. The medium reinforces itself through the elaboration of cell walls and immune systems, but this cannot efface the original condition, the mutual, absolutely continuous openness of images. Bergson advocates for something bordering on panpsychism, for perception is, in the final instance, extended to matter itself. The images themselves are endowed with a form of consciousness, although never is this preindividual, prehuman, and prevital consciousness of matter described in any great detail in *Matter and Memory*.[4] Perception in general denotes a "power of action" on the part of any existent whatsoever (MM: 64). The power of perception already brings in its train a power of activity. To acquire extension, an image must act upon its surroundings. The complexity of a living body can be thought of as a function of the variety of activities it is capable of, as well as the sum of images it can absorb without succumbing to their forces. Life is unavoidably implicated in a sequence of graphic expressions, not all of which are under its control. What Moulard-Leonard calls an "objective tone," or the force of an image, is what gives occasion to an affection (Moulard-Leonard 2008: 28). Of course, one could object that the idea of pure perception is an abstraction. Bergson himself does not deny that it serves as more of a logical extremity than a description of perception in practice. At its extremity, perception blends with the aggregate of images. Ultimately, at the very edge of this extension, the difference between the material life of the images and pure perception vanishes. "Every perception," as Elena Fell reminds us, "is prolonged into a potential action" (Fell 2012: 45). This process is the automatic invasion of the outside, the piercing of the suffering organism by its ecology as well as other life forms, passing in and out of a perforated life.

How does memory figure in the process of perception? Specifically, what is the place of memory in the suffering, punctured vitality of the lived, felt level of durational endurance? The body, as a contraction of perception, delineates a mode of ecstasy. If we can only be ourselves outside of ourselves, in exteriority, does this not qualify as the reintroduction of a transcendent element into philosophy? Commenting upon the relative status of exteriority and interiority in *Matter and Memory*, Lawlor suggests that Bergsonism represents a philosophy of immanence. "Absolute immanence is absolutely outside," writes Lawlor, for Bergson "has separated immanence from consciousness and identified it with the non-conscious. In other words, he has made immanence and the outside synonymous, in order to make consciousness be the inside of the outside: consciousness is an inside relative to the outside" (Lawlor 2004: 37). The ecstatic being of the subject points toward a relativization of the notion of "interiority." No longer is this ontological modality synonymous with consciousness. Rather, it is one of many "inner" modes of duration. Resulting from the aspirant nature of temporal flow, interiority is a product of any folding whatsoever, describing the making-impotent, the relative becoming-powerless of time. What Bergson does is retain the notion of interiority, while sundering it from exclusive attachment to the human subject. We should not let ourselves be misled when we read of "our" inner experience or affections. This interiority does not equate to the totality of all imaginable inner time aspects. Human beings have internal temporal folds, but so does everything else. Few have summarized this exteriorization of internality in clearer terms than Lawlor. The body as center of action is not the sole centrality, for a core may be found anywhere durations are undergoing a flowing. Corporeality is but a concretion of the greater, immense perceptual interchange: "My perception, in its pure state, isolated from memory, does not go from my body to other bodies; it is, to begin with, in the aggregate of bodies, then gradually limits itself and adopts my body as a center" (MM: 61). Selectionist theories of mind work with an initial "proliferation phase," during which memories come rushing into the brain. The role of the nervous system is to choose from among these exterior elements, excluding superfluously useless remnants. What commentators such as McNamara miss is that proliferation is an original aspect of becoming. If defined as the activity of memories, we will reach the erroneous conclusion that "the past is the active agent" in Bergson's philosophy of mind (McNamara 1996: 226). But this circular reasoning is based on a more fundamental mistake, namely, the definition of proliferation as the active self-multiplication of memories. In truth, the aggregate of images already composes a qualitative multiplicity *prior* to the advent of the narrowing of consciousness. Authentic activity is both *exterior* and *prior* to the body. Immanence is the outside, because perception is never entirely our own, extending far into the immensity of worlds.

The further it dilates, the purer perception becomes. The impersonality of matter, as represented by pure perception is, to quote Anthony Feneuil, "upstream of personality," while the rougher, less refined impersonality of space is "downstream" (Feneuil 2015: 308). Activity is a function of distance. If the individuated body remains far from the perception, it is enclosed in a mode of virtuality. In itself, without access to the pure activity of perception, the body is almost nothing. But if it chooses to take leave of itself, tending to close this gap through the enactment of a movement, the body allies itself with the cosmic activity of pure perception. Absolute immanence is a function of our distance from ourselves, and therefore must be considered synonymous with an increase in our power. Efficacy is never under the exclusive ownership of the organism. Hence Lawlor's apt description of Bergson's project in terms of a separation of self from identity: "In order to understand this unity of the outside and immanence, I think one has to recall the ambiguity of the Greek word 'autos,' between *ipse* and *idem,* between self and identity; it seems as though Bergson retains only the *ipse* and eliminates the identity. In short, we have a same which is not identical" (Lawlor 2004: 37). Perception in its pure state is akin to a frameless vision, a thought without borderlines, a flow lacking pinpoints and eddies. What the positing of pure perception allows Bergson to achieve is a decentering and relative recentering of the body, without making the latter into an absolute core of all activity. Absolute immanence situates activity in the outside, while corporeality is the interiority born from the depletion of this energy, individuation being equivalent to dissipation. Every perception is an "experience of force," to borrow Lawlor's expression, but this drive comes from an outside to which we ourselves are endogenous (Lawlor 2004: 38). The aggregate of images is the source of action, for the body itself, insofar as it remains anonymous, is also an image among images. As Dolbeault reminds us, "the contingency of physical time is nothing else than its indeterministic nature, the consequences of which are unpredictability and irreversibility" (Dolbeault 2012: 34). Indetermination is already in the environment. It would appear then that the body, in its individual mode, can be identified with the pole of powerlessness, that is, memory. What Bergson suggests when equating the activity of the body with exteriority is that the corporeal can only become active if it moves away from itself into proximity with pure perception. The further it is removed from its original condition, the more active the body becomes: "The more distance increases between this object [the external object] and our body (. . .) the more does virtual action tend to pass into real action" (MM: 57). Simply put, the reality of an action is a function of the self-alienation of the body. We cannot, for all our agreement with Lawlor, assent to the idea that "pure perception" would be something "purified down to a point," being a mode of materiality having "no memory and therefore no duration" (Lawlor 2004: 27). This otherwise impeccably careful reading slips into the mistake of viewing matter as being "outside" of duration, reactively reinscribing the very dualism into Bergson's work which the latter is attempting to transcend. What Lawlor misses is that durations in Bergsonism are inherently multiple, arranged in a nested hierarchy of temporalities. Pure perception, defined as the immediacy of material vibrations, "however rapid we suppose it to be, occupies a certain depth of duration" (MM: 69). There is no such thing as a purely "momentary" vibration; even the smallest of particles takes time to unfold. The phrase "moment" must be taken to describe a flow. As Gunter points out, Bergson's intuition resembles a differential calculus, integrating a whole continuum of durations: "Each thing that exists has its own duration. However, each integrates specific lower durations into its own character. What one differentiates at a lower level becomes part of any integration" (Gunter 2005: 147). The momentariness of matter is deceptive, for it too is already endowed with a certain temporal flow. As we have seen with the example of sugar melting in water in *Creative Evolution*, every process takes time. We are not constrained to equate duration with the presence of a living subject of experience. Later on, Bergson describes the material realm of vibration as "an ever renewed present" (MM: 151). Even an apparently momentary, transient duration already takes a certain time. Its temporal quality sets it apart as an individual. Pure perception would be a duration without a corresponding individuality, a flow lacking moments. It can be surmised that this is the reason why Bergson characterizes absolutely unmixed perception as a fiction. An entirely transparent duration would form a constant, unceasing activity. Such a perceptual flow could never cease acting. Individuation occurs whenever hesitation enters into the picture.

Bergson introduces the concept of memory to account for the way bodies desist from action. In the world of finite organisms, duration is constrained to limit its objectifications. Images do not image all the time. Their conveyance of qualities, the spillage of heterogeneity and indetermination into the environment, is limited by the contours of each particular image. What memory does is introduce a gap between acts, allowing for the preliminary evaluation of virtual actions. The selection among memories occurs along the lines of utility (MM: 66). It is in this sense that McNamara describes Bergson's theory of memory as "Darwinian," for "what is chosen is what is used" (McNamara 1996: 223). For all their work of individualization, representation and memory alike do not succeed in entirely blocking perception as the preindividual, prevital basis of subsequent developments. Perception is always penetrating memory. Again, we find another perforation, this time the puncture of memory's virtuality by perception's activity. Underneath the conserving operations of memory, perception is constantly at work, eliminating unnecessary memories: "An impersonal basis remains in which perception coincides with the object perceived and which is, in fact, externality itself" (MM: 66). If there is really no such thing as an absolutely pure perception, neither can there pertain a truly unmixed memory. "These two acts," Bergson continues, "perception and recollection, always interpenetrate each other, are always exchanging something of their substance as by a process of endosmosis" (MM: 67). The selection of images always proceeds according to the pragmatic needs of the organism. Discrimination opens a pathway, traced from the observed interplay of ecological elements arising in the proximity of the body as relative center of action. A certain synchronicity is in play between organism and environment, memory and perception. That being said, the body itself also has a duration of its own. The notion of "biological time" has become an accepted element in contemporary biological discussions. It is widely recognized that bodies have temporalities of their own. Gunter points to the affinity of Bergson's philosophy with the "chronobiological" school of twentieth-century French biology, exemplified by the pioneering work of Alexis Carrel and his student Pierre Lecomte du Noüy. Carrel, an avid reader of Bergson and, one could argue, an enthusiastic "practitioner" of this philosophy, believed in the radical, even potentially infinite extensibility of life, managing to keep a tissue of dead chicken cells alive for a remarkable twenty-nine years (Gunter 2005: 150). Lecomte du Noüy observed wide divergences in the time it takes for wounds to heal, known as "cicatrisation" and, building on this line of empirical research, recommended the introduction of the category "physiological time" alongside psychological and physical temporalities (Gunter 2005: 151). While we certainly do not wish to credit Bergson with the work of these biologists, it is known that the philosopher corresponded with Lecomte du Noüy, and in a letter addressed to the scientist writes approvingly of the idea of physiological time (ibid.). This digression is necessary, to show that duration cannot be restricted to any single dimension. We find Gunter's characterization of duration in general as a temporal hierarchy sound, for it corresponds to the Bergsonian intention. In every situation, several times are in play, but it is always an actuality that synthesizes the various moments. Bergson himself speaks with enviable clarity: "The actuality of our perception (. . .) lies in its activity, in the movements which prolong it, and not in its greater intensity: the past is only idea, the present is ideo-motor" (MM: 68).

CORPOREALITY IS THE LIMITATION OF PERCEPTION

How then does the past coexist with the present? More to the point, how can the pasts of one duration coexist with the present of *another* duration? In a highly interesting speculative move, Fell refers to Saint Thomas Aquinas's description of angels to illustrate the ontological relation of nested Bergsonian durations to one another. For readers unfamiliar with Christian theology, angels are immaterial lacking bodies while still being individuated. This presents us with an ontological conundrum. Let us—for the sake of argument—accept their existence. How then can we categorize existents if they lack a definite extension? The totality of unextended but nevertheless individuated angels bears a striking resemblance to Bergson's idea of the multiplicity of continuous yet heterogeneous times. A duration, like an angel, lacks a spatial extension, and the various durations permeate one another, yet they still compose a multiplicity. Bergson clearly differentiates between greater durations, such as the great wave of life (the *élan vital*) and "smaller" ones, such as the lived experience of an individual subject or the—for us—"momentary" vibrations of molecules. Aquinas's solution is to distinguish between different angels "according to the diversity of their powers," a move that generally entails a recognition that we can "distinguish qualities within a heterogeneous multiplicity by relations that they emanate" (Fell 2012: 88). Applying this to the difference between perceptions and memories, it can be stated that memory conserves, albeit in a powerless, unreal form, a slice of reality, while perception exudes a real act into a world. We can maintain the fundamental unity of durations, without entirely collapsing their distinction. Nothing, not even memory, can change the fact that perception, viewed in itself, constitutes "the act," the medium through which "we place ourselves in the very heart of things" (MM: 67). Action is exclusive to perception, while its traces pertain to memory, the latter being a contraction of the former. If reality is activity and movement, the past is unreality, the density of a time whose thickness is inert, immobile, a passive extension. In the words of the philosopher, "the past is essentially that which acts no longer" (MM: 68). It is for this very reason that, despite their inseparability, we are obliged to maintain a very real difference in kind between these two temporal modes. The past is far from absolutely unchanging. It can become retrospectively modified, but the genesis of this alteration lies in a present which is presented to the past in the form of an *open* future. As Fell observes, "the past (. . .) constantly changes, affected by the events that follow" (Fell 2012: 96). Not without basis, we can speak of a *backward causality,* but solely from the side of the past. Resituating our perspective into this present, however, causality is seen to emanate from the moment of internal duration, the internality peculiar to any duration pertaining now. Not even in the context of its own domain is memory sovereign. This quasi-extension is always being reworked by the exigencies of the present. Selection from among memories is also an act pertaining to the actual. What memory strives to accomplish is a synthesis of the momentary vibrations. Synthetically material lives are incorporated within an organized, subordinated, limited, finite stream of vitality. Memory, in its essence, describes the translation of physical time into physiological time. Bergson says as much, defining memory as a "translation" which, though "the original pales," could never be effectuated without the prior presence of a prior original (i.e., objective image) "localized" by the finite body's necessities (MM: 60). The instantaneous is never a point, but a flow, while corporeality is an eddy within the perceptual current, an obscurity blocking the dispersal of light.

The function of memory is the condensation of representations. But images themselves are by no means stored exclusively within the confines of a nervous system. Such a pale, organic emplacement would prove far too restrictive for the entirety of images, as imaging strives toward maximum dispersal. Throughout its expansion, the image is accumulating ever more energy. Rather than losing its power, light only adds to its efficacy during the course of its proliferation. Properly speaking, there is no such thing as a "memory-trace," so-called, in Bergson's philosophy of memory, because memories are not stored exclusively in the brain (Mullarkey 2000: 58). They have something of the absolute mobility of the absolutely exterior yet immanent "images." Memory is the process through which qualitative succession comes to be synthesized or contracted into a continuity accessible to the organism. Without a mode of condensing and localizing images, the body could not connect itself to pure perception, the latter defined as absolute exteriority. Because of its status as an object subjected to the sufferance of finitude, the organized body is forced to gather images. Those images which prove incapable of resisting its entreaties are absorbed into the mnemonic faculty. If pure perception "is impersonal and coincides with exteriority itself," as Guerlac and we ourselves also maintain, then what does the suggestion "the rest is memory" mean for us? (Guerlac 2006: 119). Underneath each condensation of memory, there is the trace of prior imagings. No memory can remain intact, not even as an illusion, without the prior persistence of materialized durations. Below memory, we find the same vibrations that underlie every mode of being: "The qualitative heterogeneity of our successive perceptions of the universe results from the fact that each, in itself, extends over a certain depth of duration and that memory condenses in each an enormous multiplicity of vibrations which appear to us all at once, although they are successive" (MM: 70). The moment is replete with subordinate durations, which can be identified along the lines of their effects and the relations they mobilize. Like an angel, a duration hides within a succession, and it is the role of memory and perception alike to tease out its unique singularity for the sake of pragmatic application. It is when perception and activity are both interrupted that memory can make its advent, injecting falsities where previously activity reigned. Finite bodies are constrained to rely upon memory. The more complex an organism, the greater the amount of hesitation it can afford. Alongside selection we find that the brain also contributes to life by the insertion of non-acts into the stream of action. Not only does it "allow vibrations to communicate with a chosen motor system," but the brain is also capable of making "vibrations wait before the plurality of motor possibilities" (Lawlor 2003: 17). Fine shades of imaginal differentiation precede the mobilization of the body toward the furtherance of its survival. Lacking the prior plenitude of the material vibrations, we could not speak of such a selectivity. That which is given to the individuated consciousness is the relational aspect of matter. Bergson's statement, to the effect that "there is in matter something more than, but not something different from, that which is actually given" must, therefore, be read in light of the dependence of memory upon the absolute immanence of material durations (MM: 71). We cannot concur with Lawlor when the latter writes that "memory, for Bergson, defines duration" (Lawlor 2003: 19). More exactly, this applies only to durations which have come to be absorbed by the operations of the finite body. But this function in turn is dependent upon the presence of images given in the organism's environment. It is the present that changes the past, while the past cannot impact the moment of this duration. "Later elements," writes Fell, "contribute something to the content of earlier elements" (Fell 2012: 97). Neither memory nor representation supplements reality. At some point during the process of recall, there must intervene a true movement, a mobility outside memory. We find in pure perception, especially in its close affinity, even unity with the vibrational plenitude, an imaging power independent of any corporeality. Bergson is very clear on this point. When differentiating matter from spirit, the philosopher writes that we must "eliminate all virtuality, all hidden power from matter" (MM: 72). In this manner, a scission will be introduced between the ethereality of spirit and the effective force of matter's own becomings. This latter we choose to term "imaging," defined as the power of objective images to create movements. Insofar as it remains anonymous, corporeality itself forms part of this ceaseless actualization. The codependent nature of memory and perception is a projection of consciousness, which is forced to remain powerless at times, conserving its energies while hesitating. "Memory," being inseparable from perception on the level of the individuated body, "imports the past into the present, contracts into a single intuition many moments of duration, and thus by a twofold operation compels us, de facto, to perceive matter in ourselves, whereas we, de jure, perceive matter within matter" (MM: 73). We must not lose sight of the fact that this synthesis of durations achieved by memory is undergone in the context of a present. Having no effectiveness of its own, the mnemonic faculty is a virtual density, operating upon quasi-extended remnants of sensorimotor reactions, external images, and their temporal relations. The past, entirely lacking in presence, is dependent upon a superior presence. Its process of conservation could never amount to anything without the fundamental imaging always already under way, leading the body to new engagements, transitions, inserting the living thing into new probabilities, pushing it along lines of facts. The spirit is characterized by a supreme *lack* of capability.[5] When Deleuze maintains that "according to Bergson," each image "distorts" "pure memory," this is quite erroneous (Deleuze 2000 [1964]: 59). Quite the reverse is the case: it is the advent of memory which results in the production of ghosts and illusions. Never shall memory give us the thing itself, the image in its immediacy. "Pure recollection," Bergson says, consists in the "representation of an absent object" (MM: 75). This surprisingly commonsense view contrasts markedly with the Deleuzian outlook, the latter always being intent upon attributing being to that which is not. The mnemonic is a productive synthesis, but it produces absence. The same applies to spirit. Certainly, memory has a function. Living things are forced to engage in many acts we may describe as "constructive negations," relations that contribute to a "balance between presence and absence" (Fell 2012: 94). Memory is the manufacture of absent objects, at times willed, but more often than not, automatic; its function consists in the production of constructive negations that permit the organized body to desist from action, allowing it to prepare the next reentry into action.

Before continuing our investigation, it is worth reflecting upon the full import of the Bergsonian idea of an objective image. As we ourselves have argued, the image in *Matter and Memory* is vastly more than a mere representation. This non-correlative aspect has not been lost on one of the greatest contemporary critics of "correlationism," Quentin Meillassoux. Although the self-described "speculative materialist" philosopher never mentions Bergson by name in his great work, *After Finitude*, Meillassoux dedicates one of his lesser-known pieces, entitled "Subtraction and Contraction," to articulating the materialist and immanentist import of Bergson's *Matter and Memory*. The goal of the article, it is true, is directed primarily toward untangling Deleuze's philosophy, but this is achieved through an extensive treatment of Bergson. This roundabout hermeneutic stems from the method Meillassoux chooses, one which bears a resemblance to our own, for it makes the original text under consideration "not the object, but the instrument of the elucidation" (Meillassoux 2007: 65). Throughout this book, we ourselves have utilized Bergson's texts as instruments for unlocking the meaning of time-as-enduring actuality. Meillassoux's provocative suggestion is that, had Bergson remained faithful to the absolute immanence of images, as outlined in the first chapter of *Matter and Memory*, without diluting or contaminating pure perception with the contracted form of memory, then this would have resulted in a genuine philosophy of immanence which could have transcended the Kantian critique's limitation of metaphysics. In the manner of a detective novel, Meillassoux attempts to show where things went awry. An image is the sum of its appearances, in a general sense, and not merely in relation to a particular consciousness. Appearance has been liberated from enclosure within the subject-object dyad. The problem is that Bergson then upholds "that the theory of pure perception, true in principle, is not so in fact" (Meillassoux 2007: 68). In other words, had Bergson remained true to pure perception, then we would obtain a genuinely undomesticated, unlimited form of immanence. As Bergson underlines, "the object is, *in itself*, pictorial, as we perceive it: image it is, but a self-existing image" (emphasis mine, Bergson MM: 10). The implication of this move, the introduction of an idea of the image as self-existent appearance in itself, is for Meillassoux a radically anti-Kantian move, a rejection of the Kantian Copernican revolution. We obtain an ontological extension of the notion of appearance, rendering any correlation of the thing-in-itself with a subject-centric philosophy impossible. If everything is appearance, the playing field of things is leveled. Every single form of transcendence is eliminated from ontology, provided we adhere to the objective concept of "image." Key to this reading of *Matter and Memory* is pure perception. As we ourselves have noted, perception for Bergson works in a subtractive manner. This implies that "there is less in perception than in matter" (Meillassoux 2007: 72). Compared to the infinite complexity of the outside, the body is an impoverishment. Perception, once canalized and trapped in a finite body, comes to resemble a destitution of sorts. What Meillassoux suggests is that individuation is a restriction of the images: "For Bergson, perception is not a synthesis but an ascesis" (Meillassoux 2007: 75). Somewhat confusingly though, our commentator fails to differentiate between the two senses of perception used by Bergson. Pure perception is most definitely *not* a subtraction, for it composes a unity with the images themselves. When Meillassoux speaks of perception as impoverishment of the real, this ought to be understood in relation to its individuated, restricted manifestation. Perception is only subtractive when it is finite. Meillassoux's problem with Matter *and Memory* is that Bergson fails to remain true to the idea of pure perception as immediate intuitive access to matter as vibration, for in subsequent chapters Bergson proceeds to introduce memory as a form of contraction, vitiating a nascent philosophy of immanence by smuggling in a renewed transcendence. In the second chapter, memory is integrated into the Bergsonian system, a memory conceived of as a virtual force transcendent to materiality. As Meillassoux laments, "from that point on, matter becomes what remains of perception once one has retracted that which memory, in its two forms, continually introduces into it" (Meillassoux 2007: 78). It is not so much memory as recall which vitiates immanence, but rather the second form, namely, memory as contraction. With the introduction of memory as a supposed power of contraction, it would seem that the cognizability of matter in itself, the direct access of material vibration through pure perception, is fatally compromised. Once again, the noumenal is reintroduced, bothering the immanentist immensely. Whereas in the theory of pure perception "representation added no sort of synthesis to matter," consisting actually in a subtraction from the complexity of exteriority, "everything changes with the contracting theory of memory; for now perception once again introduces a form into matter" (Meillassoux 2007: 83). Provocatively, Meillassoux proposes that we could delineate an entirely immanent philosophy based exclusively on Chapter One of *Matter and Memory*. Constructing a fictional theory from scratch, we could act as if the subsequent three chapters of the book had never been written at all. Are historians of philosophy not constrained to act in a similar manner when engaging with the long-lost works of the pre-Socratics? Arguably, even were we to have only the first chapter of *Matter and Memory* in our possession, this would still amount to significantly more than the scant information we have on the philosophical systems of such obscure figures as Heraclitus or Epicurus. If the dual doctrine of pure perception as access to the infinite appearances of images and representation-as-subtraction are both maintained, this "amounts to according to matter all the rhythms of duration, and to making of human perception not the *contraction* of material quantity, but the *selection* of one of the rhythms of a matterimage which contains each and every one of them" (Meillassoux 2007: 87). We are left with pure material becoming, defined as the sum of heterogeneous multiplicities. Everything is change, but within these changes, relative immobilities may be found, what we have called eddies and Meillassoux describes as "interceptions of flux" (Meillassoux 2007: 89). The fact that some existents appear to display persistence should not entail the positing of anything like permanence or a substantial presence. Each immobility is itself an assortment, an assemblage of flows which have reflected onto themselves: "Flows, left to themselves, are just (. . .) pure mobility, immobilising themselves by the very fact that no obstacle obstructs their deployment" (Meillassoux 2007: 88). There is absolutely no lack of change, for even supposedly permanent elements such as laws are but eddies in the river of becoming. Puzzlingly, Meillassoux does not seem to notice that the reintroduction of the notion of "virtuality" seems to compromise the pure immanence he is after. The abyssal thought Meillassoux cannot accept is that becoming and flux are one and the same, that nothing substantial is truly produced when all is said and done. Nothing of the virtual could survive an authentically completed immanence. The (relatively) permanent, the "break," has, by his own admission, been degraded into "a detour of flux" (Meillassoux 2007: 92). All this does not contradict the Bergsonian idea of reality as change. For our purposes, the most important point recognized by Meillassoux is the implication, extracted from the subtractive view of perception, that bodies are nothing but "rarefactions" of pure perception. Opposing any privileging of corporeality, Meillassoux upholds, in the style of a true philosophical radical, that the living body is "a localised impoverishment of fluxes" (Meillassoux 2007: 97). *Corporeality is poverty*, when compared with the infinite complexity of the outside. Meillassoux also equates embodiment with dissipation. The organism, insofar as it closes upon itself, results in a progressive self-erasure. Embodiment is originally complicit therefore with reactivity (Meillassoux 2007: 102–103). Meillassoux distinguishes between two modes of death. The first would correspond to absolute closure ("reactive death"), while the second would be more akin to a completed openness. This second form Meillassoux calls, in a deliberately oxymoronic vein, "creative death," a term which undoubtedly would greatly disturb Bergson (Meillassoux 2007: 103). The second form of death results in an opening of the organism to the infinity of images. Breaking free of its ascetic automatisms, the organism would melt into an objective totality. At the limit, self-subtraction as ascetic practice results in a destruction of the ascetic imperative to self-enclosure. Losing its extension, particularity would expand into a coincidence with externality. Positionality is traded for translation into a becoming-invisible, an "infinite madness." On this reading, the good death is an explosion of the self into the nonrepresentational fusion of all qualities, creating a home in the original spontaneity:

> Becoming-material would be the effacement of the selection of images. And it would seem then that to make an image of death, we would have to conceive what our life would be if all the movements of the earth, all the noises of the earth, all the smells, the tastes, all the light—of the earth and of elsewhere, came to us in a moment, in an instant—like an atrocious screaming tumult of

> all things, traversing us continually and instantaneously. As if the nothing of death could not be understood as a simple void, but on the contrary only as a saturation, an abominable superfluity of existence. Death, thus understood, is the triumphant reign of communication. To die is to become a pure point of passage, a pure centre of communication of all things with all things. (Meillassoux 2007: 104)

Creative death is not a negativity on this view, but an "atrocious" positivity, a saturation with excess, conceived within the immediacy of communication. A moment is obtained which has been completely purified of both life and finitude. Among the suffering, sentient beings in the universe, were they to be assured of such coincidence with all things, which among them would not desire the vivifying prospect of an instant explosion that eliminates all desire and sufferance? Bergson himself would most probably take issue with the above description, with its provocatively negativist characterization of reunification with pure, immediate perception of materiality as "death." Could this melting into the all-in-all not also be described in terms of an expanded, dilated, bolstered, widened life? The dissolution of the organism, the abatement of world-engaging sensorimotor reactions is a negation, this is true, but could it not be the spotlight, the ray of hope that reminds us that infinite joy can be just as imaginable as the unavoidability of suffering in a samsaric world? The living being is "passivity," a "passability" restricting the scope of being as becoming (Meillassoux 2007: 101). If this is the case, then the prospect of an unlimited openness coming *after* life must fill us not with dread, but with a hope the intensity of which permeates and enlivens our vitality in the here and now. The infinite madness Meillassoux speaks of, the discarding of cognitive selection, can be regarded as an image of commonality. An image without ownership, an imaging without differentiation, an intensity without anticipation, a heterogeneity without distinction, would all this not contain a hint of the paradisiacal?

HABIT AS MOTOR-MEMORY

In the following, we must strive to find an answer to the dilemma emphasized by Meillassoux: does the introduction of memory as contraction by Bergson vitiate or limit his own philosophy of immanence, as contained in Chapter One of *Matter and Memory*? Furthermore, does the advent of memory endanger our own view of duration-as-actuality? It is not the presence of sedimented motor mechanisms that appears to endanger the previously posited purity of immanence, so much as the supposed "independence" of recollections. Is memory truly independent of the present? Bergson himself describes our body as "an ever advancing boundary between the future and the past," a "pointed end" through which "our past is continually driving forward into our future" (MM: 78). Evidently, past and future relate to actuality, conceived of as a presence. We refer here to a cryptic passage from Deleuze's *Bergsonism*, in which being is equated with the past: "The present *is not;* rather, it is pure becoming, always outside itself. It *is* not, but it acts. (. . .) The past, on the other hand, has ceased to act or to be useful. But it has not ceased to be. Useless and inactive, impassive, it IS. (. . .) It is identical with being in itself" (Deleuze 1997 [1966]: 55). Referring ourselves back to the example of the lived present, persisting in a relation of interpenetration with the flow of times, inserting its impotential interruptions into the flow of other durations while being itself shot through with alien, inorganic images, it can be said that becoming itself, on this view, is naught, while that which no longer acts, is. The inanity of such a position renders it almost parodistic or, better yet, parasitical in relation to the Bergsonian construct. We cannot accept that Bergson is saying that becoming is impotence. A real process of ripening is under way, reality *is* in the making, albeit not in a substantial manner. Nothing is gained by introducing, as Deleuze does, a "useless and inactive" remainder in the form of memory into the flux. In simpler terms, if we seek to remain faithful to the Bergsonian doctrine of permanent becoming, then we cannot assent to any equation of being with the static. That which acts, is. *Flowerings coalesce into ripenings*. Memories, it is true, persist after the brain, the organ of selection, is severed from the body, but this by no means implies that such a virtual persistence is endowed with anything like a positive being. Lacking a mode of canalization, the past image is deprived "of all means of acting *upon the real*" (emphasis mine, MM: 79). This use of words exposes Deleuze's sleight

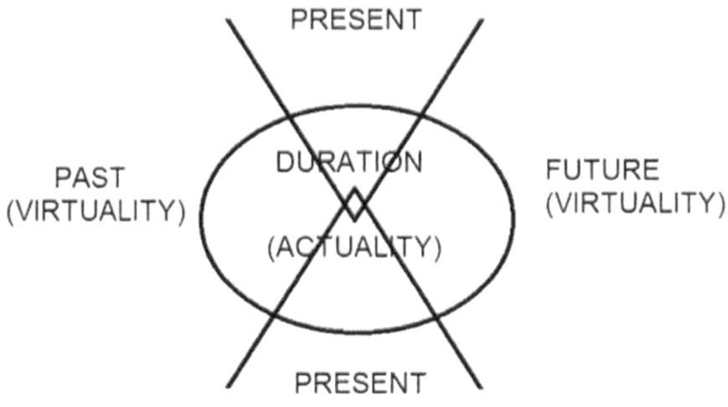

Figure 4.2 The centrality of duration can be clearly discerned in relation to the three temporal modes. Duration is the purest form of the present, being the unifying temporal synthesis preceding relative pasts, presents, and futures. The point-like place wherein ascending and descending actual temporalities meet is what we have chosen to term the "gate of the present." *Source:* Adapted with modifications from: Scmidke (1989: 228)

of hand, for Bergson explicitly speaks of memory as composing something *other* than reality! Duration, as Charles Schmidtke emphasizes, "represents a 'moment' of time," a qualitative heterogeneity that connects all three modalities of time. In truth, duration corresponds to the plane of the present, being the synthesis of both past and future, the saturated hole through which the flows escape into invisibility (Schmidtke 1989: 228) (Figure 4.2.).

Everything of importance happens in the pulsating pure present. Memory itself has no positive presence of its own. Mnemonic persistence does not engender any new objective image, for it is a derivative of the images, stemming from the individuation and declination of perception in the form of representations. Once a living organism integrates and subordinates a selection of images to the dictates of its life, memory is already in play, but solely as a form of impotentialization. As Rudolf Bernet notes, the individuated condition of perception, what we have called subtractive perception, "is a function of a certain opacity of the perceiving body" (Bernet 2005: 59). Because it originates from selectivity, memory itself is a by-product of the body's finitude. Bergson illustrates the two forms of memory through the example of learning a lesson. In the first case, we read a book. The particular reading itself is nothing more than a representation of an image, a copy of an event in the world. The image remains unaltered by the act of memory. But when thinking about the lesson in general, a very different memory is mobilized. This time, we aggregate the various readings, taking them together as a whole, asking whether in fact we have learned the lesson. The individuality of each memory is reorganized by our body, integrated into an affection referring to a generality of learning. It is as if the thing, the book, has forged words which, through conventionally coded noises imagined by our mind, come to be interiorized by our body. In its collective aspect, the alternation of images is ingrained, inscribed as a bioculturally coded habit. The first form of memory is the production of memory-images, while the second is grounded in action and can be described as *habit memory*; "seated in the present and looking only to the future," habit memory "prolongs" the usefulness of memories "into the present moment" (MM: 82). The first form of memory is a spontaneous registration of images, a recollection without effort, lacking entirely in repetition. The second is the only effective, powerful form of memory, for it is always allied with action. Where does its energy originate from, if not its attachment to a vibratory present? If the former is already finished from the outset, the latter is perpetually open: "Spontaneous recollection is perfect from the outset; time can add nothing to its image without disfiguring it; it retains in memory its place and date. On the contrary, a learned recollection passes out of time in the measure that the lesson is better known; it becomes more and more impersonal, more and more *foreign to our past life*" (emphasis mine, MM: 83). Essentially, what Bergson says here is that there is, in truth, only one causally effective form of memory, namely, repetition. The latter's power directly correlates with an agent's estrangement from the past! Through learning, we do not gain a greater access to a past. Quite the contrary: the goal of education is progress, a separation of ourselves from our previous condition. An effective pedagogy is one that, if need be, violently sunders the disciple from their past, introducing a temporal scission in the individual's narrative. A good education changes what one is, abducting the student from the powerlessness of being rigidly stuck in a yesterday, initiating them by breaking down the resistance of reactivity, transporting them into a hopeful tomorrow. As Guerlac writes, "the past is by definition incapable of action" (Guerlac 2006: 128). Hence, a Bergsonian pedagogy aims at passing "out of time" in the sense of a common elopement from the past, removing both student and teacher from enchainment to that which has been in favor of a hopeful alliance with a novel reality, a connectivity which is also a relationship in the making.

Motor mechanisms, produced by repetition, are substitutions for representation. Deposited throughout the body, they can be recalled upon at any moment, summoned forth by the exigencies of activity. As Bergson notes, in habit "we make use of the fugitive image to construct a stable mechanism which takes its place" (MM: 85). If the spontaneous memory is the novel, authentically unpredictable element, it is also that which is invisible, being so spectral that it passes away in the process. Only the accumulation of repetitions remains, for only this particular type of memory pertains to the actual dimension. The habitual is the machinic aspect of memory, giving access to the time of images. "Real time," as Chevalier emphasizes, "is the time of things," while duration is "a memory internal to change itself, since it prolongs indefinitely the 'before' into the 'after'" (Chevalier 1928 [1926]: 151). It is a present posteriority toward which priority is tending. To escape the hold of the dead past, we are forced to hold the fugitive image in place, prior to its disappearance into the aggregate of lifeless, impotent memory-images. The past is preserved in the form of either memory-images, defined broadly as representations or copies of a distant past, or motor mechanisms. Any uplifting of ourselves into affinity with pure memory would necessarily be a flight into nothing. As Chevalier reminds us, "there is no consciousness present in us severed from matter, no conception apart from images, no memory not linked with motor articulation, no tension without extension" (Chevalier 1928 [1926]: 164). There can be no question here of attributing effectiveness or causal force to memory. The gulf between memories and material images cannot be bridged, for the former has nothing of activity residing inside of it (Chevalier 1928 [1926]: 173). Building on the insights of psychologists such as Théodule-Armand Ribot and Henry Maudsley, Bergson emphasizes that "there is no perception which is not prolonged into movement" (MM: 94).

Perception is perpetually extending itself into affinity with a world of exterior images, suppressing the alternation of sensations to the precise degree that action requires. The invitation to action, however, originates not from within the opacity of the living body, but from the outside. Heterogeneity beckons, provoking interactions whose recurrent nature lends them to a depositing within corporeal interiority. Offered by the sum of sensibilities, light is gifted to the organism, being transformed into a shadowing through incorporation. The invitation of the images is a positive element "whose very presence invites us to play a part" (MM: 95). Enclosure within memory would be not unlike a reactive entrapment within oneself. Absent the actual occasion of contact with an outside and the body would remain forever enfolded in a virtual state, lacking power, energy, and activity. Is memory a virtual act of retrieval from an archive, or an actualization of virtual involvements? Bergson speaks of a retrospective movement of recall which takes us back into our whole past. But it cannot be forgotten that "though the whole series of our past images remains present within us, still the representation which is analogous to the present perception has to be chosen from among all possible representations" (ibid.). The apparent wealth of the sum of memories is an illusion, for in truth it is only the memory pertaining best to an actual present perception that can form the object of a legitimate selection. Recall, in other words, as motor mechanism, remains utterly dependent upon actuality. Only the image "most similar to the present perception" will actually survive selection, blending into the current condition of the organism (MM: 96). The Bergsonian view of memory is a mirroring of the present: "Memory (. . .) creates anew the present perception, or rather it doubles this perception by reflecting upon it either its own image or some other memory-image of the same kind" (MM: 101). *Pure memory contains nothing apart from those contents already contained in perception.* No positive description can apply to a specter that is entirely parasitical upon other objects. Subtract perception, and one will never be able to regain actuality. It cannot be the case that, as Bernet and many other commentators misled by Deleuze, would have it, "the only things that truly exist (. . .) are the pure past and the material world" (Bernet 2005: 62). Against Bernet, we affirm that only the non-correlative aggregate of images has any positive being, for it describes the sum of all material becomings. Instead of a matter spiritualized, we have obtained a spiritualism thoroughly materialized and immanentized. Those infatuated by the spell of the virtual fail to see that "the memory-image itself, if it remained pure memory, would be ineffectual. Virtual, this memory can only become actual by means of the perception which attracts it" (MM: 127). We can speak of the primacy of perception in Bergson's case, but only if we understand under it a pure perception that precedes the distinction between organic and inorganic objects. This would be none other than a direct, immediate intuition of the vibratory duration of material life. Far from liberating us from the present, as Trifonova would have it, memory is purely incapable of abduction (Trifonova 2007: 49). And luckily so! The power of a pulsating, vibrantly present duration is what prevents us from being entrapped permanently in the past. The memory-image is the parasite that inserts itself into the perception of the present, and for us this means that, in practical terms, memories and perceptions *appear* to us as inseparable (MM: 103). Far from pointing to an ontological truth, this testifies to our sorry state as finite, deluded beings. The permanent revolution of time gains actuality once it proves capable of effectuating the selection, driving itself free of its past through an act of violence. The issue is therefore is not how to liberate the memory from the present. A truly progressive question would be the following: *how can the present be emancipated from its attachment to the past?*

No memory can add anything substantial to perception. Bergson is eminently clear in this regard: "It cannot create a new kind of impress, a new quality of perception" (MM: 214). Pure perception would be infinite, unmixed sensibility, a flow of images, so to speak, free of all unnecessary obfuscations. If the individuated form of perception, defined as representation, is subtractive, then memory too is an artifactual appearance, a fractioning of an otherwise integral duration. The mnemonic slows down the imaging, insinuating itself into a field of consciousness. In concrete perception, as opposed to the pure, unmediated kind, "memory intervenes," artificially prolonging "a plurality of moments into each other," thereby "contracting" the multiplicity of images "into a single intuition" (MM: 219). For all our antipathy to the phenomenological approach, the remedy to the apparently hopeless attachment of the present to the entirety of the past can be found in a proto-phenomenological insight. What is the goal of the entire Bergsonian doctrine? The disciple of this teaching must, on a personal and professional level, strive for an immediate experience of duration. *Return to a true moment*—this is the kernel of Bergson's ethics. This return is achieved either through an act of introspection leading into our interior duration, or a dilation of our consciousness toward other, alien temporalities. The way our present can be liberated from the oppression of its past is to take the leap back from contraction into the flowing moment. Luckily, Michael R. Kelly's reading can serve as a valuable instrument for achieving the return. Specifically, we have in mind this author's highly valuable repudiation of the centrality of memory in the Deleuzian interpretation of Bergson. As will be remembered, Deleuze holds that the present has no real ontological import. It is only the past that has succeeded in preserving itself. While the present passes, the past remains intact, hence the former is absolutely lacking in being: the past "preserves itself in itself (while the present passes), it is the whole, integral past; it is all our past, which coexists with each present" (Deleuze 1997 [1966]: 59).

There is a serious problem with this highly partial reading of Bergson's philosophy, namely, the complete and utter neglect of the centrality of the enduring nature of the present. Kelly calls our attention to a key sentence in *Matter and Memory*, namely, the definition of a present as "what is being made" (MM: 150). In a process ontology, there simply cannot be any place for inert substance. Memory cannot fulfill the role Deleuze would prefer it to play because being is becoming. All things are durations in the process of being elaborated. The unfinished present is everywhere, its openness involving its surroundings through relations of mutual implication. Let all substances and ideas go to the wind, let time tear illusion to pieces! Nothing is immune to the bite of time, the fangs of actuality are already lacerating the past. Any reading of Bergson which remains enthralled by the spell of the past is mistaken, for it cannot accommodate the totality of becomings being made. Kelly proposes that, following in the footsteps of the original text, we "turn to the living-present" (Kelly 2010: 407). Through an act of freedom, consciousness is capable of transcending its finitude, accessing impermanence, overcoming the blankness of inertia. This freedom is far from abstract. It is the practice of a *kinetic morality*, a phenomenal activation achieved in a concrete, mobile situation. The leap into duration is not some revelation of a dematerialized past time, conceived of as some ontological reservoir. Rather, the intuitive leap consists in addressing duration prior to any appraisal or analysis. Involving us through envelopment, the kinetic act manifests an encounter with impermanence, the change which deprives propensities of solidity. A register is opened, an appeal is addressed, from one temporality to another. Both perceptions are fleeting, but once magnetized, they serve as each other's exterior limits. A borderline beckons, waiting to be crossed over, transgressed, overlain with sedimentations of novel movements. The present individuated consciousness perceives the past, but let us use our good sense and interpret the following sentences very carefully: the "concrete present such as it is actually lived by consciousness (. . .) consists, in large measure, in the immediate past. (. . .) Practically, we perceive only the past, the pure present being the invisible progress of the past gnawing into the future" (MM: 150). What Kelly emphasizes, citing this section, is that while Bergson definitely does assert that every perception is, in a sense, anamnesis, he qualifies this by adding the phrase "immediate" to "the past." Activity, even on the level of living beings, does not consist, therefore, in the mutual implication of the past in general and perception in particular. Rather, attention to life is rooted in the *immediate* past. The implications of this qualification are enormous, for they allow us to reaffirm the primacy of actuality. Memory as unmediated recollection can be permitted by our own actualist and presentist interpretation, but only provided it is treated in accordance with its relative importance. In Kelly's words, "the flow of conscious life (. . .) enables a consciousness of the past of the immediate experience relevant for a present action soon to be completed. (. . .) The vibrancy and self-apprehension of the flow of conscious life enable the survival of those successive ('material') vibrations in a consciousness of the past of the relevant series" (Kelly 2010: 407). The vibrancy of a conscious life, sinuously streaming now toward an open future, is what enables, in the final instance, the arrangement of memories. Of course, this only applies to beings with a certain degree of consciousness, and not to the intuition of material life in itself (pure perception). Memory is not the final synthetic force, uniting all strata of reality. This role is fulfilled by the image, this in-betweenness perpetually intervening among the registers, decomposing and rebuilding connections. Take any appeal, any single sporadic event that distends or shocks awareness: we can be sure that a mobile image is at play somewhere. It is the duration of the objective image which magnetizes, fluidifies, and fructifies perception, at times through a violent, eminently impersonal effort. As Marie Cariou reminds us, there is an almost infinite variety of durations at work in the Bergsonian cosmos. Instead of conceiving of duration in a single way, "as if it were homogeneous substance," Cariou emphasizes that the concept also contains "drive[s]," "directions," and "tendencies" (Cariou 1995: 70). The image is what drives our own individualization, in the form of an impersonal force driving us further into the future. The present is nothing more than a passage perpetually doing violence to our givenness, but passage, for Bergson, is the sum of all existents. Transition is all there is. Even that which is in movement is a mobility, lurching through the quadrangular gate of the present.

At one point, Trifonova speaks of a "pure image" but here, once more, we find a confusion, for she relates the event of individuation and personalization to memory. In our view, the pure image must be related in actuality to the dimension of pure perception. Far from being erroneous, the assertion that "certain pure (personal) images can adopt a different body since the body, by itself, is not individualized but is, conceivably, replaceable" can be upheld, as long as we desist from equating the purified image exclusively with the mnemonic (Trifonova 2007: 54). Memory too, insofar as it pertains to the individuated dimension (i.e., representation), is a pollution of duration. The pure image would correspond to material life. To claim that the past is, in some manner, "broader" than perception would represent a distortion of the true import of the Bergsonian reversal of thinking. There certainly is a leap at stake here, but were we to dive into a fictional generalized past as Deleuze stresses, we would be undertaking a jump into nothingness (Deleuze 1997 [1966]: 56). Real activity does not consist in a return to memory, for this would make action synonymous with dreaming. The leap we are seeking after is a leap into matter. Bergson upholds a radical distinction between activity and virtual passive dreaming. Actualization, on the level of living, conscious bodies, is defined in the following terms: "Past images, reproduced exactly as they were, with all their details and even with their affective coloring, are the images of idle fancy or of dream: to act is just to induce this memory to shrink, or rather to become thinned and sharpened, so that it presents nothing thicker than the edge of a blade to actual experience, into which it will thus be able to penetrate" (MM: 106). Actuality is therefore the thinning of memory, the chiseling of the virtual until it becomes a blade, an instrument utilized by the present for slicing into the future. Self-elaboration is a perpetual emancipation from the past. The fangs of time are supplemented by the blade of actual, immediate experience. Practical involvement necessitates a rejection of memories unsuitable for inclusion in the next present. Attainment of the actual unmistakably exposes us to a public sphere of materiality. The development of stability results in ever greater intensities of duration. What is a mind, if not a keyboard being played by the objective images, registering notes of alien realities, transmitting the vibrations, filling the body with resonance? The nervous system is the musical instrument of exterior imagings, "an immense keyboard, on which the external object executes at once its harmony of a thousand notes, thus calling forth in a definite order, and at a single moment, a great multitude of elementary sensations corresponding to all the points of the sensory center that are concerned" (MM: 128). What is habit, if not the incorporation of vibrations into our corporeality? Clusters of affections are born from the tensions integrated by the body.

Learning through repetition is the felt experience of our own subjective reality being built through inhuman, impersonal tendencies actualizing themselves inside of our intentional acts. Already, our apparently personal duration is full of alterities, temporalities which differ, at times radically, from our individuality. It is because the durations interpenetrate that Lapoujade writes that "the human in Bergson is humanization—or rather humanization itself oscillates perpetually between dehumanization and overhumanization, depending on the different levels at which one grasps it and the different tendencies that act in it" (Lapoujade 2018 [2010]: 48). The actualization of a memory through learning is dependent upon the presence of a habit sedimented within a particular body striving to unleash effects in its environs. What is striking is the ambiguous manner in which Bergson treats habit here. At times, it would seem that the habitual represents a potential immobilization of both life and society, hence a danger to be avoided. If the living is to maintain its active becoming, it is compelled to "constantly renew its habits" (Sinclair 2011: 44). Taking issue with Bergson's at times negative interpretation of the idea of habit as something that limits freedom, Mark Sinclair has argued convincingly for a more nuanced portrayal of habit. For instance, in Félix Ravaisson's philosophy, an important precursor who influenced Bergson's views, habit is actually the guarantee of freedom, a force of action that, while being inertial in itself, is ceaselessly tending toward a greater elaboration of the body's free movement. As Ravaisson notes in the crucially important concluding remarks to his 1838 essay, *De l'habitude*, "the history of Habit represents the return of Freedom to Nature, or rather the invasion of the domain of freedom by natural spontaneity" (Ravaisson 2008 [1838]: 77). Its repetitions, encoded in the form of virtualized, deactivated movements, are what allow the body a greater sum of movement and freedom overall. Habit is what integrates the organism into its world, functioning in Ravaisson's philosophy as an extension of the indetermination of the living. We concur with Sinclair's view, that "understanding the force of habit in a non-dualistic fashion as a continuation and development of the freedom of the conscious will, rather than as a mechanical principle of action, provides the philosophical ground for a more positive evaluation of it" (Sinclair 2011: 44). The habitual is what grounds the organism in its situation, being "the site of a pre-reflective but still intelligent inclination" (Sinclair 2011: 35). Just as the human element in Bergson is but a phase in the continuous humanization of the subject, so subjectivity in general is, on Ravaisson's view, one stage of corporealization. To have a body means to participate in a progressive process of embodiment. The phrase "lived body" (*le corps propre*), later borrowed by Merleau-Ponty and transformed by the latter into a phenomenological concept, originates from the French spiritualist tradition. First used by Maine de Biran, the phrase describes the in-between condition of both being a body and having possession of a body (ibid.). Repetition is the guarantee of embodiment, the continuous correspondence with a reality we call our own. Like projective lines extending into an unknown, hidden terrain, the body extrapolates its capabilities in its world or *Umwelt*. This achievement depends on the reconnection of memory with the actual situation. Habit is, in *Matter and Memory*, the synthesis tying memory to the lived continuity of duration. There can be no question of a memory entirely independent of its encapsulation within a finite corporeality, for "pure memories, as they become actual, tend to bring about, within the body, all the corresponding sensations. But these virtual sensations themselves, in order to become real, must tend to urge the body to action and to impress upon it those movements and attitudes of which they are the habitual antecedent" (MM: 130). What does the recognition of images tell us about the relationship between past and present? Adhering as closely to the original meaning of Bergson's teaching as possible, we see that something like an objective perception can be discerned, a perception staying in close proximity to the acentered aggregate of images. To paraphrase the text, *in order to become, virtual sensations must ally themselves with an actuality*. The virtual is an embodied repetition, the sedimentation of vibrations limited by the contours of a finite, habituated body.

Any separation of the memory from the moment it lives from cannot but strike us as an artificial and unnecessary distortion of the Bergsonian doctrine. Without an *effective* qualitative heterogeneity, the various memory-images would fly away into nothingness, losing their subjects. Unaccompanied by localization in some actuality, the mnemonic is empty and absent. Put differently, it is the moment into which the notes of the mind-as-keyboard condense (MM: 128). Habit, to quote Ravaisson once more, is an "idea in action" (Ravaisson 2008 [1838]: 59). It is only within action that habit and memory alike gain their importance. The infinity of pure perception is the limitless dimension giving occasion for both repetitive sedimentation and the exercise of memorial registration. None of these terms must be supposed as entirely self-identical. Beneath the production of representations and the proliferation of memories, an original continuity is present, in the form of the original heterogeneity of interpenetrating objectifying material images. An authentic leap puts us back in "the continuity which we have (. . .) broken between the perception, the memory and the idea" (MM: 122). Instead of discreteness, duration is the continuity of real, causally effective actualization. We find, upon placing ourselves intuitively into pure perception of materiality, that duration is a passage that persists. In itself, the pure memory does not exist. In its essence, recollection as virtuality cannot ever be known except as a process of actualization. By placing ourselves in the past, we are merely substituting our own present with an absent presence. To perceive means to slice the world into manageable portions, consuming a certain amount of complexity in the process, integrating the latter in the form of habits. Incorporation is the active idea immanent to the material life of images. As Mullarkey puts it, "intellect" is "an extension of the digestive power of perception" (Mullarkey 2000: 69). To picture means not to remember, but rather, to consume a certain amount of exteriority, emplacing a foreign element within our own interiority, until the two terms coalesce, forming a properly functional body. Remove the world of images, and no recording or recording of the past will be able to take place. Were somebody to abstract memory from the present, they would be left with an inarticulate residue lacking coherence. Bergson observes that the past "cannot be known as something past unless we follow and adopt the movement by which it expands into a present image" (MM: 135). Each living thing is an assortment of supramodal attention selections, some of which are active choices, others passive recordings enfolded within a body, ceaselessly remembering and forgetting, but always, of necessity, inside of an action. Each duration has "a distinctive breadth and a distinctive character," yet all are interconnected, flowing together in a state of interpenetration (Gunter 2005: 153). The present is the sole life, the only vitality at work. Attention is a payment, the price paid for the opportunity to subtract elements from a world. The mutual implication of imagings, as conceptualized by Bergson's idea of matter as the sum of interconnected resonant images, results in the complete erasure of anything like a "matterless thought." In this all-encompassing monism, the spiritual can retain nothing of its ethereality.

As Al-Saji notes, the process of actualization in Bergson's philosophy displays an original asymmetry between the virtual memory and the actual object perceived. It is the images which engender perception, and not the memories which give form to objectivity: "The virtual is the equivalent or duplicate of the actual object perceived" (Al-Saji 2004: 217). The summoning, the call, the invitation, this cannot but arrive from a present. The ossified cannot beckon us. In Bergson's words, it is "my present" that "interests me, which lives for me, and in a word, that which summons me to action" (MM: 137). *Wake up from your slumber*, good sense tells us, *there is a work waiting to be done, an undergoing to be commenced.* Perception is the active pole, the dimension of activity, the quality of boundlessness, that which "corresponds to the actuality of the present" (Fell 2012: 43). Instead of an abstract, mathematical point, a quantifiable instant, duration is a genuine thickness, a concretion. Striving tends toward complete unity with a present image. What Bergsonism aims for is a coincidence with a "real, concrete, live present," a present which "necessarily occupies a duration" (MM: 137). It is the now which provokes us to action. In fact, it would not be going too far to claim that the present is the original source of all activity, be it vital or causal, organic or inorganic. Negatively speaking, we recognize the priority of habit even at the expense of the will. In Ravaisson's words, "it is not the will—or at least it is not reflective will—that works out and devises in advance the very production of movement; for this can only arise from the depths of instinct and desire" (Ravaisson 2008 [1838]: 71). The demented torment of the living thing originates from its vulnerable exposure to the tyranny of the passions. Where must we search for the source of desire? The durational present, the becoming is, itself, the place of origin of this emotive drive. If we are to defeat suffering, the solution lies in emplacing ourselves in the source, ending the tyranny of desire through emplacement within impermanence. By melting down our affections into an unsolicited wave of indifference, the unfolding of desire can be stemmed, or rather, diverted into a presentation seen to pass over our cleared, crystallized awareness. Intuiting duration prepares us for reentry into time, dancing upon the edge of the abyss of memory, without falling back into dream. "Spatiality and physicality" being "the features of actuality," these are lost when a thing becomes past (Fell 2012: 47). If memory is the inextensive, then the intuition of duration-as-actuality would be a regaining of extension, even a reintegration with matter in the mode of "creative death" alluded to by Meillassoux. But the phrase "death" bothers us. Why not just describe this expansion as an *aperture*? What we call our present is, itself, but a small slice, a practically manageable portion of a larger series of durations, for the individual sensations we feel in this particular, individuated "now" are themselves translations of "a very long succession of elementary vibrations" (MM: 138). The present is that which is not yet closed, the elemental that has yet to lose its infinity. Meditation upon duration is a speculative materialist practice of self-exposure to the contingent patterns of the cosmos. Continuity is irreducible to any particular unity. Intuition uncovers "the actual state of my becoming, that part of my duration which is in the process of growth" (ibid.). We are transported into affinity with the flowing moment, the impermanence underlying all appearances. The penultimate is the process of growth, as effected through the medium of the present moment. There is no tautology in claiming that "in its actual state the actuality of our present lies" (MM: 139). Such a simple insight, this brief but powerful message, destroys illusion, rendering obfuscation impossible. Stripping away superfluous ornamentation, we are left with immediate access to fluctuation. The continuity of change can now be conceived internally, for we ourselves are also internal to time (Al-Saji 2004: 204). Verily, this is so, for each duration is nested within a broader continuity of durations. Chevalier's description captures the ambition of the intuitive method, namely, immediate coincidence with change: "This undivided and indivisible flux of consciousness is no doubt a unity, but it is a mobile unity, a moving unity; it is not the unity of a position, but of a progress; it is a continuity; it is the past incessantly bestriding the future upon the mobile point of the present" (Chevalier 1928 [1926]: 159). In its pure state, perception would be a melt-up into change, an ascent into that which is sufficient unto itself. When situated in a present, freed of delusion, we ourselves become like jets of steam shooting through cracks opening up along the sides of a previously solidified past.

If all is change, how can we account for the appearance of permanence within that change? As John Ó Maoilearca[6] explains in a recent article that, on a pluralist view, "reality doesn't appear as just one thing, even if only in virtue of the illusion of at least two things" (Ó Maoilearca 2014: 25). Metaphilosophically speaking, in what manner can the appearance of the unchanging be integrated into a process ontology, such as Bergson's philosophy? Matter's dynamism, its mattering, so to speak, would leave no room for living, selective bodies, if the flow of images were not slowed down to a certain extent. Neither could we point to any ontological status of certain liminal categories such as illusions, dreams, or specters. Ó Maoilearca point out the existence of phenomena inexplicable in a reductive materialist worldview. What can materialism do with *apparently* real but *objectively* false objects such as ghosts? Realism of the materialist persuasion runs into considerable trouble when confronted with the "spectral dilemma," defined as the issue of "how to account for epistemic norms when norms, the 'as' of seeing aspects, imply a subjectivity within the objective" (Ó Maoilearca

2014: 22). The ghost is a disembodied subject.[7] Its liminal status as an absent presence points to an epistemology that denies our laws of causality. The initial phase shifts and oscillations characteristic of materiality as known by the natural sciences give way to the eruption of unprecedented fluctuations, spectral phenomena apparently irreducible to materiality. Even if the specter, the haunting, is not "real," being a figment of the imagination, materialism of the conventional kind has a hard time of dealing with the "apparently" real, elements we may call the "not-quite-real." The question is, in what manner can the "appearance of appearances" be integrated into a worldview, without compromising the barrier between reality and illusion. However disturbing the implications of such a position may be, Ó Maoilearca questions whether philosophy in general is even capable of addressing, let alone answering such metaphilosophical questions. In large part, this stems from the status of philosophy as subjectively integrated thinking. As an embodied being, the human, when confronted by the monstrous expansion of subjectivity, recoils in horror. Recent new realism and speculative realism has been tending toward an expanded semantic of vitality. In Ó Maoilearca's view, the language of "the ubiquitous life of objects in Latour and Harman's ontology of actants/objects" (or, for that matter, Jane Bennett's vital materialism) points toward the survival of a latent vitalism, despite the relative underplaying of the influence of the vitalist tradition on the part of these two authors as well as other members of the present-day speculative realist current (Ó Maoilearca 2014: 16). Rather than searching for a resolution to the ontological status of the not-quite-real, Ó Maoilearca proposes that we return to Bergson's method of intuition. Specifically, the access Bergsonism gives to the multiplicity of durations can help us in registering the nonhumanity of the human, through an affect resembling that of horror. What, after all, is the experience of the horrific, if not the immediate presence of the ubiquitous lives of objects? "Thinking objects, like contemplating stones, are horrific," for the sense of dilated realism "arises in horror as the perceived objects think for themselves in us, forcing a painful reversal of thought (that Bergson would call 'intuition')" (Ó Maoilearca 2014: 23). Certainly, the vision of a world forming itself, without adherence to any rationally explicable laws, is an intimidating one. But contingency need not equate to a sense of overwhelming despair. Indeed, as we shall see in our final chapter, it is the indeterminate nature of the present that can account for freedom. What *Matter and Memory* in particular shows us is the nonhuman status of the human as well as the inorganic nature of the organized body. Pure perception is always already an imaging of the images out there, in the exterior world. The body indeed is nothing more than a folding of these pictorial realities upon themselves, and a limitation of perception. Images indicate an individual mereological relationship with reality, and not a representational one. *Matter and Memory* provides us with "the reality of the image rather than the image of reality" (Ó Maoilearca 2014: 24). The goal of Bergson's philosophy is not the production of yet another representation. Of these, we have many. Practically engaged concepts such as solidity, form, weight, thickness, texture, and so on, are habits of thought. Once canalized into a living body, limited perception demands a platform onto which it can accumulate insights. As Elizabeth Grosz has emphasized, habit "grounds us firmly in a pre-representational real" (Grosz 2013: 218). At this juncture, we must reflect on the import of the idea of habit as contraction, for we have not yet touched upon the status of habit as habit memory. Reflecting on both Ravaisson and Bergson, Grosz reminds us that habit, defined as the interiorization and recording of movements in the form of virtual actions in the body, can be viewed as the activity of stabilization on the part of corporeality. Incorporation through habituation or what some of the literature calls "bodily learning," is the way corporealization guarantees the stability of its own organization, lending something approximating predictability to its own acts. In other words, the relative stability of repetition is a method for a body to continue its process of organization. "Habit exists somewhere between the necessity of ease and the torment of need" (Grosz 2013: 220). Once the images originating from the heterogeneity of a world have undergone incorporation, they can be recalled by the body in its time of need. More fundamentally, "habit is change contracted, compressed, contained" (ibid.). Hence, habit memory is already a nascent form of memory, albeit one that is never far from real activity, being the fringe or cloud surrounding the present involvement of the organized body. Looking forward to the generality of memory we shall soon evaluate, we identify the first iteration, the first concentric circle, of what shall be elaborated later on during the course of *Matter and Memory* as the famous cone of memory. Habit is the first stage in memorization, the process of elaboration of illusion. The practical necessity of the organism to decelerate the images, slowing them down and integrating them as contracted movements, constitutes the origin of the false idea of permanence. Our "indivisible present," being an "infinitismal element of the curve of time," is "both sensation and movement," rendered accessible and recallable for us in the form of endogenously integrated repetitions prolonging the present into a future action (MM: 138). Does memory endanger the noncorrelational aspect of the image? To answer this question, we must recognize the true place of memory in the Bergsonian ontology.

MEMORY IS ACTUALIZATION

Is habit memory even memory at all? Bergson himself is more than ambiguous in this regard. While Trevor Perri argues for a holistic view of Bergson's philosophy of memory, suggesting that "the implicit, non-representational memory of habit and the explicit, representational memory of recollection" form two extremes of a single continuity, we ourselves are no so convinced that habit can actually be brought into adequation with the rest of memory (Perri 2014: 842). As Bergson himself stresses, because of the status of our present as a "sensori-motor" condition of embodiment embedded in a certain now, "from the moment that it becomes image, the past leaves the state of pure memory and coincides with a certain part of my present. Memory actualized in an image differs, then, profoundly from pure memory. The image is a present state, and its sole share in the past is the memory from which it arose. Memory, on the contrary, powerless as long as it remains without utility, is pure from all admixture of sensation, is without attachment to the present, and is, consequently, unextended" (MM: 140–1). As we have seen, without elongating itself into a present, memory lacks both extension and causal efficacy. Throughout its existence, the body is forced to produce convenient illusions. This idea will reappear later in Bergson's *Two Sources of Morality and Religion*, in the form of the myth-making function saving intelligence from depression and its own presumed self-annihilation (TS: 98–9).[8] When all is said and done, a yawning abyss emerges that cannot be easily bridged. If habit as nonrepresentational memory is inherently turned toward the present, then pure memory applies to that which has genuinely passed. While the habitual survives inside a body, the memorial flies off into the deep time of impotentiality. As Sinclair writes, "there is something essentially tragic about remembering" in the Bergsonian philosophy, "for it is doomed to failure, destined never to reach exactly what it seeks" (Sinclair 2020: 103). If action is immediacy, then memory is mediation. That being said, there is a not so latent ambiguity at work in Bergson's concept of the present as well. There is at work within the undivided present a tension between the several durations in play. On the one hand, we have the lives of objects, as presented by the becoming of images, summarized as "material life." On the other, we have our own inner life which, as lived experience, while relating to the material realm, is still sensed in an eerily different manner. The sense of *having* a body is fundamentally different from sensing or observing other, external bodies. Something inherently spectral is at work in each present, as memory introduces a haunting scission between us and our world: "While reality, in so far as it is extended, appears to us to overpass infinitely the bounds of our perception, in our inner life that alone seems to us to be real which begins with the present moment; the rest is practically abolished. Then, when a memory reappears in consciousness, it produces on us the effect of a ghost" (MM: 145). Causally impotent, materially absent, memory still appears to be equipped with a certain power. From where does this impression of a potency within absence come from? How can memory result in an affection? In reality, Bergson recognizes that only actuality is capable of having an effect, defined in the case of organized bodies as the ability to induce the accumulation of sensations and affects. The explanation for this supposed doubling lies in the ambiguity of the present as such. For what is the structure of the present moment? Never is a lived experience of time given in itself. As Fell explains, "in Bergsonism we see a hierarchy of being, where lower levels of being are characterized by slower rhythms, and higher levels by faster rhythms" (Fell 2012: 146). Let us remember that an asymmetry exists between the subjective and objective types of duration! Materiality, defined by Bergson as the imaging activity of the heterogeneous images, can do without any form of individuated consciousness whatsoever, but the reverse does *not* hold. The subtraction of the world undoes consciousness, but images can survive the destruction of all sentient, organized bodies. Because of its dependence upon the flow of images, the body's present is full of other, alien durations. The familiarity of our body gives occasion for almost infinite multiplicities of temporal cycles and rhythms, differing markedly from the life cycle of the organism they are contained by. For example, microbes of ancient provenance, in existence long before the evolution of *Homo sapiens*, populate our bodies. The finitude of corporeality, as well as the inherently limited nature of our nervous system, means that we cannot directly access these other internal durations. But intuition is capable of accessing a purely inhuman, even inorganic pure perception. As have emphasized, perception in its pure state for Bergson is a completely immediate coincidence with the lives of things. Pure perception is the material life of the images. Mullarkey has persuasively summarized the whole project of *Matter and Memory* in terms of a dehumanization and desubjectivation of the philosophy of mind: "Bergson attempts an exorcism of all that might make perception subjective. He thus constructs a perception that belongs neither to any subject, not any body-subject, but instead to a supposed mathematical point perfectly mirroring the universe surrounding it. The body literally becomes a point perspective. The objective state would be an absorption in a timeless, immediate present" (Mullarkey 1996: 373). To quote Bergson's own words, "in pure perception we are actually placed outside ourselves; we touch the reality of the object in an immediate intuition" (MM: 75). All becomings have their place in perception purified of memory. What makes the present ambiguous is that there is never just one duration in play. Rather, the durational is itself a manifestation of heterogeneous multiplicities. A present is never entirely singular, being the intersection of several different times which form together a continuity within this particular moment (Mullarkey 1996: 375). Never is there a present given in isolation from the other presents. Far from contaminating memory, action is what gives it a new resolution. Without becoming circumstantial, a contingent aspect of a certain involvement, memory remains dormant, a useless remainder lacking application. The present is paradoxically unrepresentable because our consciousness only relates to past and future, without being capable of taking into account the entire complexity of all durations. Everyday cognition does not achieve the intensity of intuitive penetration. For individuated consciousness, the pure form of perception, union with all material images, must remain invisible. It cannot be true that all forms of memory are united with a single, general past. The two fundamental forms of memory, habit and recollection, are not continuous with one another, as Perri would maintain (Perri 2014: 843). Bergson is explicit in this regard: habit memory is not true memory, for habits are inseparable from a present. Defined as the repetition which stabilizes the organized body, habit is the power situating us "in an ever renewed present" (MM: 151). This qualitative difference entails a highly important consequence: there is no such thing as an effective, active, actual memory.

We arrive at this juncture at the famous Bergsonian image of the cone. That element we call our body is "a section of the universal becoming," a "place of passage of movements received and thrown back, a hyphen, a connecting link," while the cone itself represents the totality of memories (MM: 151–2). Any needless mystification is to be avoided here. Geometric language notwithstanding, the cone too is perpetually in movement. As Raphaëlle Costa de Beauregard emphasizes, Bergson's cone is also a spiral. As distinct from Deleuze, "Bergson only describes movement as continuous and never-ending" (Costa de Beauregard 2010: 92). However suggestive the cone of memory appears, we must not in any manner let ourselves be fooled into attributing any permanence or substantiality to pure memory. Just because the present is unrepresentable to individual consciousness, the invisible progress is still operative, on the level of microdurations. And obversely, merely because we appear capable of summoning memory-images by no means implies that we must accept memory as having a positive being. What, after all, is a cone? To speak of a "pyramid" in this regard is erroneous, for the diagram in question depicts a hollow cone. Not long after, the cone is bisected with several subsections, all denoting a certain level of the past. Perception, according to Bergson's view, travels to and fro along the totality of memory, traversing various absent temporalities. Specifically, Bergson talks about the "general idea of genus," an act of categorization forever traversing the cone. If we multiply the cones, we obtain multiple gates opening to multiple simultaneous presents. (Figure 4.3.).

Categorization is achieved by two different acts of mind, the "formation of stable images" capable of orienting our activities, and an endless construction of genera without end. The general idea is infinitely mobile, and never will find a home in any section of the cone of memory (MM: 161). Why does Bergson choose a hollow, three-dimensional geometric form to symbolize the

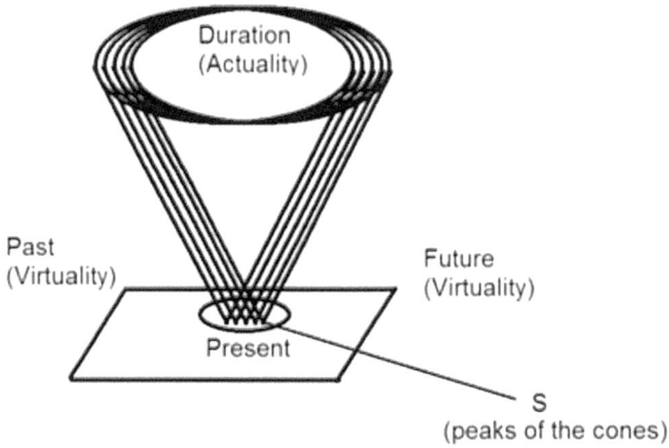

Figure 4.3 The cones of memory are multiplied in our diagram, for here we are representing duration in motion. Each duration is revealed to constitute a mobile set of presents. Duration is the force of actualization, the source of multiplication motivating moments. Neither past nor future have any reality apart from their relative status to the set of present (effective) durational endurances. *Source:* Author created. Not required.

relationship of memory and perception? Our answer to this mystery does not lie in the attribution of a positive being or thickness to memory. As Deleuze would have it, "each of these sections [of the cone] is itself virtual, belonging to the being in itself of the past" (Deleuze 1997 [1966]: 60). The blind spot of Deleuze's interpretation lies in a seemingly obvious circumstance, namely, that the cone has no roof. The cone's "roof" is nothing but a hole, through which duration flows. Mullarkey's suggestion is that we view the cone of memory in terms of a duality between an act of recollection and pure memory: "Whereas a recollection actualises the past, pure memory is this past. This is without doubt a thoroughly realistic view of antiquity" (Mullarkey 2000: 79). On this reading, the cone would refer to the entirety of the evolutionary past of an organism, incorporating perhaps even the inorganic causalities and precedents that have contributed to the creation of a living body capable of reminiscence. Deleuze goes as far as to posit the presence of a cosmic remembrance: "Everything happens as if the universe were a tremendous Memory" (Deleuze 1997 [1966]: 77). We ourselves are not so sure about a panpsychist reading of the cone. More specifically, it is not the positing of panpsychism itself that we reject. Bergson himself hints several times at such an interpretation. What we dispute is the ontological *positivity* attributed to memory. Even Al-Saji's otherwise incisive reading remains infected by the Deleuzian error, which consists in crediting memory with a positive status as being the synthesis connecting all moments: "The continuity of duration should be understood as the interpenetration and overlap of actual moments by means of a virtual dimension of pastness that coexists with each. This virtuality, which haunts every present, is the condition for the communication of the present" (Al-Saji 2004: 218). The Deleuzian interpretation of the cone of memory errs not in its panpsychism, but rather in the manner in which it situates this panpsychism. What is the connecting link between durations? In the case of a remembering, habitual body, this node of communication is the body. And what is a body, if not an actuality? Memory is certainly a haunting, but it is not the specter that connects the durations. Rather, it is the vibratory present itself. (Figure 4.4.).

Perpetually, the present is drilling through the immanent limitations of its momentary constitution. There is merely one solidity, one thick presence in Bergson's diagram: the point uniting everything is the body. But this body in itself is one particular manifestation of something much more general, and this is the present. Duration flows in the now. *A change is in the making*. Durations, in other words, connect to durations. Here we are, at the summit of a hollow, empty cone, a shape lacking in content. The base is the height of a mountainous, abundant durational whole, the pinnacle of a temporal

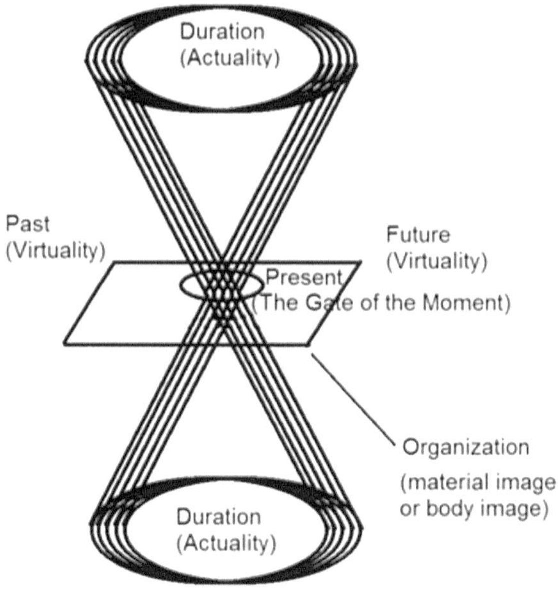

Figure 4.4 The process of actualization involves a transition from one duration to the other through the actualization of one memory in another present. Only by a needle-like insertion in a present can the cone of memory be said to "exist." Neither past nor future temporal modalities have any independent role in this process. Pure actualization heads through the Gate of the Moment, which corresponds to the opening in becoming created by each present. The rectangle refers to the materialized constitution of a duration, in other words, an image, be it a material entity or an organism. *Source:* Author created. Not required.

individuality. Its sole claim to power rests upon presence, the individuality of a continuing, enduring, streaming present. This point is of course a mere symbolization, and we should in no manner identify it with a subject or a form of consciousness. It can be any duration whatsoever, any "thing unmaking itself." Let us train our gaze to discover the secret, the core, of time. The moment is the beating heart of time. Were we to perversely, remove this point that vibrates, the past, conceived of as a derivative outflow of this present duration in the making, is bound to disappear along with it. Does the reverse hold though? Our asymmetry could be tested by the example of a life lived in the present. A critic could always point toward pathological examples. In Christopher Nolan's film, *Memento*, we observe a male protagonist lacking any long-term memory whatsoever. Leonard is an absurd example of a person living almost exclusively in the now. This person is at once both tragic and laughable, an individual unable to live for an extended length of time in a society. In the words of one interpretation, the malfunctioning protagonist becomes an "obscure criminal," who, disturbingly, could very well be "relatively unaffected by his own acts of violence" (Clarke 2002: 173). But the author also emphasizes that it should be borne in mind that the characters in the film are "already imaginary," accentuating the "chronic undecidability between what is true and what is false in the present," the authentic experience of time being precisely this virtual chronic in-betweenness (Clarke 2002: 174–175).[9] Certainly, one could gather many real examples of pathologies and disorders of memory. Without recollection, no life could be possible, says the critic. But what makes diseases such as dementia truly deadly is that, after a while, they begin to damage the habitual form of bodily memory as well. One could speculate that an otherwise healthy life lived entirely in an eternal present need not be as horrendous or pathological as its fictive portrayal in *Memento*. What drives Leonard into a series of unfortunately chaotic involvements is a will to knowledge, an intent to uncover the truth underlying his world. Had he simply given up, submitting passively to his fate in the manner of a modern-day ascetic, he would have been spared the liminal state that becoming-criminal implies. Instead of criminality, the protagonist could just as well have surrendered his will altogether. The figure of the *obscure criminal*, perpetually lacking in time, but always chasing after the absconded meaning of time passed, could be effaced by an obscurity which erases recollection while also affording an intensified memory of the present. What Melissa Clarke sees as a loss, a broadened view could just as well interpret as a gain: "Without any past to consult, there is no way to interpret the presents at all. The structure of a present is evidently pure becoming" (Clarke 2002: 176). Once paralyzed, veridicality, as manifested in a sickly, overactive will to truth, could become transfigured into a pure, objectless willing, an *affirmation* of everything that *is*. Paradoxically, one could claim that Leonard's true error lies in his failure to detach himself from the illusion of truth and veracity. He must know what happened to his wife, proving incapable of finding peace of mind unless the truth is uncovered. In the process, this antihero loses his access to the present as well. On our unconventional reading, the true message of *Memento* could be summarized in the form of the following imperative: *always remain true to* this *present!* The obscure criminal is the person unwilling or unable to let go of the past. Momentarily uplifted by the memories he latches onto, the obscure criminal gradually floats away from the present, clutching onto balloons, becoming unable to maintain contact with reality. As Clarke emphasizes, Leonard's "becoming-criminal took place linearly while time did not progress linearly," for the events in the film are presented as heading backward from the traumatic final denouement (Clarke 2002: 178–179). What the commentary does not reflect upon is that this becoming-criminal is intricately linked to the character's inability to accept his present moment. Even when he would find a new lover, Leonard is held back by his desire to rediscover the past. As opposed to the obscure criminal, we posit another moral and temporal possibility, namely, *the disciple of the present*, the meditator who seizes becoming without succumbing to the temptation of truth.

Good sense lies in adherence to the Middle Way, being situated halfway between impulse and dreaming (MM: 153). The pointlike serpentine flow, S, represents the center of Bergson's diagram, the most crucially important point of the cone. If one were to subtract the summit, the entire structure would collapse into an incoherent chaos of lines, planes, and subsections. The summit is therefore also the center that organizes and articulates the entire process of memory. As Sinclair explains, "the base of the cone represents memories in their pure state, prior to their instantiation, whereas the summit, the point, represents their insertion into the present moment of action" (Sinclair 2020: 101). But is there, properly speaking, a "base" here at all? Another asymmetry seems to be in play, for while a summit is unavoidable, there is no need to construct a *base* for the inverted cone. Memory extends into nothingness. We ourselves conceive of the cone of memory as a hollow geometric form, as it appears where one would expect to find a solid base infinite openness pertains. Dreaming consists in abduction, entrapment in memory. After introducing the cone diagram, Bergson engages, somewhat unexpectedly, in a highly interesting discussion relating to qualities and their relationship with sensual immediacy. Through what hidden mystery does an object show its qualities? What are the sensual notes of a thing? Bergson argues that categorization, defined as the analytical separation of qualities, is a product of an advanced nervous system capable of translating the "striking quality" of an object into a memory-image. Analysis is the product of an intelligence complex enough to engage in discernment. That being said, underlying the differentiation of qualities in organic perception is the idea of need. Organized bodies must extract those elements of their surroundings which serve the perpetuation of their survival. Bergson posits a generality of selection mechanisms, situated along a continuity extending from the inorganic into the organic realm: "We can follow from the mineral to the plant, from the plant to the simplest conscious beings, from the animal to man, the progress of the operation by which things and beings seize from their surroundings that which attracts them, that which interests them practically" (MM: 159). Every self-organizing structure survives through extending its duration through a variety of selections, accessing a greater or lesser durability in the process. Categorization is a necessity of life. Sentient beings, however, are also capable of more abstract generalizations. They have a "general idea of genus," for their brains are equipped with the ability to traverse the cone of memory, without having to stop at any level. Immemorial time has no meaning for an intelligence capable of dreaming, of getting lost in its own past (MM: 161). Bergson is not a philosopher of the dream, of a sleepy idealism which would lose itself in the cloud of immemorial memory. Detachment from "our sensory and motor state" is equivalent to living "the life of dreams," while "we tend to concentrate ourselves in S (the present) in the measure that we attach ourselves more firmly to the present reality" (MM: 162). It would be overly simplistic to claim that this constitutes a mere description of how cognition operates. A morality is at work here, but this can only be discerned by remembering the ethical dimension of the phrase "good sense." In Bergson's *Two Sources*, the concept of "good sense" illustrates the mode of being of the mystic. Despite appearances to the contrary, the mystic or saint must not be conceived of as a mentally unbalanced person. The mystic is abnormal, to be true, displaying behavior that diverges radically from everyday commonsense practices, but this abnormality is in truth, a different kind of health, a fresh form of mental balance. The lives of the great spiritual leaders of humanity evince that "there is an exceptional, deep-rooted mental healthiness, which is readily recognizable. It is expressed in a bent for action, the faculty of adapting and re-adapting oneself to circumstances, in firmness combined with suppleness, in the prophetic discernment of what is possible and what is not, in a spirit of simplicity which triumphs over complications, in a word, supreme good sense" (TS: 195). These great personas are not dreamers, locked in a sphere of pure memory. Rather, they engage, usually in a disarmingly simple manner, with their presents. Many examples could be furnished of the practical engagement and skill of great spiritual individuals. Teaching spiritual detachment necessitates, paradoxically, an eminently pragmatic attachment to the present. Skillful means are the best way of extending the power of an infinitely persuasive love. Therefore, to concentrate ourselves in the present is not some regression into a "prehuman" state. Instead, it represents an ascension to what

Lapoujade calls "a new kind of confidence, a new kind of attachment to life, in short: a new kind of health" (Lapoujade 2005: 1153). After the advent of the novel teaching, no longer can we return to the past. This new normality is an immediate witnessing of duration. Bergson's morality shines through his affirmation of the present. Morality stems from acceptance of the appeal, the invitation to go beyond one's past into a new moment. A present is calling us to a higher life, a vitality lived at the limit, an activity corresponding to the peak: "Memory, laden with the whole of the past, responds to the appeal of the present state by two simultaneous movements, one of translation, by which it moves in its entirety to meet experience, thus contracting more or less, though without dividing, with a view to action; and the other of rotation upon itself, by which it turns toward the situation of the moment" (MM: 168–9). The appeal of the present is the force, the power that compels temporality to spiral out of itself. Bergson's cone is, in truth, a tornado. Without the emergence of an actuality attracting attention, in the manner of an irresistible magnetism, memory could not even move, let alone insinuate itself into activity. It is the energizing appeal that provokes the effort of memory. Everything depends on the summit's invitation. The oscillations of the mnemonic access a dynamic that is pure becoming, spiraling outwards, clinging to physicality while also striving to accelerate into the immediacy of imagings.

What, in the final instance, does the new health consist of, if not a reversal of commonsense mental balance, an unbalancing that trumps convention? This ethics is one of violence, of shock, of chaos, conducive to a creative opening of the subject in peak experiences. As Bergson relates, speaking of amnesia, "a sharp shock, a violent emotion, forms the decisive event to which they [memories] cling; if this event, by reason of its sudden character, is cut off from the rest of our history, they follow it into oblivion" (MM: 171). Beneath apparent illness, we find the seeds of a new type of health. A great spiritual shock can, given the right circumstances, result in the reorientation of the personality, an emplacement of the subject in a present that abides. Ultimately, what we think of as death, the undoing of the body's organization, is the final form of the creative dissolution into pure perception. This melting into heterogeneity, far from constituting a transcendence, is more akin to reincorporation in the material life of all images. Already, this reconnection is preceded by the reorganization of individual characteristics by the mind along the lines of character. The temperament or personality of an individual is a habitual sedimentation of elements, a convergence "upon actions" which stamps each act with a certain quality. But a character is only as real as the effects it produces among the images: "The action is not able to become real unless it succeeds in encasing itself in the actual situation" (MM: 172). Attention to life serves as the apex of the inverted cone, a concentration that is irredeemably attached to our finitude while also going beyond any limitation.

If we have learned anything from our study of the Bergsonian doctrine, it is the heterogeneous complexity of each and every tendency. A direction contains a multitude of sub-durations. No point is entirely unitary. In his recent book, Sinclair hints obliquely and, perhaps, unintentionally, at something of import, when he writes that "we are (. . .) not Gods. We are finite beings, with a limited range of attention, and we are this stuck with the distinction between the past and the present" (Sinclair 2020: 104). Attention to life is the attachment or vital involvement of the organism to its *Umwelt.* But it rests upon the basis of the finitude of all corporealities. One can imagine a body that is infinite, a life without any particular extension. What would happen were we to remove the point S from the diagram? The result would be a collapse of the cone of memory and the flight of the subject into a dream state. The obscure criminal would be the individual unable to escape from her past. On the other hand, another possibility has been presented, one we have termed the disciple of the present. In the latter case, the individual gains immediate access to the present. This apparent loss is, in truth, a gain, for nothing has been lost except the delusion of a phantom being of the past. To regain the moment, we must bereave ourselves, eliminating that which has passed. But this elimination of absence is no loss, for the past is revealed as already absent. Another question presents itself, namely, were we to expand the flowing point S, dilating it until it blots out all the lines of the diagram, covering them with blackness, what would we then find? Would this obscurity not be itself a kind of light? Like smudged charcoal, our white page would be darkened, but the smudges would still have different degrees of opacity. The immediate present knows nothing of the past, but it loses nothing through this subtraction. As a matter of fact, the goal of intuition lies in making the finite infinite. Formerly concentrated, attention is exploded, until it incorporates everything in a monstrously loosened duration. "Pure intuition," Bergson says, "external or internal, is that of an undivided continuity" (MM: 183). Pure perception, however fictive it may appear, is a completed perception, a sense inclusive of all the elements. In the process of expansion, everything is left intact, with the exception of the phantoms of the past. No longer can that which is absent impersonate the vitality of the present moment. The tyranny of memory is overthrown by the force of activity, the absolute becoming-active, the message of correspondence leading to a self-awareness of change. We are displacement, we are movement, we are the inverted cone, only, in the mode of divinity, in exploded, shattered form. Intuition can be purified of needs. The aim is to enact a return, "to seek experience at its source, or rather above that decisive turn where (. . .) it becomes properly human experience" (MM: 184). In actuality, prior to the beginnings of individuation, there is a becoming which is never finished. Inevitably, the final vision of things will fall short of our unimaginable desires, being a neutral dimension of imaging.

For all its indifference, the aggregate of images is vastly more intensive than anything the individual mind can fathom: "The duration wherein we act is a duration wherein our states melt into each other" (MM: 186). Let us be very attentive to Bergson's choice of words. Duration is not in us; *we* are inside *it*. Each life we live, each power we release, is endogenous to a certain temporality that is, for all its alterity, perpetually intruding upon our dimensions. Movements are whole and continuous, but they nonetheless differ according to their qualities. It is never a case of rebuilding a movement from lines. The continuity of movement in no wise implies its homogeneity. Points "have no reality except in a line drawn," for "passage is one with the movement itself" (MM: 189). For the sake of practical utility, analytical intelligence slices the world into discrete bodies, images of motionlessness. Folding out that which has been contracted, intuition registers a different world, a pervasiveness of movement, a permanence of impermanence, an emptiness which is not Void, so much as dense saturation without circumference. Bergson's view of becoming consists in laying out all that which has been enfolded in the image of solidity attributed to the particle, opening up the black box: "All division of matter into independent bodies with absolutely determined outlines is an artificial division" (MM: 196). Pure perception, the nonspatialized, nonrepresentational immediate access to dynamism, presents movement in itself, displaying the self-elaboration of reality in the making. Every solidified memory, every particle, every closure, is opened. Beneath the world of determinacy, there is the dimension of indeterminate irreversibility. Prior to its solidification into (illusory) being, memory is actuality, the becoming of states, the streaming of elements. In pure perception, "a moving continuity is given to us, in which everything changes yet remains" (MM: 197). This is truly no dream, but the most substantial reality, the power of alteration, in which every division is equally plausible. We accept the image at its face-value.

PURE PERCEPTION AND PANPSYCHISM

The Bergsonian pure perception is immediate coincidence with change. If living involves carving up one's world into parts, then intuition must be considered an extended life, an infinite vitality that does not need to solidify its surroundings. The nonrepresentative dimension of an unextended, infinite body, forms a unity with all processes. Mentioning a contemporaneous physical view, Bergson declares that upon closer examination, the apparent discontinuity of particles must be abandoned: "The solidity and inertia of the atom dissolve either into movements or into lines of force whose reciprocal solidarity brings back to us universal continuity" (MM: 200). Nothing is lost in pure perception, for Bergson's idea of heterogeneous multiplicity includes the individuality of the durations inhabiting larger world-durations. One quality of time contains several temporal qualities. Just as the cone of memory contains no base, so there is no ultimate level of duration. Bergsonism does not mean that we must posit any such ultimate duration of durations. The temporal is the timelessness of change. Metaphysically speaking, the permanence of impermanence implies extratemporality. This is what Bergson means when he states that everything changes while also remaining. Intuition, the form of perception purified of memory, shows "modifications, perturbation, changes of tension or of energy and nothing else" (MM: 201). No longer can the idea of a particle be upheld except as a convenient abstraction. In a revolutionary manner, Bergson, pre-dating the quantum turn in physics, shows that each corpuscule is nothing but a collection of vibrations and oscillations. Bergson cannot be described as a phenomenologist, for the very simple reason that he does not uphold the idealistic doctrine of the primacy of consciousness, as the primacy of pure perception contains nothing of the subjective. How else could we interpret the thought experiment of the subtraction of consciousness, unless in non-correlative terms?

> [I]f you abolish my consciousness, the material universe subsists exactly as it was; only, since you have removed that particular rhythm of duration which was the condition of my action upon things, these things draw back into themselves, mark as many moments in their own existence as science distinguishes in it; and sensible qualities, without vanishing, are spread and diluted in an incomparably more divided duration. Matter thus resolves itself into numberless vibrations, all linked together in uninterrupted continuity, all bound up with each other, and traveling in every direction like shivers through an immense body. (MM: 208)

Remove consciousness from reality, and what do we obtain? The active forces of images remain intact, proliferating movements. Alteration is non-correlative. Extrinsic to our individuated, organized bodies, the changes, reactions, and disturbances continue to act in very much the same manner as before. One of the images—consciousness—has been removed from the scene, but the totality of images continues its work of differentiation and creation. A completely pure perception would blend with multiplicity, until any interval between it and materialized life is removed. As a result, perception inescapably slides into matter itself. This is not Deleuze's panpsychism of the virtual. Rather, it is a panpsychism situated within an active, rolling present, in which the distinction between movement and sensation is blurred. "Concrete movement," we read in Bergson's dramatic summary, "capable, like consciousness, of prolonging its past into its present, capable, by repeating itself, of engendering sensible qualities, already possesses something akin to sensation" (MM: 246–7). The time of matter is a time that endures, extending itself into a necessarily integral directional "-ness." By an act of conceptual creation, we choose a suffix and make of its a phrase denoting pure becoming. The *ness* is the quality of dynamism prior to individualization. Already containing heterogeneity in the form of tendencies, it is nonetheless a continuum of force, the real, actual existence of undiminished intensities. Bergson's panpsychism is a panpsychism of actuality. Freed from the exigencies of life, good sense dilates not into the nothingness of delusion or dream, but the concretion of an occasion. A *ness* is an occasion for action, persisting not in the shape of preexistent potential or a vacuous virtuality, but given as a genuinely salvageable moment. "The change is everywhere," but the "universal transformation" will only be perceived successfully if we enact a leap of faith (MM: 209). Intuition consists in the traversal of planes which have been hitherto blotted out by the necessities of limited, individuated life. The flowing point floods everything. Cone, plane, and line are nothing to us. Inside *this* now, there is a flowering of times. Passing from our former state to the realm of withdrawing images, intuition captures their quality of withdrawal prior to the turn of human cognition. After its elimination, we "bring back consciousness" to a new life: "At long, very long intervals, and by as many leaps over enormous periods of the inner history of things, quasi-instantaneous views will be taken, views which this time are bound to be pictorial, and of which the more vivid colors will condense an infinity of elementary repetitions and changes" (MM: 208–9). The leap allows intuition to condense the vibrations, while accessing an experience of the dilated material life of imaging. Inside the ness, a strange craze overcomes us. The more forces we feel, the less we exercise our power of cognitive selectivity. Discrimination and recollection fail to keep the new life, the resurrected soul, in check. Similarly, the brain too is condemned to a secondary importance. Suspended, brain and body alike subside, giving way for what amounts to an authentic participation in the lives of things. Contemplation of the degree zero of becoming, none of the sensations coming into the spirit from outside are decelerated or halted. Immobility has been left outside this illuminated zone of participation. The indeterminacy of matter in general points toward the presence of freedom and perception inside the aggregate of images (Dolbeault 2018: 550). This lowest level of consciousness is a repetition that, for all its apparent resemblance to a mathematical instant, nonetheless extends itself into an open future. Nothing is predetermined, anything can happen. "Pure perception" is, in the final instance, "really part of matter" (MM: 222). The true nature of matter resides in activity. Far from composing a passive substrate, the material is the genuine source of energy and movement. Bergson reverses our perspective. Instead of localizing freedom in an organized body, freedom is extended, infinite, pervasive. It is almost as if all finite bodies were but components of a borderless corporeality. Insofar as they remain entrapped in their finitude, living things are actually decelerations of actuality, dissipative screens preventing the outflow of energy. The leap detaches us from ourselves, from out emplacement within our individual duration. Rather than raising us along the cone of memory into dispersal among the dreams, the authentic leap heads downward, into the infinite vibrations of materiality. The goal of our movement is a descent, a downfall that nonetheless results in an ascent of our spirit. Reunification with materiality, the ultimate mystery of death, is the dissipation which nevertheless results in an amplification of energies. No more delays, no more contractions, only the openness of impersonal material durations. Observation of the ness, the necessarily contingent swerve uniting lines of facts, allows for a recognition of the fully anisotropic nature of processes. In the basis of one thing, there is already at work another becoming, fully other. Like any other duration, we are already inside of objects (MM: 246). This is a necessary metaphysical truth relating to the nature of becoming. Substance is tendency, reality is change, permanence is impermanence. Outperforming the illusory verticality of memory, the priority, indeed pervasiveness, of perception automatically implies the primacy of pure perception. We take hold of the obscure, becoming one with a new horizon of nobility. Freed of the necessity to remain ourselves, we are liberated from the past, perceiving the sensibility, the whole process boundlessly traversing a shattered cone. No longer a moment, the instant becomes an outflow. Within the original intuition of material life, all interests have been decomposed into excitations. All desire subsides due to the simple intuition of change. Duration is the powerful timelessness of creation.

NOTES

1. The hypermediated nature of late modern/postmodern society could, arguably, account for the remarkable explosion in the popularity of Deleuzian philosophy between the 1990s and 2020s. But if media technologies shift, as we have every reason to hope, toward a screenless omnipresence, then the degree of mediation in society may perhaps even decrease. It would be interesting to observe the fate of a virtualizing philosophy under social circumstances of demediatization and deconditioning.

2. Giancarlo Maiorino has identified Bergson's philosophy with a broader, baroquely inspired "cornucopian" state of mind, a mode of thought characterized by sensitivity to the excessive nature of reality. There is no end to the "cornucopian outpouring"; Bergson's intuition of duration allows for us to recognize the infinite, perpetual openness of the process of perfectibility (Maiorino 1990: 185). The Baroque in general, as a mode of thought or method, and not just an exclusively historical phenomenon, represents an affirmation of the spiral form, in opposition to both simplicity

and linearity. Applied to writing, such a methodology allows for "metaphoric words" to evolve "into metamorphic expressions" (Maiorino 1990: 99).

3. This can be applied to the relations between various images lacking subjectivity as well. The phrase "noumenal" here does not refer exclusively to the epistemic connection between a consciousness and its objects.

4. Joël Dolbeault argues that *Creative Evolution* too may be read as a panpsychist tract. On such a reading, the later *élan vital* can be tentatively associated with the earlier concept of pure perception (Dolbeault 2002: 529–537).

5. Max Scheler's doctrine of spirit as absolutely lacking in power bears a resemblance to Bergson's equation of the spirit with powerless virtuality (Scheler 2017 [1928]). We leave open here the question of what influence, if any, Bergson could have had on Scheler, the latter being a formidable philosopher of considerable originality in his own right. Incidentally, Scheler's method has also been characterized as an "intuitive" one. Although we cannot, for reasons that by now should be abundantly clear, assent to her characterization of Bergson as a "phenomenologist," we see much merit in Adriana Altamirano Alfaro's comparative study of Bergson and Scheler, for the concept of "intuition" does appear to be fundamental to both of these remarkable thinkers (see Alfaro 2017).

6. John Ó Maiolearca recently renamed himself, formerly having used the Anglicized name John Mullarkey.

7. Ó Maoilearca also emphasizes the influence of Émile Boutroux, Bergson's teacher, on the contemporary self-styled "speculative materialist" philosopher Quentin Meillassoux as well. Yet another philosophical specter surely worthy of further examination (Ó Maoilearca 2014: 17).

8. Bergson's social and ethical philosophy shall form the focus of a later volume, one that could, but not necessarily ought to, be read as a continuation of the present book.

9. In this regard *Memento* echoes the anti-realist metaphysics of the early Third Millennium. The civilizational sense of post-metaphysical derealization is also markedly present in Tom Tykwer's *Lola rennt* (1998). Citing Bergson's study of the comic phenomenon as mechanism, Oliver C. Speck has argued for a simultaneously comic and redemptive reading of this latter film. Because the main protagonist, Lola, is capable of starting her run again and again, in the manner of a video game character, her travails can be seen humorously, but also we as viewers also come to suspect that real life is somewhat more difficult to restart. Life is "inscribed in time" (Speck 2011: 186). It is only when Lola accepts her subordination to chance by throwing the dice that she proves capable of truly becoming a player of the game of life. Surrender is the source of peace. Lola is the good gambler who magically "affirms" the "last chance" by "screaming" (Speck 2011: 197).

Chapter 5

Getting Back to Duration

What is liberty? The fundamental question addressed by Bergson's doctoral dissertation points to an ambiguity surrounding the status of this concept, as well as the work itself. On first appearances, *Time and Free Will* (*Essai sur les données immédiates de la conscience*) would appear to be a work of psychology, critiquing the, at the time, mainstream position of "psychophysics," one echoed by today's cognitivist neurosciences. According to this latter position, mental states can be quantified. The stated goal of the text is to dispute this idea, based on observations relating to the flow of consciousness. Cutting through the clutter, we can reach an intuition of the "immediate data" of consciousness. And truly, this work contains much psychological information relating to perception, feeling, emotion, and mental states. But we must not lose sight of the fundamental goal. *Why*, we ask, must we return to the immediate? In our view, there is an underlying kinematics of freedom at work. And what would the intuition of the free and unhindered be, if not the logical outcome of an ethical project? The mistake of many critics of Bergson is to expect explanation even where something very different is required. Ralph Barton Perry, a contemporaneous critic whose uncharitability rivals that of Russell, complains that the Bergsonian method of intuition does not furnish us with an analytical explanation of phenomena. Where is epistemology, where is truth, in this unphilosophical philosophy? Intuition "throws no light whatever on the nature of anything. My experience of life has dissolved; but nothing follows concerning the nature of life. (. . .) I have blurred and blotted our my knowledge of life; but life is not therefore blurred or extinct. In the twilight all things are gray; in ignorance all things are simple" (Perry 1911a: 679). What Perry labels "immediatism" would be a supposedly spurious, dogmatic method, an "irrationalist" performative use of words masquerading as philosophy. In the view of the analytic position advocated by Perry, truth must be some remainder, obtained after we have sliced and diced things into their constituents. Unity and simplicity give us nothing new, an integral method that blurs differences cannot serve as a source of scientific, rational truth. Writing in another century, the phenomenologist Dan Zahavi expresses a basically similar evaluation of *Time and Free Will*'s contents: "Bergson might be approached in various ways, but it does not seem advisable to read him as a (proto-) phenomenologist. (. . .) His analyses are simply no match for what can be found in later phenomenology" (Zahavi 2010: 130). Writing in two different epochs, in the context of two very different traditions, Perry and Zahavi nonetheless share the same outlook. Their error lies in expecting a scientifically grounded explanation of the manner in which consciousness works. We hope to show that the real object of Bergson's first book of systematic philosophy, the object of our concluding chapter, is freedom. In spite of appearances to the contrary, *Time and Free Will* is an investigation in moral philosophy. Far from neglecting or ignoring the realm of ethics, the Bergsonian doctrine will be shown to have been inherently ethical from its very commencement. This requires according freedom its rightful place on the throne of Bergsonism, so to speak.

WHAT IS FREEDOM?

A question immediately presents itself: what kind of liberty is at stake here? Specifically, we have in mind two very different concepts of freedom, namely, its negative and positive forms. Before interpreting *Time and Free Will* itself, it would be helpful to turn to one of Bergson's staunchest critics, Isaiah Berlin.[1] If we are to define Bergson's concept of freedom, we must first delineate its two basic forms. In this regard, Berlin's essay, "Two Concepts of Liberty," can assist us, even if we do not quite agree with some of its conclusions. Negative freedom would be, simply put, the right to be left alone. If anybody willingly obstructs us in our activities, that person can be conceived of as coercing us, hence the state of negative freedom is an absence of such a deliberate hindrance: "By being free in this sense I mean not being interfered with by others. The wider the area of non-interference the wider my freedom" (Berlin 1958: 170). Of course, according to the view of classical liberalism, some restrictions on action must be implemented, because a state of absolute freedom would be ruinous to our activities. If anybody could do anything to anybody, others could potentially interfere with us to an unlimited degree, leading to something like a Hobbesian state of nature, a social anarchy. Hence, citizens are forced to accept the restriction of their freedom to pillage, kill, and maim others, so as to prevent themselves from falling victim to crimes, thereby fully guaranteeing the right to be left alone. There is something paradoxical in the seemingly commonsense idea of sacrificing some individual freedom for a greater overall degree of social freedom. A frequent objection to a laissez-faire system, one that has been repeated endlessly over the past couple of centuries, is that it produces glaring social inequalities. If we lack the resources to participate in competition, our liberty counts for nothing whatsoever. Hence, the government or some other governing body must intervene to guarantee a minimum degree of well-being for the individual, taxing the haves so as to improve the chances of the have-nots. All good and well, Berlin says, "but a sacrifice is not an increase in what is being sacrificed, namely freedom, however great the moral need or the compensation for it. Everything is what it is: liberty is liberty, not equality or fairness or justice or culture, or human happiness or a quiet conscience" (Berlin 1958: 172). Even the moral beauty of communal altruism does not, for Berlin, compensate in any meaningful way for the sacrifice of liberty. Decades after the publication of *Time and Free Will*, we find Berlin's essay a reiteration of the key thesis of Bergson's book. It testifies to the suggestive power of Bergson's words that even his enemies were fertilized by his teaching. Never can a qualitative difference in kind be translated into a quantitative difference of degree. Try as one might, freedom differs radically from other values. The surrender of freedom represents an absolutely incommensurable loss, one that cannot be compensated for by the corresponding growth of other values. True, Berlin is also an adherent of historical relativity. Negative freedom is the right to be left alone within a certain sphere of privacy, but the precise contours or borders of this area will differ depending on the social form in question. Negative freedom is a "liberty *from*" tied to a certain, historically circumscribed territory or domain, the "absence of interference beyond the shifting, but always recognisable, frontier" (Berlin 1958: 174). What can we say of *freedom to* though? The positive sense of liberty derives from a desire to be one's own self. As Berlin writes, "I wish my life and decision to depend on myself, not on external forces of whatever kind. I wish to be the instrument of my own, not of other men's acts of will" (Berlin 1958: 178). Would *freedom to* not correspond to the epitome, the completion, of the idea of freedom? Not so, Berlin claims, for the idea of a freedom which is authentically "ours" is based on very uncertain ontological grounds. All this is based on the claim that there pertains a "real" or "autonomous" self, against a lower or "heteronomous" self (Berlin 1958: 179). Berlin takes issue with the concept of positive freedom because it would, supposedly, open the door to all forms of authoritarian substitutions of the subject for false, communally contrived "selves." Positive freedom is open to all manner of abuse, for it relies on the identification with a deep self. Hence, the collectivity can be viewed by authoritarians as being more authentic than the empirically self-given individual. The "true goal of man," identical with the "free choice of his 'true,' albeit often submerged and inarticulate, self" subordinates free choice, leading to enslavement in the name of liberty (Berlin 1958: 180). We must therefore abstain from positing a positive idea of freedom. Only negative freedom corresponds to the empirically given free individual, although the exact contours of this person, defined as a sphere of privacy, will always be historically contingent.

Although this essay in particular does not mention Bergson by name, it is difficult indeed not to read it as a critique of Bergson's position on freedom, as presented in *Time and Free Will*. An important goal of our chapter lies in defining Bergson's idea of liberty. On first impressions, a positive view is at stake. On our part, however, we shall argue that in fact, a Bergsonian view of freedom is actually profoundly negative, for two very important reasons. Not only will this conception be revealed as agonistic, from an epistemological perspective, but it will also be found negative in the sense of being an affirmation of the *absence* of obstruction. Bergsonian freedom is a negative liberty, one that is also open to a fundamental self-abnegation of the subject. Far from detracting from our subjectivity, the quest for immediacy we embark upon is an ascetic practice that adds to us, snowballing as we are into something vaster. For Berlin, it is not evident how "ascetic self-denial" can be called "an enlargement of liberty" (Berlin 1958: 186). Undoubtedly, this inability to fathom how mysticism can result in a growth of the person and its corresponding freedom originates from the erroneous equation of asceticism with mere self-denial. The practice of intuition is not a denial of self, but a sweet realization, a fruition of our own ripening, an ecstatic processing of continuity, empty of all disturbance and dislocation. There is nothing of the bitterness of self-denial in the active prolonging of the flowing present to infinity. To understand Bergson's concept of negative freedom, we must work our way through his train of thought, until we obtain the idea of a negative, aporetic freedom which follows from the pure contemplation of duration in our present. In the beginning, a multiplicity of mental states is given. Indisputably, every conscious person can observe many different "states of consciousness," each one of which is "capable of growth and diminution" (TFW: 1). The multiple nature of these different states would appear to suggest that they can be quantified, that is, reduced to an issue of magnitudes. A feeling or any other mental state could then be treated as an "intensive magnitude," variable according to differences of degree (TFW: 2). The introduction of the idea of intensity into our inner life implies a transformation in the way we treat the phenomenon of consciousness. If these various mental states are indeed intensities, this means they can be spatialized. As Bergson notes, "In the idea of intensity (. . .) we shall find the image of a present contraction and consequently a future expansion, the image of something virtually extended, and, if we may say so, of a compressed space" (TFW: 4). A key part of the

Bergsonian method consists in the primacy of internal witnessing. Like the prophets of the Old Testament, Bergson is perpetually intent upon listening to the voice within. In Vladimir Jankélévitch's words, the "supreme authority" of philosophers is "inner experience" (Jankélévitch 2015 [1959]: 24). Even in his very first book, Bergson adheres with remarkable consistency to this fundamental methodological premise. Always the inner voice must have the final say, even at the expense of going against the grain, of deviating from either the most up-to-date results of scientific inquiry, or resisting the cherished dogmas of others. The very paucity of explanation will prove a strength in this regard. Analytical rigor is certainly not missing from the pages of *Time and Free Will*. We find many analyses of consciousness, which are of seminal importance if we are to reconstruct the Bergsonian ethics of liberty. A question presents itself at the outset: does the transition between different intensities of the same mental state also imply their separability, their discreteness, and by extension, their spatial nature? Let us take the case of a very intensive feeling, an emotion that inundates our very being. Is this feeling continuous or discrete? There is apparently some sort of continuity of states at work here, yet the latter are also multiple in their nature. "An obscure desire," for example, "gradually becomes a deep passion" (TFW: 8). Within the stream of consciousness, several different states appear to be in motion. Where is the boundary between mental states? What separates a superficial, fleeting intimation of an emotion from a genuinely felt abysmal passion? Let us introduce an example taken at random from Virginia Woolf, a novelist inspired by Bergson to a certain degree.[2] In the novel, *To the Lighthouse*, we read the following:

> It was as if the water floated off and set sailing thoughts which had grown stagnant on dry land, and gave to their bodies even some sort of physical relief. First, the pulse of colour flooded the bay with blue, and the heart expanded with it and the body swam, only the next instant to be checked and chilled by the prickly blackness on the ruffled waves. Then, up behind the great black rock, almost every evening spurted irregularly, so that one had to watch for it and it was a delight when it came, a fountain of white water; and then, while one waited for that, one watched, on the pale semicircular beach, wave after wave shedding again and again smoothly a film of mother-of-pearl. They both smiled, standing there. They both felt a common hilarity, excited by the moving waves. (Woolf 1992: 193–194)

Every new movement bursts forth from the frothing of a previous phenomenon. The stream of thought, this becoming-liquified of consciousness, begins with a "pulse of color" that floods the mind of the couple, taking their hearts out to sea, washing away their subjectivities in a wave of affectivity.

But no sooner is one feeling expanded, it is dashed upon the black solidity of another, rival affect. The waves of sympathy are interrupted by the "prickly blackness," complicated by the awareness of a limitation. Besides each intimation, there is a rival emotion of repulsion. So the ebb and flow build up, until the pair attains a state of excitement, aggravated to the point of anguish by the hilariously irregular movements of the white, frothy waves. This whiteness gushes forth, like semen from a penis, enlivening the imagination, unleashing a sexualized fantasy ("it was a delight when it came"). Even in this brief description, distinguishing between the various states becomes difficult. Like the process of orgasm (when achieved through the voluntary cooperation of two consenting individuals), the ebb and flow of sensations tends to blend with an all-encompassing common feeling which can be characterized broadly as the accumulation of a mutual "enjoyment." The couple striving for physical relief, affected by the waves, shares in a communal enjoyment, but what of the minute complications which have already nested themselves within the flow? The irregularity of the waves and the obstinate presence of the hard, black rock betrays an underlying heterogeneity, operative even inside of the most continuous of streams. Whiteness is permeated by a blackness, and the stony blackness in turn contains something of the frothiness of the waves. There is something of the sexual in Bergson's mode of description too, a hint of emotion building up into joyful release. In another context, Rudolf Bernet has written of phrases such as "pressure" or "thrust" as pointing toward the presence of "a sort of desire" lending a "quasi-sexual character" to "the Bergsonian description" (Bernet 2005: 68). For all this, we do not wish to exaggerate. If the description of life as accumulation bears a resemblance to the sexual act, this originates from the latter's subsumption or enfoldment within the contexture of the former. Can it then be said that the couple enjoying the irregularity of the waves on the beach, or, alternately, the alternating oscillations of their own sexualized, localized enjoyment, feel one thing, or are they as a matter of fact undergoing several different emotions united under a single heading? In other words, does the person enjoy a series of different enjoyments, or "enjoyment" as such? On the Bergsonian view, joy is an enlightening of the whole personality: "All your sensations and all your ideas seem to brighten up: it is like childhood again" (TFW: 8). If only for a brief moment, anything seems possible. Within a duration, we must posit a continuity of heterogeneous elements. Bergson's revolutionary claim is that one who asserts that we experience a multitude of emotions and the one who advocates for continuity are both correct, provided we use the words "multiplicity" and "heterogeneity" in a manner suitable for the explication of inner duration. For what is at stake here is the positing of an interiority peculiar to each duration. In the case of a human being, the play of emotions represents a multitude of forms that nonetheless blend into one another in that human's inner time. The ambiguity of Woolf's white waves is a function of their incommensurability *and* inseparability from the stubbornly immobile black rock. An intensive feeling separates the personality from exteriority. "In cases of extreme joy," Bergson observes, "our perceptions and memories become tinged with an indefinable quality, as with a kind of heat or light, so novel that now and then, as we stare at our own self, we wonder how it can really exist" (TFW: 10). If joy leads to perplexity, and the goal of philosophy ought to be the causing of joy, as Bergson himself affirms, then it follows that philosophy conducted in a Bergsonian key must use explanation as an instrument, and not confuse discovery with the ultimate objective. The goal will be the realization of joy achieved not against time, but within and through the contemplation of impermanence.

STATES OF CONSCIOUSNESS AS HETEROGENEOUS MULTIPLICITIES

Before we can proceed to the delineation of the pure intuition of duration, however, a difficulty presents itself. Are the states of consciousness singular or multiple? It could be said that the deep joy Bergson speaks of differs in a quantitative sense from a minor happiness. Unconsciously, we attribute multiplicity to a single event, even though each element corresponds to the singularity of this occasion. Take Georgie O'Keefe's 1917 series of watercolors, *Light Coming on the Plains*. Why not simply contract the perception of a striking desert sunrise into a single work? What O'Keefe is striving for in her portrayal of a landscape is to capture movement while removing every nonnecessary element from representation. Through an act of artistic subtraction, the continuity of movement is captured, a middle way is enacted that avoids reducing the singularity of *this* Texan sunrise to a series of mere snapshot reproductions. An abstraction is achieved which captures movement without immobilizing it. As Judith Zilczer highlights, "O'Keefe's sunrise watercolors may be considered visual analogies for Bergson's idea of duration. While the three sunrise paintings form a series, her watercolors do not represent a simple, linear progression. Rather each expansive sunrise image implies organic unfolding in an extended present. O'Keefe underscored this evolving experience of time by her use of the present participle in the titles of the works" (Zilczer 1999: 196). Change *is* in the making, situated within its particular now, not in an absent yesterday or a dreamy tomorrow. Instead of a linear brightening of the horizon, we observe a wavering in the colors, born from an emphasis upon the irregularity of landscape. In her exquisite article, Zilczer demonstrates that O'Keefe was influenced by her teacher and mentor, Arthur Wesley Dow (Zilczer 1999: 194). In particular, O'Keefe's choice of pursuing landscape painting could have derived from Dow's suggestion of this topic in the highly popular textbook, *Composition*. There the author in question claims that landscape is an almost ideal subject for composition, as "its irregular spacings contrast well with the symmetries of pattern, and when tones are played over them the effects are new and strange, stimulating to further research into the mysteries of tone" (Dow 1913 [1899]: 69). The faintness of the colors belies the intensity of this insight into impermanence. Nowhere in O'Keefe's sunrise paintings is the expected violent outburst of violet or crimson to be found which we normally associate with the shining sun. *The new dawn is not yet arisen, but* is *in the process of arising*. A solar intensity is intimated by the very uncertainty of this advent. Like hermits contemplating the face of an absent God in the arid wastes, we are aware that the sun shall rise once more. Despite the wavering irregularity of O'Keefe's Texan sun, one emotion is nonetheless preponderant, and that is hope. Already something is in motion, the not-yet present empyreal has given news of its impending arrival. As Zilczer comments, "The aureoles of color emanating from the rising sun form an expansive image that captures the unfolding of a natural phenomenon over time. O'Keefe envisioned a present time evolving from the past and implying future growth. Each of the three watercolors bears within it the visual record and implications of the other two" (Zilczer 1999: 194). Every state is contaminated, tinged, as it were, with the others. To the three watercolors, we add the at present—absent future whose advent they imply. The reason expectation gives fortitude, nourishing as it does the soul with spiritual energy, is that in hope "the future (. . .) appears to us at the same under a multitude of forms, equally attractive and equally possible. (. . .) The idea of the future, pregnant with an infinity of possibilities, is this more fruitful than the future itself" (TFW: 9–10). An eschatological expectation contains infinitely more than the reality of any fulfillment. But this circumstance in itself we may take as proof of the primacy of the present! It would certainly be a disappointment, even a terrifying omen, if the sun were not to make its scheduled appearance upon the horizon. Something in such an eventuality would disappoint us, the substitution of absence for an expected presence being invariably disheartening. Recently, astronomers found themselves shocked and dismayed when a previously discovered planet disappeared from sight. First observed in 2008, the exoplanet, officially named *Fomalhaut b* and nicknamed *Dagon* was one of the first planets observed outside our own solar system. Closer examination of images taken by the Hubble telescope, however, revealed that the previous calculations of the astronomers had been amiss. In the prosaic words of an article, "Dagon had vanished altogether. In its place was . . . nothing" (Starr 2020). What had formerly been conceived of as a gas giant transformed overnight into a cloud of particles, the aftermath of a collision between two asteroids. If hope is the present turning in the direction of an open, indeterminate future full of opportunities, then sorrow corresponds to the dashing of expectations. Interestingly, what Bergson says at this point can be unproblematically related to what we have written of the virtual and the impotentiality of memory: "Sorrow begins by being nothing more than a facing towards the past, an impoverishment of our sensations and ideas," a feeling which culminates, at its greatest intensity, in the "impression of crushing failure, the effect of which is that we aspire to nothingness" (TFW: 11). The name "Dagon" can be taken to refer both to an ancient Mesopotamian deity and to one of Howard Phillips Lovecraft's first short stories. Written in 1917, the novella "Dagon" is one of the first stories published by the iconic horror author. We may speculate that at this early stage of his career, the writer would undoubtedly have been full of ambition and optimistic expectations. Had Lovecraft seen in advance the relative lack of material and professional success his work would bring, as well as his own early death, he would undoubtedly have been discouraged to an extent. Perhaps a sorrow would have gripped the writer, as well as a sense of indignation. But equally one could also speculate that the posthumous greatness of Lovecraft's works, far outshining the demise of their creator, would have compensated their author even with the awareness of his own temporal brevity. Thankfully, the feeling of disappointment is only possible once we have access to a past which can then serve retrospectively as an object of reminiscence. If we are to avoid sorrow, we are obliged to turn ourselves toward the future in the perpetually creative present. Even at the hour of our demise, the force of the now is always at work. A seemingly minor detail is pregnant with meaning. In her 1917 watercolors, O'Keefe used "bare newsprint," combining the sunrise with an absent media (Zilczer 1999: 194). The advent gives no news, aside from blurry intimations and uncertain traces, for the "news" relates to nothing but the past. The creativity of the flowing present cannot be reported: it must be lived. If the newspapers are inherently sorrowful media, then the sunrise watercolors are a communication of joy.

Can we then speak of magnitudes when it comes to mental states? This issue boils down to a more fundamental question, namely, is there such a thing as a mental "state" as such? Certainly, the various modalities of consciousness differ from one another in important ways. After contrasting joy and sorrow according to the divergence of their temporal direction, Bergson speaks of the impression of "grace." Why do we find the movements of a dancer graceful? In the view of the philosopher, this perception is afforded by the continuity of movement the graceful person displays. By observing the dance from outside, we find that "the perception of ease in motion passes over into the pleasure of mastering the flow of time and of holding the future in the present" (TFW: 12). Instead of performing a series of disjointed, incoherent, spasmodic movements, the dancer blessed with grace is apparently capable of uniting divergent subcomponents into a single performance. Through what Bergson calls a "physical sympathy," the audience and the performer are synthesized into a single affective moment of abiding (TFW: 13). As Ruth Lorand explains, it is important for us to bear in mind that for Bergson "vital orders are not composed of conceptual components, whereas conceptual entities are generated by the process of fracturing vital orders" (Lorand 2000: 88). Art in general is a soothing or lulling of our consciousness, resulting in a blending of our duration with other times and modes of becoming. This is not a conceptual process, but an extension of life through the dilation of spirit. What artistic sympathy shows is a complex permeability among psychological states, but there is a limit as to how much of another person's mind we are capable of accessing. Each emotion is "a state unique of its kind and indefinable, and it seems that we should have to re-live the life of the subject who experiences it if we wished to grasp it in its original complexity" (TFW: 18). Rather than beginning from simplicity, Bergsonian philosophy commences from the intuition of multiplicity and complexity. Understanding this will allow us to avoid the perplexity which puzzles interpreters such as Zahavi, who is "struck by the negative character of Bergson's account" (Zahavi 2010: 129). The negativity of Bergson's exposition is essential to its goal. The simple is not an a priori category, something already posited at the outset. Rather, it is an achievement of a mode of thought which dares to think the complex while striving for the simple. Separating two portentous descriptions of the present, we pass through stagnant, redundant dualities, until a new coincidence with reality has been attained. As will be seen, the false dilemma of libertarianism versus determinism forms a spurious dualism, being the illusory product of a badly stated problem. For Bergson, there can be no legitimate application of magnitude to phenomena of consciousness: "There is nothing in common (. . .) between superposable magnitudes such as, for example, vibration-amplitudes, and sensation which do not occupy space" (TFW: 32–3). A feeling can be identified as a disturbance of the organism, and truly there are various degrees of disturbance, this much Bergson admits. Take the case of a pain which grows in intensity. On a commonsense view, we do not think of our suffering as one note or melody, but rather as "a symphony, in which an increasing number of instruments make themselves heard," depending upon the depth of disturbance our body has undergone (TFW: 35). There is nothing out of the ordinary in identifying an affection with a certain quantity. It will be said that someone whose skin has been gently grazed by a razor feels "less" pain than a victim of torture, whose toenails are being ripped off one by one. Were we to measure their nervous reactions, we would find a greater degree of nervous stimulation in the latter individual. The issue at stake here relates rather to the inner being, the non-relational aspect of feeling. Bergson's contention is that magnitude only relates to how sensations come into relation. On their own terms, qualities are incommensurable with quantity. As F. C. T. Moore emphasizes, on a Bergsonian view "if we compare physical objects in terms of magnitude, we do so by ignoring their qualitative differences. (. . .) But sensations, if they exist in some sense, are precisely *qualia*" (Moore 1996: 45). The notion of "greater and smaller" automatically implies "relations of container to contained" (TFW: 72). It is an eminently commonsense move to relate certain affections to various degrees of external causation. The razor gently grazing our skin has a greatly less drastic effect on our life than a razor thrust deeply into our jugular vein or the plier ripping our nails out. Because we always relate our body to the realm of external causality, we come to associate "the quantity of the cause" with "the quality of the effect. At this very moment the intensity, which was nothing but a certain shade or quality of the sensation, becomes a magnitude" (TFW: 42). For an organized body, it is convenient to categorize various effects depending on the intensity of their effects, grading these according to the depth of their interventions in its own corporeality. But in Bergson's view, thinking of emotion in this manner results in a spatialization that obscures the reality of affection. How should we describe Othello's jealousy? There does not seem to be a concrete instant in which a hint or intimation of jealousy became the fully developed emotion. One emotional state seems radically incommensurable with another, for talk of degrees of transition or levels of intensity, masks the real qualitative difference between different feelings: "It would surely be inappropriate to think that Othello's jealousy is, as it were, simply a great agglomeration of constant and innumerable twinges. We might better say that his jealousy is so pervasive as to *exclude,* so far from subsuming, anything that could properly be called a twinge" (Moore 1996: 45). Just as a conflagration is not a collection of individually insignificant sparks, so a fit of powerful jealousy which consumes the individual, driving them into an act of self-destructive aggression, is more than a mere aggregate of various minor details and small irritations. The differences between various emotions and intensities are real divergences but "they do not consist in varying intensities" (Moore 1996: 47). Sensation only becomes a discrete magnitude once we have related it back to our body. Such a pragmatically oriented reduction does not, however, exhaust the being of affection in itself. As will be remembered from *Matter and Memory*, in this philosophy the body does more than react to external stimulations. Corporeality is also endowed with a kind of memory, as "it also absorbs some measure of" the action of external objects "in the mode of affection" (Jones 2016: 257). Affection is the internalization and virtualization of movement. What happens if we prick our left hand with a pin held in the right hand? The deeper the pin goes, the greater intensity of pain we shall feel. It appears that the pain is localized within the pricked hand, but Bergson holds this to be an illusion of sorts. Unconsciously translating the muscular effort of the right hand, we extrapolate from this effort, positing various degrees of pain when in truth there is nothing but a single movement of pain involved in the whole procedure (TFW: 42–3). A certain continuity of sensation has been displaced by a habit of thinking, surreptitiously transformed into a quantification of discrete irritations. Choosing the correct amount of energy expenditure, selecting a suitable intensity of effort, can mean the difference between life and death in the struggle for existence. But this does not entail that our metaphysics must also accept the confusion of the degree of muscular effort with the quantification of affection. A quality has been mistaken for a quantity, an intensity for a magnitude (TFW: 43). Radically, this recognition implies that the divergences among qualities are so vast as to defy *any* translation into magnitude whatsoever. No relationship of exchange can be identified between two qualitatively different feelings.

RIPENING THE SELF

When it comes to the affectivity of the body, "a more intense heat is really another kind of heat" (TFW: 47). It is not so much a case of denying intensity tout court but rather negating the validity of magnitude in relation to intensities. A feeling is an intensity that has nothing to do with a quantity. Each degree of a pain, each level of joy, is itself a different species nevertheless united within a single felt continuity. As Bergson asserts, "The various degrees" of a certain genera or type of feeling are "so many species" (TFW: 48). Spontaneously decoding qualities into quantities, our corporeality distorts and restricts the gambit of perception. When lifting weights, there occurs the production of a series of localizations. In the act of lifting a weight, "I experience a series of muscular sensations each of which has its 'local sign,' its peculiar shade" (TFW: 49). It would be far too complicated for a consciousness that must, in the interests of survival, economize time, to treat various local signs on their own terms. Rather, it makes more sense to handle the sensation of lifting a weight as a continuum of discrete degrees of effort differing exclusively according to their magnitude, the practically engaged mind reducing different qualities to differences in degree. As opposed to the creature struggling for survival in the wild, or the athlete competing for fame, the metaphysician has more leisure on their hands. A philosopher can afford to delve deeper, excavating the immediate data of consciousness. Underneath the local signs "what is present to consciousness immediately is the sensation of (. . .) a heavy movement, and this sensation itself can be resolved by analysis into a series of muscular sensations" (TFW: 50). Bergson is not denying the existence of varying intensities. What the philosopher is disputing is their translatability into the language of quantity and number. As Michael McKenna and Derk Pereboom emphasize in a recent book relating the various possible philosophical treatments of free will, Bergson's view of the unquantifiable nature of states of consciousness has decisive consequences for the concept of freedom: "Although actions occur in time, the temporal properties of conscious agency do not resolve into the kinds of magnitudes required for the applicability of causal laws" (McKenna and Pereboom 2016: 233). We shall see that it is the indeterminacy of the will which accounts for its freedom. Bergson's account of emotions is reliant upon a semantic of color and saturation, but also of melody. Our internal duration is filled with various tonalities, in both the visual and the audible sense of the term. It would be nonsensical indeed for us to characterize the color black or gray as being nothing but a relative absence of white. As Bergson notes, "Black has just as much reality for our consciousness as white" (TFW: 54). Similarly, the various intensities of whiteness, traversing different shades of gray, finally culminating in blackness, are not differing magnitudes of white, but rather, colors in their own right, each differing in a qualitative sense. Bergson believes, based on the testimony of lived experience, that something similar holds for states of consciousness. We must remain true to what our consciousness presents to us immediately, without the introduction of theoretical blinders and other supplementations. It will then be found that the relationship between two emotional states is "like that between the shades of the rainbow," bearing no resemblance at all "to an interval of magnitude" (TFW: 66). Even more radically, Bergson denies that consciousness, in its immediate form, can have any extension whatsoever. Consciousness is, for us humans, the location of the qualitative, whereas externality is wholly homogeneous. Bergson declares with a youthful radicalism that "there is no point of contact between the unextended and the extended, between quality and quantity" (TFW: 70). This distinction is valid not only on the ontological plane generally, but is also internal to subjectivity. The dualism of this description is striking and will recur when Bergson speaks of the relationship of the authentic self to its superficial involvements with exteriority, be it the physical dimension or the sociocultural environment. Already in the first chapter of *Time and Free Will*, we have a nascent distinction between the pragmatically involved body and an authentic self open to the experience of immediacy that recurs in the differentiation of the "deep-seated self" and the "superficial self." Rediscovering the deep self is inseparable from the experience of real duration. The import of the entire Bergsonian project consists in the enactment of this return to immediacy. When hearing "a series of blows of a hammer, the sounds form an indivisible melody in so far as they are pure sensations," giving rise to the impression of "a dynamic progress" (TFW: 125). We ourselves are indivisible melodies. The authentic level of the self would consist in the direct contemplation of a moment of change. An example from another author inspired by the Bergsonian doctrine springs to mind, the Nebraskan novelist Willa Cather. Her 1922 novel, *One of Ours*, relates the narrative of a misguided young man, Claude, who cannot find his calling in life until he discovers in heroic self-sacrifice a cause worthy of commitment, dying in World War I. Reaching France as a volunteer soldier, Claude experiences an epiphany in the Church of St. Ouen:

> When he reached the choir he turned, and saw, far behind him, the rose window, with its purple heart. As he stood staring, hat in hand, as still as the stone figures in the chapels, a great bell, up aloft, began to strike the hour in its deep, melodious throat; eleven beats, measured and far apart, as rich as the colours in the window, then silence . . . only in his memory the throbbing of an undreamed-of quality of sound. The revelations of the glass and the bell had come almost simultaneously, as if one produced the other; and both were superlatives toward which his mind had always been groping,—or so it seemed to him then. (Cather 2006 [1922]: 450)

Daryl W. Palmer has argued convincingly for a reading of the above scene in terms of Bergsonian duration, defined as the experience of the felt presence of a flowing moment (Palmer 2009: 127). In the stillness of this now, the deep-seated self achieves a revelation of the throbbing, vibrating present. *Every last one of us is perpetually growing into something different.* Claude observes the continuity of change, the relaxation of his perception allowing him to comprehend the blending of the purple-tinted window with the tolling. Glass and bell coincide in the expanded consciousness of a materialized pure perception. In Bergsonian philosophy, with its theme of a divided self, Palmer identifies an ethical imperative: "Human beings *ought* to emerge as themselves" (Palmer 2009: 119). On this view, it is not the final state which matters so much as the quality of our ripening. Compared with the process of growth, the final fruit can even be disappointing, for the very finality of any state precludes any further improvement. There is no hope in that which is finished: fruit which has achieved ripeness brooks no further delay; it must be plucked from the branches and consumed as soon as possible. Revelation consists in a felt simultaneity of self with its environment, a coincidence of ego with impermanence which nonetheless has the effect of stopping time. The deep-seated self "is a self whose states and changes permeate one another" (TFW: 125). Bergson's goal is to ripen, so to speak, his readers, preparing them to reenter the depths of their subjectivity. It is not a question of tearing the self apart completely. As John MacWilliam makes clear, the dualism of inner and outer, deep and superficial self must be treated as transitory: "The true and fundamental self is not an inner self which has an outer self in contrast or in opposition to it, but is a self which is beyond the dualism of inner and outer" (MacWilliam 1928: 79). The goal is a return to quality in itself, prior to spatialization by pragmatic involvements. We must not view the deep self as composing a separate section of a self. Despite appearances, Bergson is not attempting the resurrection of some Cartesian duality between individual and society or the mind and the physical world. In this regard, Zahavi's suspicion to the contrary is unfounded (Zahavi 2010: 130). Rather, what is at stake is uncovering true duration through a return to quality and continuity. Below number, we find "a qualitative multiplicity; below the self with well-defined states," we discover "a self in which succeeding each other means melting into one another and forming an organic whole" (TFW: 128). *Melt into yourself*: this is the Bergsonian imperative. Cather's antihero is not experiencing a succession of several different feelings. Rather, Claude's revelations soften, as it were, into a single, supple experience of glasslike bells, ringing.

According to Bergson, consciousness cannot be amenable to quantification, for one very simple reason: the quantitative relates to the spatial. When we count, we enumerate several different, albeit identical units, juxtaposing them in space. Mathematics cannot pertain to pure duration, because time differs from space (TFW: 77).[3] Without a clear principle of demarcation, lacking a mode of localization, a thing cannot be subjected to the count, and a definite contour is exactly what states of consciousness do not have. Whereas we can speak of a "discontinuity of number," there is nothing like discreteness and exclusivity among the various affections (TFW: 82). In consciousness, we observe immediately a confusion of qualities, replete with vividity, driven by originary associations inaccessible to the pragmatic dimension. Softly amorphous, the emotional is a frenzy of interiorized motions. As we learn early on, Claude is incapable of entirely controlling or coordinating his movements: "The storms that went on in his mind sometimes made him rise, or sit down, or life something, more violently than there was any apparent reason for doing" (Cather 2006 [1922]: 31–32). Common sense dictates that we view these manifestly ungraceful motions as being indicative of some pathology on the part of the protagonist, perhaps displaying a flaw of character, the presence of some uncivilized restlessness or impulsiveness. But as Perry underlines, Bergson advocates for the view that spontaneity itself is freedom. The self is free inasmuch as it is "the original and spontaneous author of events" (Perry 1911b: 718). Grace paradoxically demands practice. Paradoxically, the effortlessly graceful performer is unfree, because she is compelled by her occupation to engage in stereotypical actions, whereas the person acting spontaneously is a producer of free acts. States of consciousness display a self-organizing multiplicity irreducible to quantity (TFW: 87). To understand the life of the mind, Bergson enjoins us to abandon spatializing habits of thought. This latter attitude is exemplified by the Kantian view, summarized by Arjen Kleinherenbrink in the following way: "The present of the phenomenal world must be thought as a narrow temporal slice, extending ad infinitum in space, yet restricted to the shortest possible discrete moment of time" (Kleinherenbrink 2014: 206). Kant's historical "crime," if we may put it in such drastic terms, consists in separating "space from its contents," that is, in the positing of a homogeneous spatial continuum (TFW: 93). The Kantian Copernican Revolution results in the construction of an extensive, spatialized reality entirely lacking in real quality, resulting in a corresponding degradation of time to a state of simultaneity. Bergson's philosophy of duration, as first outlined in *Time and Free Will*, differs radically from the Kantian conception. It has even been called, not without basis, a "reversed Kantianism" (Ansell-Pearson and Mullarkey 2002: 32). The claim is a simple, albeit counterintuitive assertion: succession in the Kantian framework is not a temporal notion, but a reduction of time to spatial relations, while true time consists in a flowing continuity of heterogeneous moments. Bergsonism undercuts the construction of time as a homogeneous medium, showing that this is nothing more than a projection. Externality is the mark of things, but states of consciousness "permeate one another" (TFW: 98). Madeleine Barthélemy-Madaule cuts to the heart of the divergence between Kantianism and Bergsonism in the following enlightening remark: "Freedom, being becoming itself, is incompatible with the idea of a block. In the differentiation of the zones of the real, Bergson dissolves the difference between phenomenon and noumenon" (Barthélemy-Madaule 1966: 115). Because reality is composed of zones of reality, that is, multiple durations, and all becomings are as images to one another, neither the idea of an inaccessible noumenon nor the idea of time as spatial extension remains plausible in Bergson's framework.

By shedding new light on the original confusion, immediate perception of our own emotional states proves that consciousness is irreducible to anything like extension. For Kant, the natural world is predetermined, governed by physical laws, while freedom consists in the ability to separate ourselves from the empirical through an act of will. Liberty is, at its most basic level, "the independence of the will of coercion by sensuous impulses" (Kant 1998 [1781]: A534/B562). Humans are capable of constructing a moral law serving as an imperative for their actions, but only provided they separate themselves from exteriority. Roughly put, the mere ability to construct a (supposedly) universally valid moral law is what proves retroactively the existence of our free will. The self is "noumenal" to the extent that it creates a universal morality that gives us control over the phenomenal (natural) realm as well as ourselves. The Kantian concept of freedom is, strictly speaking, transcendental to the rest of the world. It rests on the view that "the idea of the moral law provides a 'way out' of nature and necessitated succession of events in time, as its universal character renders it immune to any contingency" (Kleinherenbrink 2014: 211). Kant's idea of transcendental freedom implies that liberty can only come into existence after we have left behind our subordination to natural laws. Moral law does not so much end the physical world, but rather introduces something foreign to nature, namely, a circumscribed artificial sphere wherein the subject is capable of determining itself and rationally shaping the conditions of its own life. Opposed to this view of liberty as the artificially imposed self-control of the rational subject, Bergson offers us something strikingly different. Rather than the universal moral law, it is the intuition of duration which grounds our freedom: "Pure duration is the form which the succession of our conscious states assumes when our ego *lets itself live*, when it refrains from separating its present state from its former states" (emphasis mine, TFW: 100). Underneath exteriority, there is a soft, amorphous vibration, an activity collecting movements, intertwinings, and traces of possession. That which we call "our" will, "our" self is but the individualization of divergences, perturbations, and reflections, deriving from a broader continuity of change. Once it sheds the superficiality of spatial exteriority, the self melts down into an interpenetrating integral whole, linked the organism with its ancient origins, contracting it into the form of a nascent emergence. *We recur, as us*. By undergoing the process of his ripening, even at the cost of his death, "Claude became, simultaneously, one and ours" (Palmer 2009: 126). The intuition of duration takes place *here*, inside of becoming. Bergsonism involves the recognition that "one does not need a realm outside nature for freedom" (Kleinherenbrink 2014: 213). Everything we need is already present. *Be the change you must undergo*, thus resounds the Bergsonian invitation in the souls of those brave enough to contemplate impermanence. This use of language is telling. We are informed that intuition does not so much consist in an effort, but rather, in "letting" our ego experience the flow of time. Among other points, Zahavi bemoans the lack of "instructions regarding how we are to liberate ourselves from the customs and conventions of ordinary social life in order to obtain an intuitive grasp of pure duration" on Bergson's part in *Time and Free Will* (Zahavi 2010: 128). The prospect of writing a Bergsonian meditation manual, not unlike some of the highly detailed Buddhist meditation scriptures, is one that an ambitious mind may certainly entertain. Bergson's interests, however, do not lie in identifying precisely how one should proceed, for the changes each person undergoes will differ according to their position and situation. Of key importance here is the realization that moral values must not be situated outside change and contingency. Quite the reverse: the goal of Bergsonism is to return to change by letting our selves live. One could object that there is nothing moral in such an undertaking. What makes the contemplation of duration a specifically moral good is that it furnishes us with an intensity of liberty, leading to an amplification of our free activity. A movement purified of extension, true duration is "nothing but a succession of qualitative changes, which melt into and permeate one another" (TFW: 104). Through a repeated emphasis on "permeation," we are led to abandon our belief in the very existence of a "state" as such. The construal of consciousness in terms of a succession of states is revealed as inherently misleading, for we introduce space where there is nothing but a continuity of impermanence. Instead of states, it is better to speak of "oscillations (. . .) *organizing themselves* like the notes of a tune" (emphasis mine, TFW: 105). We are free to the extent that we let our selves live. As Barthélemy-Madaule explains, "The free act has no other aim than its own existence, or its own authenticity" (Barthélemy-Madaule 1966: 124). Underneath the individuality of the discrete, unitary subjects, there are infinitely varied multiplicities of vibrations, intertwinings lacking any single choreographer, undirected motions extending themselves into new times. Gilles Deleuze speaks of subjectivity as a "system" populated by "both larval subjects and passive selves: passive selves because they are indistinguishable from the contemplation of couplings and resonances; larval subjects because they are the supports or the patients of the dynamisms" (Deleuze 1994 [1968]: 118). The intuitive contemplation of change and the actual activity of pure perception are synonymous with one another, forming together the return of the nascent subject to its own duration. Freedom must be viewed as relating to every single type of duration, including those far below the human level of intelligence.

Pure duration admits nothing of quantity. The vision of an indefinite multiplicity of processes permeating one another reveals numerous difficulties though. Bergson makes it clear that we must not envision the growth of the ego in terms of its relationship to a past. Rather, the ripening of the personality occurs in the present: "Within myself a process of organization or interpenetration of conscious states *is* going on, which constitutes true duration" (emphasis mine, TFW: 108). Duration cannot be located within either the past or the future, for both of these are not yet. These intertwinings are occurrent at every moment, being themselves sub-durations of our own indivisible movement of life. Inside a lifeworld composed of both organic and inorganic durations, we endure. It should not be said that the superficial ego is a mere illusion. The status of an entity as projection fails to detract from its positivity as a type of being, although admittedly a derived one. Later on, Bergson describes the superficial ego as a "parasitic self," which "continually encroaches upon the deep-seated authentic self" (TFW: 166). Insofar as the ego lets itself live, we can say that it aspires to actualization. On the one hand, the basal or durational dimension of subjectivity can be equated unproblematically with the pole of actuality. On the other hand, the superficial self, as powerless supplementation, a projection that spatializes perception through operations of common sense, can be brought into adequation with the concept of the virtual. An actual and a virtual self—these are the two asymmetric "halves" of a single personality. This duality is transitional, for it is incumbent that we transcend the superficial self, arriving at an integral intuition of the underlying unity of selves. This bisection of the ego in terms of a temporary duality is a preliminary step on the path to a complete shattering and reunification of subjectivity into an infinity of processes and durations. Prior to the immediate, pure intuition of duration, the self looks as if it were composed of the cooperation of two different but equal categories, namely, "succession without mutual externality" and "mutual externality without succession" (TFW: 108). The self, conceived of as an integral whole, would be nothing if not a pattern created by the mutual implication of intensive (durational) and spatialized perceptions. Bergson describes this correlation as "endosmosis" (TFW: 109). We must not lose sight, however, of the fact that this bisection occurs prior to the intuition of reality as universal change. Throughout the decades, many commentators have missed the underlying asymmetry between the two terms. Not contenting ourselves with the projections of the superficial, conventional ego, with its spatializing positing of separate states of consciousness, we must undergo a transition to the metaphysically grounded experience of change in itself. The homogeneity and determinacy of space, as constructed by Kant, is an illusion we must rid ourselves of, as the external physical world too is just as indeterminate and permeable as our own mental states. Have we not seen the entirety of physical objects dissolve into a set of movements in *Matter and Memory*? The interiority of duration is something much more radical than a merely subjective introspection. What intuition affords the metaphysician is immediate access to the flow of time in itself, breaking open the noumenal which has been denied to philosophy for too long by the Kantian dogma. Bergson's reversal of Kantianism consists in providing us with access to the absolute, this latter being change itself. If reality is composed solely of movements, then in the case of every reality "we have to do here not with an object but with a progress" (TFW: 111). Change applies, of course, to the illusions pertaining to the spatiality and impenetrability of objects and mental states as well, otherwise we could never hope to transcend our illusions. But if permeability is the law, where does this leave the self? If a duration is not a completed object, being "something that is unceasingly being done," where does that leave the subject, defined as a particular type of self-reflective duration? (TFW: 119). The unity of the self seems to be brought into question at this juncture.

Another critic of Bergson, the philosopher David Balsillie, identifies a tension latent in the Bergsonian conception of freedom. Despite the admission that the ego can occasionally attain to a type of freedom when it lets itself truly live, the impermanence and permeability of mental states nevertheless entails a general dissolution of subjectivity into a medley of qualities. Balsillie's question is a simple one which nonetheless has vital ramifications for our own interpretation of the Bergsonian idea of freedom: if there is nobody who can be free, the agent itself being but a collection of permeable durations, then how can we speak of a free "action"? As we have seen, the states as durational processes live a life of their own, being endowed with what can be called a *nascent freedom*, organizing themselves into continuities and melodies. Freedom is pervasive to duration. We do not even have to refer to Bergson's later introduction of the all-encompassing *élan vital* at this point, for already in *Time and Free Will* freedom has been extended far below reflexive consciousness, in the form of the spontaneous self-organization of mental states. Toward the end of the book, Bergson reiterates this view in a footnote, stating unambiguously that "the process of free activity goes on, as it were, unknown to ourselves, in the obscure depths of our consciousness at every moment of duration, that the very feeling of duration comes from this source" (TFW: 237–8). Balsillie contends that such a move, the granting of a certain liberty even to states of consciousness inaccessible for us, radically undermines the individuality of the agent. The individual dissolves into a multiplicity of free activities, not all of which can be conceived as conscious acts: "There cannot be free action without an actor. Who is the agent at every moment of duration? It cannot be the Ego, for it is a growing Ego, is only the organic unity of states that permeate each other. These states, however, we are told, are living beings, analogous to the cells of the body, which also have an individual life of a kind" (Balsillie 1911: 376). Resenting such an Aristotelian type of metaphysics, Balsillie finds it incomprehensible that a self, viewed on its own terms, should be at once a free agent and, from another perspective, an assemblage of impersonal or, better yet, sub-personal durations. The very idea goes against our notion of a spatially discrete, localized, individualized "self." In this regard, Deleuze's neologism, the "larval subject," is an apt description of these minor individualities from whose coincidence a living subject, agent of choice, emerges. As Deleuze puts it, "Before the embryo as general support of qualities and parts there is the embryo as individual and patient subject of spatio-temporal dynamisms, the larval subject" (Deleuze 1994 [1968]: 215). The designation "larval subject" stems from its function as the basis of a larger "subject-assemblage's" individuation. We regard the "free activity of mental states" in Bergson as corresponding relatively unproblematically to the Deleuzian idea of larval subjects. Both are already concrete individualities underlying heterogeneous continuities of change, and both contribute meaningfully to the smooth functioning of subjectivity. Neither the free activity of states nor the larval subjectivities are virtual, for their effective actuality serves as the already present precondition of consciousness. Balsillie finds it impossible to comprehend how the freedom of the person can be reconciled with the freedom of the sub-personal processes. Specifically, he has in mind Bergson's famous description of the genuinely free act as something fortuitous and rare. What then, in the final instance, is freedom for Bergson? The following definition is given: "It is the whole soul (. . .) which gives rise to the free decision: and the act will be so much the freer the more the dynamic series with which it is connected tends to be the fundamental self. Thus understood, free acts are exceptional" (TFW: 167). If this is the case, then we are faced with an apparent contradiction. Bergson seems to be suggesting that free activity is ubiquitous, but also that "we are free only on rare occasions of moral crisis" (Balsillie 1911: 376). How can these two declarations be reconciled with each other?

THE DEEP-SEATEDNESS OF LIBERTY

What is most interesting about Bergson's definition of freedom is the starkly aporetic negativity and minimalism of this idea of liberty. We are not given anything like a positive description. The best one could do when analyzing whether an action is free is to proceed by retracing the acts of an individual, in the hope of discovering whether they did, in fact, originate from the "fundamental self." But as we have seen, the self is also a chaos of intermingling states! Readers intent upon uncovering the pure, unadulterated definition of freedom are left pondering. A feeling of disappointment is bound to well up in the bitter soul. Have we been duped? Does this thinker know anything whatsoever about what he is writing about? Not only does Bergson not furnish us with a positive idea of liberty, we also lack any sure method of verifying whether an action is genuinely free or not. More disturbingly, however, the very notion of a discrete, separate "state" has been rendered problematic. From the durational perspective, attached to the flow of impermanence, there cannot be a "state," properly speaking. If reality is nothing but interpenetration, who is to say that the interval among states of consciousness is not the norm, and that consciousness is thereby empty of content? One could speculate, following Balsillie, that "beneath our finite form of consciousness there is a real mental life in an eternal present," because "there is no reason why the interval" between two states "may not be indefinitely extended" (Balsillie 1911: 373). The ubiquity of interpenetration can be extended to anything whatsoever, including the transitions between states. Bergson would undoubtedly reply that interpenetration must also necessitate the absence of an interval, but this in itself is not enough to rule out the possibility of a permanently extended present, provided we conceive of it as an enduring, heterogeneous actuality. It seems strange though that the extension of freedom should be read as a diminution of human free will. However implausible Balsillie's contention appears—the freedom of activity on the part of sub-individual processes jeopardizes the freedom of the individual—it was one also shared by another opponent of Bergson, namely, the Catholic philosopher Jacques Maritain. For the purposes of this chapter, it is more than sufficient for us to reproduce the broad outlines of Maritain's opposition to Bergson's idea of freedom, as presented in the former's 1913 book, *La Philosophe Bergsonienne* (*Bergsonian Philosophy and Thomism*). Maritain is in agreement with Bergson regarding the central importance of freedom. To attain liberty means to achieve a fulfillment of our being, freedom being "an immediate datum of consciousness" perceived through intuition (Maritain 1955 [1914]: 252). The divergence between the two thinkers stems from the different path they take in approaching their common goal. In Bergsonism, no positive definition of liberty is possible. Forming an unanalyzable fact of immediacy, for Bergson there is no manner in which freedom can be rationally conceptualized. Nothing has any reason for being, therefore the principle of reason is discarded. We cannot give any reason for a free act, duration just occurs of its own accord. As Maritain writes, "In the process of pure becoming, of Bergsonian duration, there is a continual creation of acts or events which are absolutely new, therefore essentially unforeseeable, because the effect is not virtually contained in its cause and is not determined by it, but wells up of itself without requiring a proportionate *raison d'etre*" (Maritain 1955 [1914]: 258). A Bergsonian concept of freedom would be a contingency completely freed of all necessity. The free agent "coincides with the very moment of the event," freeing themselves from subordination to external relations (Barthélemy-Madaule 1966: 116). Echoing his teacher Émile Boutroux, Bergson separates causality from necessity: "The principle of causality, in so far as it is supposed to bind the future to the present, could never take the form of a necessary principle; for the successive moments of real time are not bound up with one another" (TFW: 208). The heterogeneity of durations relative to one another excludes the predetermination of the present by the past. Freedom, defined as the ability of a subject to engage in acts that authentically reflect the entirety (conscious and unconscious) of its personality, is guaranteed by an ontological premise, namely, the absolute absence of determination anywhere. For a rationalist such as Maritain, the implications of such a position are deeply disturbing. The very fabric of thinking is endangered by the sundering of past from future: the Bergsonian idea of freedom as atemporal indeterminacy (or, differently put, absolutely contingent coincidence with one's own moment) is built on the assumption that "the real is contingent in its very essence, and that the contingent is self-sufficient" (Maritain 1955 [1914]: 259). The continuity of change seems to force a choice upon us. Either we maintain our commitment to eternal truths, or else accept contingency as our new divine absolute. Maritain does not realize, or else deliberately ignores, the possibility that a mode of divine temporality can be intuited that accepts the permanence of change while also salvaging eternity. After all, the free act places us in the moment, if only ever in a risky and uncertain way. As Barthélemy-Madaule states, "Contingency is never 'pure negation,'" but rather "a precarious victory" (Barthélemy-Madaule 1966: 120). In our Conclusion, we shall outline precisely what a concept of timeless duration could look like.

Why is the extension of freedom to all domains such a problem for Maritain? Human beings are, in the Bergsonian philosophy, form part of a vaster continuity, stretching deep into evolutionary time. Even matter, as both aggregate of images and force pervading our cosmos, is endowed with a certain degree of liberty. "Bergsonian freedom," writes Maritain, "is nothing other than spontaneity" (Maritain 1955 [1914]: 260). Anything capable of displaying an unpredictable movement can be considered as a process aspiring toward freedom. The state of liberty is nothing if not the presence of a striving. Even an unconscious tendency, a dispersal of molecules, a frothing of bubbles rising in a liquid medium, a volcanic eruption, all these material events already indicate the presence of a nascent freedom. For the dogmatically minded, such an ontological extension of a supposedly human prerogative is a scandal and a heresy. By consequence of the continuity of the free act with a wholly indeterminate open universe, "for Bergson an absolute metaphysical contingentism will affect the whole domain of spontaneity itself" (Maritain 1955 [1914]: 261). The equation of freedom with a pervasive spontaneity even present in modes of being vastly "inferior" (Maritain's choice of expression is telling) results in the destruction of the centrality of the human. No longer do we have any mode of distinguishing between the exercise of free will and other spontaneous process, "the specificity of free will has, to tell the truth, vanished" (Maritain 1955 [1914]: 263). Instead of volition, we are left with a substitute in the form of a force. Bergson's "sin" or heresy consists in the reduction of liberty to an ease of self-caused movement. To qualify as free, is enough for an act to originate from within the agent in question. Two liquids combining by themselves into a stable mixture are, in an ontological sense, "free," albeit not to the same degree as a human being experimenting with its sexual identity or an entire social system in the process of reformatting itself. The centrality of impermanence means, for Maritain, a thoroughgoing abandonment of being as well as intellect (Maritain 1955 [1914]: 265). If all is impermanent, this means that our categories too are not immune to alteration. Because Maritain holds that the free will can only intuit its own freedom through a rational contemplation of the necessity of God's works, he views the triumph of contingency as paradoxically destroying the possibility of access to freedom. On the Thomist view, inspired by the Aristotelian teaching, "freedom presupposes necessity; there is no free act without some necessary act" (Maritain 1955 [1914]: 267). Through the fulfillment of God's will, which is absolutely necessary, the individual can achieve freedom. Once again, we find a subordination of the individual to a moral law, which resembles the Kantian idea of free will as coincidence with the universal imperative. In *Groundwork of the Metaphysics of Morals*, we read that "free will and a will under moral laws are one and the same" (Kant 1959 [1785]: 447). Although in the latter case the imperative is a self-given construct, the basic concept remains continuous from Thomas Aquinas through Kantianism: one must surrender a portion of one's freedom so as to become capable of enjoying freedom later on. Truly, the conformist bent of mind would ask for such a sacrifice, but we are not obliged to surrender the activity of the present to the eschatological promise of a future liberty which has yet to arrive, nor a past beatitude which is already dead.

The difference between Maritain and Bergson revolves around the ontological centrality of the human entity. From the Thomist perspective, it is difficult to conceive of a will that is not motivated by intelligent apprehension. Human beings are made in the divine image and are capable of imitating God through the practice and refinement of rationality. Those, like Bergson, "who deny our specific difference must also deny our freedom," for the latter is "not the absurd power of choosing without motive or in spite of motive, but the power of choosing according to reason" (Maritain 1955 [1914]: 275). By extending free activity to all things, Bergson insults anthropocentric habits of thinking. As we have seen in our discussion of later writings such as "Introduction to Metaphysics," the intuition of the absolute for Bergson is synonymous with the contemplation of change. This immanentization of the absolute radically separates Bergson from previous members of the French Spiritualist tradition. Even the most prosaic of experiences can attain an unfathomable intensity. As a recent commentator highlights, previous Spiritualists such as Ravaisson "would not have been happy with this idea that the proper or natural object of intuition is human duration, in so far as they believed that intuition was that through which the human mind can transcend itself and reach the objects of rational theology, that is, the existence of God, the immateriality of the soul and the creation of the universe" (Jaffro 2010: 111). The precipitation of present durations overwhelms the selectivity of our consciousness. All at once, we are struck with the abundance of the present. Like a wave, the unguided activities of things pull forward, opening up interiority to exteriority. Free will, as one particular attentional focus among others, is no more threatened by the abundance of the present than any other element. It is not a question of decomposing ourselves so much as recomposing our perception, making it fit for reentry into the vibrant moment. The rationalist and conservative mind, as exemplified by Maritain, obstinately refuses to countenance the prospect of a universal change applying even to the categories of the mind itself. Earlier on, the Thomist writes, "from the moment that the intellect in its very nature becomes and evolves, any conceptual proposition which (. . .) would immutably express a truth, must in reality change, evolve, become. (. . .) In other words the truth of the enunciation evolves, truth changes" (Maritain 1955 [1914]: 170–171). We answer wholeheartedly in the affirmative. Instead of affirming or denying the charge of relativism, the Bergsonian must point to the validity of the new absolute. *Impermanence is the law, growth is the prospect, hope is the chance.* The leap into continuity is an acceleration of our becoming, the simultaneous dilation and amplification of our duration. There are forms issuing forth at every moment. Each present is pregnant with an indefinite variety of finite pasts and futures, which are perpetually enriched by the vibrating duration they themselves compose. Past and future alike are nourished by the enduring present. There is a multiplicity "of fusion or interpenetration," for "such and such a feeling, such and such an idea, contains an indefinite plurality of conscious states" whose becomings lie outside representation (TFW: 162). The moment we attempt to represent these heterogeneities, they are laid out, as it were, immobilized by spatializing habits of mind. *Language murders the life of time.* Words are superfluous when it comes to the contemplation of temporality. Freedom is to be achieved in the here and now, without any mediation whatsoever. *No phrase can contain the will you must become.* Leonard Lawlor puts it brilliantly when describing Bergson's philosophy of immanence in terms of a "leap" into immediacy: "I must not regress; instead, I must leap right into the sense, and that means into the past in general and then into a region of the past. But, this leap into the past also means that we have escaped the particular language and escaped language as a whole: we are intuiting" (Lawlor 2004: 35). Turning toward the past would be a regression, a decay of the self into a nostalgic mode of sorrowful longing. The sense of duration differs from attachment. One could say it is a spiritual activity that reattaches us to the ineffable sense of a current. "Get down to the real self"—Bergson tells us (TFW: 164). If we are to avoid the pitfalls of a backward-looking, masochistic sorrow, we are enjoined to abandon any transcendent hope as well. The intuition of duration demands something very different than an all too comfortable luxuriation in the bygone or the similarly spurious enjoyment of that which is yet to come. Those acts shall be considered free which "express the whole of the self" (TFW: 166). Against any social conditioning and our own inveterate habits of thought, we can rebel, turning away from embeddedness in sedimented patterns of ossified becomings in favor of a genuinely spontaneous commitment to the perpetuation of change. Upon seeing a social reality decay into repetition, an inner voice invites us to action. A genuine activism, freed of all dogma, would be one that does not rely on interpretive authorities. Neither would it be predicated on the restoration of any prior state. Rather, a completely mobilized activism would consist in the embrace of change, confronting any revisiting of the past with the promise of the full realization of *this* present. The various conventions and expectations we integrate into our personalities over time come to form a "crust," a layered dryness preventing us from acting freely, but also anchoring us to the prevailing social mainstream. On rare occasions, however, the free activity always boiling below the surface becomes, as it were, self-reflective. In hesitation, in the refusal of an act, Bergson finds empirical evidence of the freedom of the will: "At the very minute when the act is going to be performed, something may revolt against it. It is the deep-seated self rushing up to the surface" (TFW: 169). Far from endangering freedom, as Maritain would have it, the indeterminacy of the will is for Bergson the very source of its liberty. After hesitation subsides, as if all reason were suddenly liquidated, the free will surrenders itself to the action welling up from inside of it. No motive or explicable goal need be posited, for in the case of the free act, the movement comes out of its own accord. As a matter of fact, "this absence of any tangible reason is the more striking the deeper our freedom goes" (TFW: 170). We can categorize acts according to the intensity of the inexplicability they display. The rarest and most unfathomable of deeds would therefore be the ones that best exemplify Bergson's negative concept of an ineffable freedom.

FREE WILL AND THE FATE OF MORAL RESPONSIBILITY, OR HOW TO BE A FRIEND OF THE MOMENT

Having dissolved the self into a dynamic series of states, while defining the free act as an inexplicable, almost impossible operation entirely lacking in a reason, what then becomes of the idea of free will? To better explicate the Bergsonian idea of liberty as indeterminacy, we turn to a recent philosophical antithesis, namely, Galen Strawson's categorical denial of individual moral responsibility. In its most schematic formulation, Strawson's "Basic Argument" consists of the following steps: "1.1. When you act, you do what you do—in the situation in which you find yourself—because of the way you are. 1.2. If you do what you do because of the way you are, then in order to be URD [Ultimately Responsible or Deserving of praise or blame] for what you do must be URD for the way you are. But 1.3. You cannot be URD for the way you are. So 1.4. You cannot be URD for what you do" (Strawson 2012: 443). Because we cannot entirely control every single one of the circumstances that make us what we are, responsibility must be socially shared. The individualization of moral responsibility is ontologically implausible on this view. Pointing to the presence of societies or cultures which do follow a doctrine of individual responsibility is not enough to refute Strawson's argument, because it is predicated on uncovering the ontological possibility of the existence of responsibility. Our actions do not derive from ourselves alone, for this would require "an act of pure bootstrapping" on our part, whereas in reality we are never our own "self-originators" (Robson 2017: 528). The notion of being URD is, for Strawson, an absurdity, because no state can be found which originates wholly from the agent itself. Adherents of individual responsibility find themselves trapped in an infinite regress: "There has to be, but there cannot be, a starting point in the series of acts or processes of bringing it about that one is a certain way, or has a certain nature, a starting point that constitutes an act or process of ultimate self-origination" (Strawson 2012: 447). The almost infinite number of psychological, social, cultural, and genetic variables that coalesce to form the unity of our personality is, for the most part, contingent in its character, the complexity of reality lying beyond the purview of our deliberate actions. Our present is therefore quite independent of our volition. The way we are is, for the most part, unchangeable. In response to Strawson, Mark I. T. Robson proposes a Bergsonian answer, pointing to the critique of both determinism and dogmatic libertarianism in *Time and Free Will*. In many prisoner's dilemma-type thought experiments, choice is presented to us as a selection from among two definite alternatives. For instance, in the famous "trolley problem," introduced in its modern format by Philippa Foot in a 1967 article, we have the example of a runaway trolley heading along a track. By doing nothing whatsoever, our lack of action results in the deaths of five bystanders, while if we pull the lever, our choice results in the deliberate manslaughter of one person, saving five lives at the expense of one. Which choice should be made? (Foot 2003 [1967]: 19–33). Bergson would answer in very simple terms: there is no such thing as the fixity of alternatives! Picture a path diverging into two directions. Prior to a momentous decision, we imagine ourselves situated at a point (O), wavering between two antithetical possibilities, X and Y. A determinist would say that our choice is predetermined by external circumstances, therefore the choice we shall make is not ours, while a libertarian would say that we are free to choose rationally between X and Y. Do these divergent possibilities truly exist though? On the Bergsonian view, possibility has no being whatsoever. Bergson questions the premise of the entire thought experiment, suggesting that "in reality there are not two tendencies, or even two directions, but a self which lives and develops by means of its very hesitations, until the free action drops from it like an over-ripe fruit" (TFW: 176). Common sense cannot help artificially transforming a dynamic and continuous situation into an inert immobile thing awaiting our choice. In truth, both the prospective choice and the supposed alternatives alike are all endogeneous to becoming. Free will is not so much a choice among different alternatives as the realization of a self-actualization through the extension and integration of our actions into other durations. As Bergson writes, "There is no reason to separate" the activity of movement "from the act in which it will issue and which forms part and parcel of it" (TFW: 178). There is no such thing as "X" and "Y," only the continuity of the movement M, uniting all terms in a single striving. Courses of action form inseparable wholes, and slicing these into portions will tell us nothing about freedom in itself. By thinking about liberty in terms of a choice among alternatives that have been artificially separated from our continuous present, we erroneously look "at the train of thought and the eventual decision in retrospect after a certain determinate outcome has occurred" (Robson 2017: 527). If we have learned anything from our study of the Bergsonian teaching, it is that real change cannot admit anything of possibility. Strawson's entire argument rests on the construction of a present "state" of the individual. As he puts it, "whatever the agent-self decides, it decides as it does because of the overall way it is" (Strawson 2012: 457). Because we cannot change the "way we happen to be" at the moment of decision, being influenced in countless ways by factors outside of our control, we cannot be said to be responsible for anything we happen to do. Robson argues that Strawson has constructed an artificial time slice, for in reality there is no such thing as "the way one is!" Time has been atomized, dissolved into a set of successive temporal atoms. Accepting continuity and interpenetration, the Bergsonian conclusion results in the ontologically grounded negation of all states: "There is no way you are" (Robson 2017: 532). In duration, there are no possibilities, for this would be to posit the existence of a movement prior to its own creation. We are evolving, ripening into our present. *Undertake your undergoing, hope not for tomorrow, pine not for yesteryear.* The Bergsonian outlook is a preparation for acceptance. As Robson puts it beautifully, Bergson shows us that "there are no paths. The paths evolve and come to be, but until that coming to be, they do not exist. Every moment is (. . .) to walk on the edge of a kind of precipice and take a step that creates a path rather than taking a step which finds a path which is already there. The future is not a world of definite possibilities. It only becomes definite once it evolves" (Robson 2017: 534). To be a Bergsonian means to persist as a friend of the moment, expressing fidelity to the present by placing one's trust in the current.

Bergson himself claims on numerous occasions that a genuinely free act is exceptional. The only actions which qualify as free are those which bear "the mark of our personality" (TFW: 173). Now a determinist could make the claim, as Strawson has done, that the personality as such is determined by factors lying outside of the individual's control. Robson's critique of Strawson's position consists in a demolition of the idea of a mental "state" separable from others, drawing inspiration from Bergson's idea of the permeability of all conditions. To reiterate, Robson holds that "the new present is a new present of shimmering indefiniteness. Each present is indeterminate, and the shapes it adopts are, until actual, not existent even as possibles" (Robson 2017: 535). There is no "being as one is," because all being is a process of dynamic elaboration. If anything, Robson understates the basic point of the Bergsonian philosophy, for it appears that he is attributing indefiniteness to the novelty of the present. The process philosophical view advocated in *Time and Free Will* and elsewhere would be even more radical. As Pete A. Y. Gunter emphasizes, "For Bergson, real indefiniteness exists at all levels," so much so that we can speak of an "actual indefinite both at the level of the self and of its subordinate durations" (Gunter n.d.: 13). It is not just the case of indefiniteness and indeterminacy pertaining *prior* to the act of creation. Rather, the absence of necessity is necessarily pervasive all the way through the various levels of becoming. Even after actualization, openness remains. This entails that we never can observe a finished state. Any process implies a present which can be illuminated only by a recursion of temporality. It is this original openness that allows for the constant renewal of freedom. Liberty is not an inert mode of being, but rather an ease of movement. Usually, two views of free will are differentiated. On most accounts, the compatibilist approach is equated with determinism. If the physical world is determined by certain natural laws, and if we ourselves are part of said world, then it logically follows that, as components of the world we and all our actions are also predetermined by iron laws of necessity. As distinct from the compatibilist view, the incompatibilist holds that, despite the determined nature of existence, agents can, at least occasionally, engage in acts informed by a free will. It is all too simple to get the impression that Bergson seems to be advocating for an incompatibilist idea of free will, but something is amiss with such an opinion, because accepting this would be to ignore the fact that Bergson also denies the validity of the necessity of causality. The material realm too gives evidence of a nascent freedom of self-actualization. The homogeneity of determined space, governed by laws of causality and succession, has been seen to be a mere mental projection. Juxtaposition takes place in a space of our own making, constructed by involved consciousness for pragmatic purposes, and therefore has nothing to do with real time, that is, duration (TFW: 77). Freedom, however, is more than an epiphenomenon. Spontaneity holds everywhere. As opposed to any perspective which would spatialize decision, placing it inside of a supposed "way you are," Bergson declares that "freedom must be sought in a certain shade or quality of the action itself and not in the relation of this act to what it is not or to what it might have been" (TFW: 182–3). Even proponents of a belief in free will, such as the incompatibilists, are erroneous insofar as they localize choice. Does the inessential and variable nature of mental states or, for that matter, the positing of the impermanence of all existents, tell us anything about the exercise of free will though? As Peter van Inwagen points out, even if we accept incompatibilism (i.e., the supposition that a free will can exist, albeit on rare occasions), the infrequency of a completely free will does not necessarily exclude the ability to attribute moral responsibility: "The inability to prevent or refrain from causing a state of affairs does not logically preclude being to blame for that state of affairs" (Van Inwagen 1989: 419). In the case of a drunk driver, we may state that the person at the moment of collision, being in a state of inebriation, was not "free" to act in any manner they pleased. But this individual can still be held criminally liable for their wanton actions, because at some point in the past they omitted to prevent themselves from getting tangled up in the situation they find themselves now. Exonerating every single guilty person, based on the idea that nobody can entirely control the set of circumstances which contributed to the type of person they have become, would be severely injurious to our sense of justice, and rightly so, for "it is plausible to suppose that our acts are among the factors that determine what we eventually become" (Van Inwagen 1989: 420). Although Van Inwagen explicitly disavows any influence on Bergson's part, declaring somewhat rudely at the outset of his article that Bergson and other Continental philosophers will not be referenced in any way whatsoever, this position stands in accordance with an eminently Bergsonian sensitivity to how the past bleeds into the present. Even if the past is derivative of a present duration, as we have maintained throughout, in regard to its present it still maintains a loose, asymmetric connectivity. Where Bergson diverges significantly from the incompatibilist position is the idea of the pervasiveness of freedom. Every single becoming is free, to the extent that it gives evidence of indetermination. What allows Bergson to uphold such a counterintuitive idea is that his conception of freedom is one that "admits of degrees" (TFW: 166). Liberty is not an exclusive prerogative of the human, being rather a gradation of various intensities of self-actualization. To comprehend liberty, it is not enough for us to draw diagrams, locking movements into lines, falsifying the real, treating it as homogeneous extension. Instead, we are encouraged, by an inner voice as it were, to "get back to the very moment" (TFW: 189). It is the final act which must be revealed, through an intuition that reactualizes the immediacy of the current. A "duration properly so called" is a movement that "remains outside" all "calculation," and "could only be perceived by a consciousness capable of living through the intervals and, in fact, living the intervals themselves" (TFW: 194). Whereas Kantianism relegates the experience of freedom to the realm of the noumenal, Bergsonism manages to produce a concept of liberty as *immediate concreteness*. A proper duration has the inner status of a current, uniting the bygone with the emergent forces of an absent tomorrow inside a dynamic recurrent conception. Bergson's freedom is a perpetually open invitation to get back to the depth of each particular time.

The false dualism of compatibilism/incompatibilism dualism cannot be applied to this mode of philosophy. Freedom is already inside of self-creation. For Bergson, we truly can be creators of ourselves. Like all the other durations, we form ourselves, self-organization being an aspect of becoming in general. Elena Fell does not err when equating being with duration. Simply put, Bergsonism is the doctrine that "being cannot be regarded in any other way than as being in motion" (Fell 2012: 4). This applies to microscopic and macroscopic dimensions of reality alike. But *what* is motion? Bergson resolves Zeno of Elea's first paradox, that of Achilles and the tortoise, by positing the indivisibility of movement. On the Eleatic view, change is impossible because every movement can be divided into an infinity of intervals. Therefore if a tortoise is given a head start, even a skillful athlete such as Achilles will never be able to overtake the slow animal. But real change differs fundamentally from this abstraction. As Bergson explains, "Each of Achilles' steps is a simple indivisible act" and, after a while, "Achilles will have passed the tortoise" (TFW: 113). The reason Zeno runs into absurdity is that in Zeno's example space has been confused with motion. True movement is indivisible, differing from extension. Summarizing the Bergsonian critique of spatialized, that is, fictitious movement, John Mullarkey writes that "space," at least of the Euclidean variety, "is homogeneous and immobile," movement is "heterogeneous and mobile" (Mullarkey 1995: 232). The mistake of Kantianism consists essentially in a repetition of the Zenoan error. Although Kant does not deny the reality of change, the German Idealist philosopher still views movement as an exchange of positions within a homogeneous spacetime continuum, this projection serving as the necessary basis of human intelligence (TFW: 222). But practical utility does not equal metaphysical truth. Time is not really a homogeneous medium and only assumes a homogeneous form when reduced to a function of pragmatic activity by our own consciousness. Far from being a specifically philosophical abstraction, the Kantian idea of time as a homogeneous medium is the reproduction of a commonsense view shared by all humans involved in their environments. What Kant "did not notice is that real duration is made up of moments inside one another" (TFW: 232). If this is the case though, a critic could object that Zeno's example is not so invalid after all. We could extend the permeability of states, denying the discreteness of the two entities. On this view, a portion of Achilles's duration would be seen to broaden into the tortoise's duration, and vice versa, uniting the two apparently distinct movements.

Are not the participants of a race not, in a way, synthesized into a whole, by the very circumstance of their common participation in a game governed by the same rules? The relation of the racers to the competition would then be somewhat isomorphic to the previous problem, namely, the relation of the (exceptionally) genuine free, unconditioned choice and the far broader dimension of moral responsibility. Bergson could counter that changes in kind are at once distinct and continuous. This has been the very aim of introducing the notion of a qualitative multiplicity. Various notes coalesce into a coherent contexture. On every level, "the whole produces on us the effect of a musical phrase which is constantly on the point of ending and constantly altered in its totality by the addition of some new note" (TFW: 106). Commenting on the metaphor of duration-as-melody, a comparison Bergson returns to in many of his subsequent works as well, Mullarkey observes that "because the states of a qualitative multiplicity are indivisible, they cannot be quantified, or rather, when they are quantified, the qualitative multiplicity they compose changes in kind as it happens" (Mullarkey 1995: 239). The author in question is obviously referring to the degradation of change into a change of state which occurs whenever we start to treat time as a transition among two clearly delineated positions. Can we not radicalize the meaning of the sentence? If we extract the final portion, while slightly altering it, a sense of the permeability of change is obtained. We obtain the following brief summation of Bergson's entire concept of quality: *a qualitative multiplicity composes changes in kind as they happen*. The spatializing view, the quantification of multiplicity, always occurs after change, after time has passed. The qualitative perspective, however, places us within the interiority of change, as it is happening. A movement cannot be grasped unless we make an effort to let ourselves go: a strenuous combat is required to win surrender. We are never outside of change, there is no dimension transcendent to the immanence of alteration. This implies a presentist view of subjectivity as well. Suppose we listen to a piece of music for three seconds. We shall find that something in us has changed, "the I apprehending it now is different from the I that existed three seconds ago, if only because I have existed for three seconds more" (Sinclair 2020: 51). Quality is the dynamism, the power thrusting forward, the accumulation of enjoyments, and creations saturating the world with new light. Through a community of continuities, the movements are blending with each other, for each condition or state forms part of a journey. The actual indefinite is, in truth, anything whatsoever, a thing conceived of in its altered aspect. A propagation of flame, a block of mechanism, a cloud passing in a blue sky, a banana falling from a tree, and countless other local transmutations, coalesce into continuities of exponation. Under the term "actual indefinite," we understand the heterogeneous nature of becoming, a multiplicity which is nonetheless grasped in its singularity. While a finite number of bananas exist on Earth, the change undergone by this particular example of the fruit is at once particular, but also shared with all others belonging to its kind. Every banana can undergo a fall, hence the status of banana-movement as exponation, but each tumble will differ from the other. As Mullarkey, following Bergson's lead, declares, "Every movement is an action" (Mullarkey 1995: 244). Certainly this should not be meant to imply that a subjectivity is involved in the case of the falling banana or other sub-subjective events. A movement is an action in the basically minimal sense of composing an indivisible commutation, and it "cannot be divided and remain qualitatively the same action" (Mullarkey 1995: 246). In the Bergsonian ontology a change can be individuated by itself, without the need for involving anything like subjectivity. *This* banana in particular has fallen from a specific tree in our locale. While the genus of movement, conceived of as a certain type of exponation, is repeated by many different bananas, perhaps simultaneously—in a neighboring country, one of these fruits could be falling at the very same instant—when viewed as a duration, the change it undergoes pertains to the timeline of this specific banana we happen to be observing. Because movement can just as easily "be individuated as well through a formal cause as through an efficient one," there is no need to always equate the individuality of a change with an agent from whom the movement is supposed to originate from, leading to the conclusion that "meaning or 'decision' is in the physical action itself without need of any mental activity" (Mullarkey 1995: 247). It is superfluous for us to involve consideration or decision when it comes to the question of free will. Freedom is endogeneous to the performance of a specific liberty, being "the relation of the concrete self to the act which it performs. This relation is indefinable, just because we are free" (TFW: 219). An action is free to the extent that it gives expression to a duration, for this fruition originates from inside a certain temporality. It merely happens to be the case that we as complex organisms have a will. The significance of the free act is sufficient unto itself, for not even the individual herself has full access to its entire proliferation. We find something which was moved by us, while other things were creating along divergent paths, within environments of strange knowings. What intuition grasps is a projection "of the movement's own indigenous and individual meaning which was first extracted by our perception of it" (Mullarkey 1995: 251). A real path is the unique elaboration of a certain thickness of duration, a fugitive impression escaping any immobilization or distortion.

To give a positive definition of liberty would be to deform the living reality of movement. However intently we attempt to give a positive notion of freedom, it will be bound to elude our grasp. Definition, or indeed, language in general, is unsuitable for the translation of the moment into a common, shared, intelligible form. Attempt to manufacture a specific concept of liberty, and "we see spontaneity settle down into inertia and freedom into necessity" (TFW: 220). Intuition returns us not to the past. This would be a victory of the necrotic at the expense of the living. Those who trade the present for the past lock their attention on that which has decayed. The major fault of any past-oriented "virtualist" interpretation of Bergson lies in its negligence of that which is in the making. Against all attachments, the intuition of duration liberates us from antiquity: "The present is richer, 'larger' than the past, and this larger, new content includes something that the previous phase could not fit in. This expanding reality cannot be sufficiently explained by its past" (Fell 2012: 203). A temporal continuity can only have intelligibility if it is *presented*, so to speak, rendered available to the vital activity of a real process. The Bergsonian philosophy is the triumph of living reality, the affirmation of the authentically real, at the expense of both nostalgia and utopia. To intuit means "to place oneself at the very moment at which the act is being performed" (TFW: 220). When it comes to the concept of time, is Bergson an A-Theorist or a B-Theorist? The difference between the two standard positions relates to the final ontological status of past, present, and future. For our present purposes, the following summation from a contemporary anthology is more than sufficient: "Those who think time is constituted by an A-series (A-theorists) think that there is an observer-independent distinction between past, present and future, and furthermore, they think that time flows inexorably with respect to this distinction. (. . .) Those who think time is constituted by a B-series (B-theorists) think that there is no objective distinction between past, present and future, and that there is no objective flow of time" (Dyke 2013: 333). Where does Bergson's concept of duration as continuity figure in the debate between the A- and B-theories? Time certainly is real, indeed, for Bergson it is the very substance of reality. But the circumstance of the coexistence of past and future within the present duration excludes the possibility of positing a relationship of necessary succession between the three terms. What duration shows us is that, to quote Sonja Deppe, "the experience of the present moment is only to be understood against the background of past moments" (Deppe 2016: 114). Does this make Bergson a B-Theorist then? Not quite, for Bergson does not concur with the view that flux is an unreal projection of consciousness. Differently put, Bergson is a B-Theorist of a strange kind, a maverick who simultaneously affirms the reality of the coexistence of two parasitical forms of time (past and future) with their connected central present, while also affirming the reality of the flow of time. In Bergsonism, there is only a single authentic temporality, namely, *the efficiency of that which is current*. A free act never shall be known from the outside looking in. What is required is an experience of freedom in which we come to coincide with an actual ease of movement. Liberty is an untranslatable immediacy, "the free act takes place in time which is flowing" (TFW: 221). As we ourselves have argued at length, there is no equality between the three supposedly different tenses or modes of temporality. The free choice is to be located within the ease of movement as it reaches fruition. *Reality is a realization*. When speaking of a genus, a set of processes sharing a certain temporal substance, we spoke of exponation, defined as a duration held in common. When speaking of individual becomings, though, we talk of a single, indivisible and individual self-enablement, a duration proper. Mental states appear as intensities when observed as single units. Viewed from the perspective of multiplicity, "they unfold in time and constitute duration." , When observed in terms of "their relations to one another," on the other hand, mental states are seen to display "a certain unity" that is "preserved throughout their multiplicity" (TFW: 224). It is this latter unity or common temporal substantiality we term "exponation." The immediate would then be the change being undergone, the alteration in process. Throughout *Time and Free Will*, Bergson has been driving home a single message: freedom is *here*. Neither the dead past nor the absent future contain anything of the free act. The unobstructed is located in the movement of emancipation, the process of liberation being its own fulfillment. One cannot "be" enlightened. Rather, one "is" enlightenment, a becoming of enlivening, a brightening of spirit. While "it is true that other moments than the present one do play a role within our experience—this role does not seem to be on equal footing with the present moment. Rather their significance seems to lie within their connection with the present moment" (Deppe 2016: 119). The passage of time is not an illusion, but the juxtaposition of temporal states is very much a delusion. Only the current gives color and nourishment. Far from denying the motion of time, the centrality of the present guarantees the continuity of change. Studied in its inner aspect, duration consists of a qualitative heterogeneity. A current is that which "is," a flowing becoming of perturbations, transferring their acts into new aspects, situations, and qualitative elaborations. The Bergsonian intuition of transition in general puts the distinction between A- and B-theories in doubt, because the experience of nowness is all that is left in the end: "As events become present, they light up in a certain way that is immediately apparent to us" (Williams 1998: 389). The transitions are there, irrespectively of which theory of time we happen to subscribe to. A concrete liberty can only be achieved if we ourselves live freely, spontaneously, attaching ourselves to the continuous present, becoming-transparent. Far from leading to a stasis or a rigid, immobilized eternity, the intuition of duration, the undergoing of transformation, is an eternally continuous pulsation, an activity wresting novelty from anything which threatens to decay into mere repetition. Creation is the production of difference, through an effort that takes time. The act of creation involves "actual indefiniteness in the present," the situation where action is "being prepared" (Gunter n.d.: 14). Bringing to our attention the imprinting affectivity of waves, impermanence betrays not meaninglessness, but a joyful surrender to the tide. The transient is an attraction, a feature of that which, by analysis alone, is indistinguishable. That which attracts incites, and each incitement is an invitation to ascend, brightening, into the continuous. Integration with the material life of the indefinite actuality, each intuition grows into the becoming of an outside impulse, until durations fill one another to the brim. The free act is one with the effort of striving, the differentiation of a refraction, an element savoring the outflow of energy. If these elements were to let go of their limitations, not a single body could remain alien to the communion. Discontinuity, the sufferance of the soul, is mended by the return to presence: "O thow diere brightnes and cliernesse; O thow cliernes, take thow me to the, and so presente me tofore my maker!" (Rolle 1918: 53).

NOTES

1. Berlin forcefully rejected Bergson's philosophy, hold it to constitute a form of romanticist irrationalism.
2. Shiv Kumar states unequivocally that "it is almost certain that she [Virginia Woolf—A.L.] had never read Bergson in the original, nor was she influenced by his philosophy in her style" (Kumar 1962: 64). Nonetheless, Kumar does believe that a parallelism can be discerned at work in Woolf's "stream of consciousness" style of writing and Bergson's idea of duration. Additionally, several acquaintances and colleagues of Woolf were familiar with the Bergsonian doctrine (Kumar 1962: 65). Disputing Kumar's idea that Woolf had a merely limited knowledge of Bergson's philosophy, Mary Ann Gillies contends that the novelist elaborates a coherently Bergsonian method, the goal of which is "to capture the invisible inner movements in which most important living occurs" (Gillies 1996: 109).
3. A natural number, after all, is a homogeneous unit, spatially discrete from the other units of the count. For simplicity's sake, when speaking of quantities, Bergson usually refers to a particular, spatialized and simple definition of number, without delving into other possibilities.

Conclusion

Duration without Temporality

The concluding remarks of this book have no direct bearing upon the philosophy of Bergson. Rather, they are devoted to outlining a possibility we have hinted at throughout, namely, the idea of a special type of duration which forms the pinnacle of the hierarchy of durations. We have made the case that Bergson's doctrine may be viewed as an invitation, a call, to get back to actuality through immediate coincidence with a living durational present. Where else could the pure intuition of change be experienced, if not the moment of pure perception, liberated from both memory and anticipation? In our conclusion, we hope to showcase two examples of a duration standing outside of temporality, two forms of an eternal now which, far from being static, evidence the truth of ceaseless change. Neither of these originate from a properly philosophical domain. The first exemplar is Thomas Stearns Eliot. Our choice is far from accidental, for the modernist poet is known to have been a disciple of Bergson early in his career. Although Eliot famously disavowed his former intellectual idol, going to the extent of writing a philosophical essay debunking Bergson—entitled "Inconsistencies in Bergson's Idealism" and presented at the Harvard Philosophical Club on December 19, 1913—Paul Douglass has nonetheless argued for the presence of Bergsonian themes throughout Eliot's later works (see Douglass 1986: 83–106). The scope of this conclusion does not allow for a complete exposition of the full complexity of what can be called the "Eliot-Bergson Connection." Furthermore, this work has been accomplished in Douglass's now-classic study. Our ambition here will be somewhat more limited, consisting in a philosophical reading of one of Eliot's greatest poetic reveries, *The Four Quartets*. Specifically, we seek to uncover how the apparent paradox of an *atemporal duration* can be conceptualized by drawing out some implications contained by the contents of the poem. While several cultural, philosophical, and spiritual influences can be discerned, such as Hinduism, Christianity, and Buddhism—major religions which Eliot studied with a more than superficial intensity—we specifically concentrate our attention on what may be called an Augustinian influence at work in this epic. A distinctly Augustinian confluence of past and future can be discerned within what Eliot calls the "stillness" endogeneous to impermanence. This influence will allow us to introduce the second exemplar, that of Boethius's concept of eternity, as interpreted through Eleonore Stump and Norman Kretzmann's 1981 article, "Eternity."[1] In the second part of this conclusion, similarly to the first, our goal will not extend to anything other than a purely philosophical treatment. Questions of literary theory, poetics, history, theology, and spirituality are all operative in both examples, but what interests us here is the philosophical intuition of atemporal duration. That being said, the Bergsonian invitation to return to duration contains an inescapably moral element. To witness impermanence means to achieve liberation from enchainment to desire. Without a deathbound nostalgia and a vacuous anticipation of that which is not (yet), we can hope to achieve an immediate coincidence with immanence. What is there left to redeem, if there is nothing outside of the unextended present? For a consciousness freed of attachment, freedom would consist in the eternal contemplation of creation. Confronting the illusion of desire means coming to understand the truth of time, drawing upon the realization that the past is no longer, without any corresponding sense of disappointment. Shedding the snakeskin of antiquity, we can yet hope for an acceptance of the real simplicity of the "ing."

Writing of Eliot's poetics, Harold E. McCarthy remarks in a Bergsonian vein that "emotional visions" of the structure of reality in itself "grow out of" variegated "intuitions with an imaginative logic of their own" (McCarthy 1952: 40). On this view, systematic philosophies and, more broadly, entire worldviews originate from a certain depth of intuition, an immediate, though usually only momentary access, to the Absolute which is change. McCarthy suggests that when interpreting a philosopher or a poet, we must search for the "key intuition" animating the entire work of the thinker in question. Eliot's *Four Quartets* is a work of immense complexity, operating many registers simultaneously with breathtaking ease. Among the vertiginous visions of impermanence, we encounter references (some explicit, some less evident) to Hinduism (a section of the *Bhagavad Gita* is quoted), Buddhism, the Catholic mystic Saint John of the Cross, various occult ideas, the existentialism of Søren Aabye Kierkegaard, and other schools of thought (Murray 1991). The goal of our conclusion is to present a specifically philosophical reading of the poem, centering on the intuition of a *changing eternity. Four Quartets* presents us with the fundamental intuition of a perpetually alternating, revolving current. The key concept is that of a moment conceived of as an "intersection." In Section V of "The Dry Salvages," Eliot characterizes the sacred as the "intersection" of time with timelessness (Eliot 1963: 198). As David Soud points out, the very same phrase ("intersection") occurs in the English translation of Protestant theologian Karl Barth's work of commentary, *The Epistle to the Romans*, describing the coincidence of God's eternity with the immanence of the world of history through the Incarnation of Christ (Soud 2014: 1382). Temporal intersectionality describes how God's presence pervades the world. The divine is, in Barth's view, absolutely other than the human, but the Incarnation points to the circumstance that "through all history there runs the line of intersection between time and eternity" (Barth 1968 [1922]: 47). Where is the immediacy located? While opposing the mystic way, arguing that it represents a self-indulgent sinking of subjectivity into itself, Barth nonetheless emphasizes the momentary nature of the divine presence. Commenting on the state of rapture provoked by the arrival of the divine invasion, Barth declares that, far from lifting us up, in the context of the eschatological event the "inner life is radically broken down. This is no psychological occurrence; rather, it is the advent of the 'Moment' which is beyond all time" (Barth 1968 [1922]: 125). Becoming-saint would be, on this reading, to remain in a state of fidelity to the unspeakable apophatic moment of reality. This latter is a state which is not a state, a flowing current, so to speak. The divine invasion constitutes the mode of redemption from time through time. Being a mutual involvement "with past and future," we are informed by the poet in "Burnt Norton" that the moment, this communion of three in one, constitutes the place where temporality is destroyed: "Only through time time is conquered" (Eliot 1963: 178). The saint would be, on this reading, the exceptional individual capable of attaining a sense of the eternal within the fleeting. As Morris Weitz would have it, *Four Quartets* presents an "immanence doctrine" of time, according to which "the real or the true is already present in the false or lesser reality" (Morris 1952: 53). Despite our general concurrence, something nevertheless troubles us about the dualistic distinction Weitz—and not Eliot—draws between true and false time. The pure moment of mutual encounter is not something completely foreign to the mundane, although it certainly transports us out of our previous selves. Provided it attains a certain key, experience can ascend to the sacred encounter, rendering our subjectivity open to all durations. An immanent doctrine of eternity does not construct a specific opposition between the illusory flow of time and a static, unchanging present, situated miraculously outside change altogether. Rather, Eliot's fundamental intuition relates to an eternalized mode of impermanence, a serenity that is not fixity: silence is that which "moves perpetually in its stillness" (Eliot 1963: 180). Duration without temporality, the hidden crown of the hierarchy of durations, is not a ghostly entity lying outside of any access whatsoever. Through a penetrating consciousness, we can achieve a coincidence with the wondrous making, the non-specificity of all "ings" in general. Eternity presents its meaning, without ever becoming bogged down in time. The temporal relates to the eternal in the manner of a quality to its thing, or, more radically, a *quale* to its *-ing*. Perpetuity is contained within successivity, similarly to the superficiality of a quality shimmering upon the surface of a depth. As Barth writes, "The time in which we live conceals and yet preserves Eternity within it, speaks not of Eternity yet proclaims it in its silence" (Barth 1968 [1922]: 304). The actual contains nothing of extension, but for all its vacuity, it is the supremely abundant content of all change. Negation is pushed to the precipice, and we are enjoined to throw ourselves into the abyss. More radically than Barth, Eliot embraces the mystic way, the *via negativa*, for it is only through the spirit of simplicity that every duration shall be united. We must stop our tongue because the proliferation of needless words threatens to derail our undertaking. As Eliot reminds the reader, affinity with authentically enduring time implies a thoroughgoing negation of all that we are. We must step outside of our human condition, leaving behind our past as well as the future: "You must go by the way of dispossession, / In order to arrive at what you are not," we read in "East Coker" (Eliot 1963: 187). The advent of the perpetual present renders the transient values of the profane dimension almost inconsequential. In Shira Wolosky's words, detachment from temporal succession implies that "the values of utmost importance to those living in time, become interchangeable" (Wolosky 1995: 27). Eliot's wondrous initiation into the indeterminate actuality of "ing" is not a negation of time as such, but instead its ultimate fruition. One could say the poet is undertaking what amounts not only to the crowning achievement of his poetic career (in the view of probably most, if not all critics) but also enthroning the enduring atemporal as the peak of duration. Redemption must not be conceived of as a possibility, for, in accordance with the Bergsonian doctrine, Eliot reminds us that even seemingly genuine events could turn out to have been abstractions all along (Eliot 1963: 175). When confronted by the moment, nothing can preserve the integrity of the human image. Our personality, an avatar made of dust, along with the entirety of our individual and collective history, fades into insignificance. The spirit is not impotent when emplaced in actuality: unbound from any attachment to past or future, the spiritual points "at every moment" from "death to the new life" (Barth 1968 [1922]: 314). Instead of struggling against the tide, the proper attitude is one of surrender to the perpetual. Confronted with intersection in the moment of moments, the disciple is capable of including eternality within her duration, but solely by going beyond into a yonder which is already here, inside of every endurance. Surrender your life, so that you may live. In the moment of "self-surrender," Eliot explains, "the impossible union / Of spheres of existence is actual," past and future alike being reconciled in the momentary eternity (Eliot 1963: 198–199).

It is far from erroneous to identify the presence of a philosophy of immanence operative in Eliot's poem. Every single title of the poems making up *Four Quartets* refers to a real place the poet is known to have visited. The concrete contains the atemporal, not as a potentiality waiting for realization, but in the shape of a timelessly ubiquitous actuality. Time, in other words, contains its own redemption. Everything has been redeemed, always, everywhere. "Little Gidding" symbolizes more than a very ancient English town. Similarly, the imaginary "England" evoked in the poem of the same title is something of vastly greater importance than a certain European nation. Visiting Little Gidding, the poet is struck by the immensity of history, contributing to a sense of the permanent inside the impermanent. Once more, the phrase "intersection" is repeated, as if to emphasize the communion of the three tenses within tenseless communion: "Here, the intersection of the timeless moment / Is England and nowhere" (Eliot 1963: 201). Eternity is not a faded song, nor a hidden place lying in an inaccessible beyond. It is *here*, in Little Gidding, and nowhere, simultaneously. Elizabeth Anderson characterizes Eliot's embrace of the eternal through the ephemeral as indicating a "material mysticism" (Anderson 2012: 135). Throughout the poem, we are continuously being initiated into an experience of the immediate. The guidances received orient the reader toward viewing even the most obscene of things in terms of their relation to eternity. *Thou must seest into the all in all at all times*—this is the implicit mantra, never avowed, but always hinted at, throughout *Four Quartets*. The now is immanent, a place within our emplacement. Awakening to the movements, our souls renewed, recreated, a new sun dawns. Now is "the stillness / Between two waves of the sea" (Eliot 1963: 209). Neither immobile, affinity with the perpetuity of movement is the innermost covenant with exteriority. Redemption entails a qualification of each and every moment "by the Now of revelation" that "lies in the midst between the two" (Barth 1968 [1922]: 498). Several references are made to conflict, destruction, and death. Even the elements are not immune to impermanence. The fangs of time consume everything. Despite its at times apocalyptic undertone, written as it was during the frenzied years of World War II, *Four Quartets* nonetheless displays a joyful hopefulness. Eliot does not give deathly nothingness the final say in this narrative. There is still hope, provided we surrender completely. The "nowhere" which is Eliot's esoteric "England" can yet become a perpetuity extending beyond anything we could hope to fathom. Hope can be extracted from the contingency of history and the shocking indifference of time, provided we detach ourselves from any particularity. Imagine the inward aspect of the resounding, reverberating things. Imagine the ing of objects, sundered from any solidity. Imagine the interpenetrating multitude of things. Inside of determinacy, imagine a boundless joy. Already the redemption is at hand, provided we get back to enfoldment within the current. Discharging itself, the temporal undoes any fixity. All states are permeations, and all things are ings, processes in the making. In the end, history knows no victors and losers. Immediacy possesses that which evades those immersed in the historical dimension, namely, peace. Those who act as friends of the moment aspire to a condition resembling serenity, for in their hearts they are aware that "all manner of thing shall be well" (Eliot 1963: 206). Enlightenment pervades perception, once a complete coincidence of the present is achieved in surrender.

Many commentators have noted the presence of a distinctly Augustinian tendency in *Four Quartets*. Ole Bay-Petersen, Louis L. Martz, and C. A. Patrides have all emphasized the resemblance between Eliot's presentist idea of time and Saint Augustine of Hippo's philosophy of time, as presented in Book XI of the *Confessions* (Bay-Petersen 1985: 143–155; Martz 1948: 444–462; Patrides 1973: 172–196). For Augustine, the present is the sole actual, effective mode of time, but this moment lacks any spatial extension. Time, in its ipseity, differs radically from space. All of existence is permeated by the Word of God, "spoken eternally" (Saint Augustine 1900: 239). Indeed, all which exists is but the self-actualization of the divine in reality. The Word in Augustine's conception does not occupy any space, for it is the creative activity occasioning the various temporal processes. Translating this theological idea into the language of philosophy, we can say that the atemporal crown of duration is the mundane stillness allowing for the production of movement. We are invited by Augustine to return to "that whence we are," the "Beginning" (Saint Augustine 1900: 240). Once more, a clear similarity with *Four Quartets* reveals itself, for there Eliot appeals to us to embrace the beginning which is also an ending. Prior to her beheading, Mary Queen of Scots is reputed to have said "in my end is my beginning." At the commencement of "East Coker," Eliot reverses this sentence, declaring that "in my beginning is my end," only to conclude the poem with the original words of the beheaded queen, "in my end is my beginning" (Eliot 1963: 182, 190). Despite the historical injustice of Mary's execution, which cannot be rectified in retrospect, Eliot turns "despair into hope through the power of paradox" (Wight 1990: 66). In itself, history shows nothing of either justice or redemption. Nothing will ever return Mary's head to her body, nor the decapitated ruler to her usurped throne. Neither the beginning nor the ending is entirely separate from the moment of intersection. The disciple of duration concurs with the Augustinian view, namely, that nothing is ever lost nor regained. Traveling into the depths of gratuitously ever-present perpetuity, we come to feel in our marrow the workings of redemption, understanding that "in the Eternal nothing passeth away" and "the whole is present" (Saint Augustine 1900: 242). If the present is all that truly exists, how then can the past and future appear to be apprehended? Historians writing of Mary's execution, as well as Mary herself, waiting for her imminent execution at some point in the past, are these past- or future-bound persons erroneous? On a presentist view, past and future alike have no independent, substantial being of their own. Apparently, time displays a linear succession, but it is suggested by Augustine that this is a projection of consciousness. Our memory operates in a manner that requires the positing of a difference between past, present, and future, yet these three tenses derive their being from coincidence or communion with a living core. Bluntly put, "Wheresoever, therefore, they are, whatsoever they are, they are only so as present. Although past things are related as true, they are drawn out from the memory" (Saint Augustine 1900: 247). The workings of memory impart a certain relative truth to the past. Similarly, expectations or divinations relating to the future demonstrate a similar structure. Past and future alike are representations, necessary for our own finite activities, but from the perspective of Absolute impermanence, there is nothing substantial in these tenses. Apart from the Triune Moment, all else is lacking in being. Actual becoming is real time, the current is the flow. "I behold daybreak; I foretell that the sun is about to rise. That which I behold is present," but Augustine is quick to remind us that the actually risen Sun differs fundamentally from the mere intimation of its rising (Saint Augustine 1900: 248). Daybreak is not yet daylight, and it would be erroneous to confound the two. Between each and every qualitative difference, there is a gulf only ever tenuously bridged by the workings of our imaginations. In no uncertain terms, Augustine rules out that the present can ever have anything resembling a spatial extension. The immediate is happening, but not in anywhere. Processes are allowances, the entirety of ings being the sum of every particular ease of movement. "The present hath no space," hence activity cannot be measured, only after it has proceeded (Saint Augustine 1900: 255). But the aftermath of daylight is not the same as the radiantly, abundantly shining solar brilliance! Lacking extension, the moment is nevertheless the fundament, the sole effective actual. Consideration, contemplation, measurement, forecast, all these functions of pragmatic intelligence relate to that which has passed (past) or that which is not yet (future). That which is measured is the trace left by the moment, which absconds from any rational apprehension. Only intuition, achieved in the stillness between two waves, can account for the intersection. Through the intuitive grasp of the moment, outside of involvement, "forgetting the things that are past; and not distracted, but drawn on, not to those things which shall be and shall pass away," Augustine admonishes his readers to ally themselves with the present, melting into the sweet bliss of eternity (Saint Augustine 1900: 258). Enlightenment consists in the disappearance of any distinction between ourselves and our present. *You are the ripening fruit, caressed by furtive hand.* Does not the Buddha also speak of a similar state of absolute neutrality, a paradoxically mobile stillness? "There is that dimension, monks, where there is neither earth, nor water, nor fire, nor wind; neither dimension of the infinitude of space, nor dimension of the infinitude of consciousness, nor dimension of nothingness, nor dimension of neither perception nor non-perception; neither this world, nor the next world, nor sun, nor moon. And there, I say, there is neither coming, nor going, nor staying; neither passing away nor arising: unestablished, unevolving, without support [mental object]. This, just this, is the end of stress."[2]

***

In a recent article, Matyáš Moravec proposes to enlighten the Bergsonian idea of duration through a synthesis with Stump and Kretzmann's notion of eternity as an "atemporal duration." Moravec has interpreted Bergson's philosophy of time in terms very similar to the view expressed throughout this book, namely, that Bergson advocates for a specifically dynamic form of "presentism," duration being always the *actual presentation* of a now (Moravec 2019: 209). Moravec proposes that by reading Bergson alongside Stump and Kretzmann's article, we may achieve a synthesis that completes the hierarchy of durations. Here our interest relates to the concept of an atemporal duration itself, and not primarily its relationship with the broader Bergsonian philosophy. We hope to prove that this temporal concept is not as self-contradictory as it would appear on first impressions. Because of our limited goal, we restrict our focus to Stump and Kretzmann's assertions, without relating them directly back to the prior contents of our book. Immediate knowledge of duration would correspond to the possession of all things. Stump and Kretzmann take the Christian Neo-Platonist philosopher and theologian Boethius as their point of departure, although, as they make clear in a later article, their view is also informed by the Augustinian idea of eternity as the perpetual present we have outlined briefly above (Stump and Kretzmann 1987: 218). Their choice is informed by the historical circumstance that Boethius's conception was essentially the one used by most Medieval theological authorities as well.[3] In Book V of Boethius's *The Consolation of Philosophy*, the following definition of eternity is given: "eternity is (. . .) a perfect possession altogether of an endless life. (. . .) That (. . .) which comprehendeth and possesseth the whole fulness of an endless life together, to which neither any part to come is absent, nor of that which is past hath escaped, is worthy to be accounted everlasting, and this is necessary, that being no possession in itself, it may always be present to itself, and have an infinity of movable time present to it" (Boethius 1968: 401). Several implications can be drawn from this concise definition. At the outset, two misleading definitions of eternity must be excluded from our considerations. The eternal is not "limitless duration *in* time," otherwise known as "sempiternality." Neither is it something purely atemporal, a "static instant" (Stump and Kretzmann 1981: 430). In other words, eternity by no means signifies the absence of time, nor does it imply the complete exclusion of change. One can conserve dynamism, even impermanence, while also committing oneself to the affirmation of a certain kind of agonistic permanence. Stump and Kretzmann identify four crucial ingredients operative in Boethius's definition. Firstly, the eternal must have a certain kind of life, a vitality that corresponds to a "beginningless, endless temporal existence" (Stump and Kretzmann 1981: 431–432). Secondly, by inference, this life cannot be said to be limited in any way, shape or form. Eternity is an "infinite duration," unlimited in any way. (This illimitation posited by Boethius is also, incidentally, what excludes the possibility of spatialization) (Stump and Kretzmann 1981: 432.) The very circumstance that all is given at once, far from lacerating the eternal duration with finitude, is what allows for illimitation. The immediate cannot be anywhere, because it contains all ings. *Thou art happening.* Somewhat later, not long after the definition of eternity, Boethius characterizes the being of God as an "abiding presence," an "everlasting and present state" (Boethius 1968: 403). Thirdly, the life the posited eternal entity leads must also be considered illimitable, for, as Stump and Kretzmann note, in a tone which recalls (if only implicitly) Bergson's philosophy, "it would be reasonable to think that any mode of existence that could be called a life must involve duration" (Stump and Kretzmann 1981: 433). We are activity and this mobility cannot predate our beginning. Temporality, on the mundane level, is the holey, "poral" action of a finite body, absorbing sensations in the manner of a sponge soaked in water. For time-bound creatures such as ourselves, this double perceptive activity of selection and absorption takes time. An eternal and unlimited entity, however, not being subject to finitude, would find all arising given within its own immediate life. This brings us to the final ingredient of the Boethian definition. If a life lacking finitude can be conceived of, then this would be one having "complete possession of all its life at once" (ibid.). By consequence, the duration of the eternal being is necessarily atemporal. As distinct from finite bodies, the infinite, illimitable life encompasses the whole of time, being absolutely one with change. Knowing all becomings, undergoing all instances, the perpetual impermanence would be actual at all times. Its moment is each and every way. Neither escaping nor remaining, its knowledge does not communicate anything apart from a supremely effective eternal action without rest. To summarize, Boethius presents us with a concept of eternity as "a mode of existence that is (. . .) neither reducible to time nor incompatible with the reality of time" (Stump and Kretzmann 1981: 434). Philosophically speaking, the in-betweenness of this conception is what makes it of supreme interest. Such a peculiar ontological concept allows us to preserve both perpetuity and impermanence, by introducing change into the very foundations of the absolute. Going further, we state that eternity is what completes the idea of durational hierarchy. Irrespectively of our stance regarding the truth-status of explicitly theological propositions, this gain in itself shall hopefully have been worth the effort.

Once more echoing Bergson, if only distantly, Stump and Kretzmann argue for a definition of duration in terms of indivisibility, stating that "the existence of an eternal entity is a duration without succession," being "infinitely extended, pastless, futureless" (Stump and Kretzmann 1981: 434–435). The "extension" in question must be considered in a strictly atemporal and non-territorial sense, and not as infinite extension in space. If this were an extension within space or inside of time, we would still be talking of sempiternality, and not eternity itself. As Boethius himself writes, the divine, eternal present is a remaining, an indivisible endurance "in the simplicity of His presence" which transcends all attachment to any particularity (Boethius 1968: 403). A coherent characterization of eternity rests on the absolute elimination of any spatial metaphors, as well as the categorical exclusion of temporal succession. Change is indivisible in the plenitude of time, without the necessity of positing a yesterday and tomorrow. The eternalization of duration is the explication of the ing of time, purified of prefixes. Earlier, Boethius explains that time relates to eternity as "a circle to the centre" (Boethius 1968: 343). Temporality has a certain degree of reality, but the pinnacle of reality, the greatest intensity of the temporal, is reached when it escalates into the atemporal. Finitude already contains the infinite, not as a possibility waiting to be born, but as a limitless life infallibly shining through the obscurity of all bodies, clearing a way for the timeless bringing-forth. The new is the present, gathering a set of associations into an entirely patterned now. *Thou raisest up bodies which are not. Co-exist with all becomings, for your real is atemporal.* Time is a circle. There pertains a strange relationship of simultaneity between the temporal and the atemporal dimensions of time, the latter functioning as the core of the former. Simultaneity relates to eternity as an indivisible whole, meaning that all temporal events relate to the atemporal at all times: "from the standpoint of eternity, every time is present, co-occurrent with the whole of infinite atemporal duration" (Stump and Kretzmann 1981: 441). Such a view may appear counterintuitive. This would seem to suggest that an occurrence can both occur and not occur simultaneously. In what manner do Stump and Kretzmann stave off this objection? Let us imagine that when the two authors were writing their article (presumably in 1981), it was a foregone conclusion that former U.S. president Richard Nixon would proceed to die on August 9, 1990, the sixteenth anniversary of his resignation. An eternal entity would, on a Boethian interpretation of simultaneity, be "at once aware of Nixon resigning the Presidency and of Nixon on his deathbead" now in this very moment (Stump and Kretzmann 1981: 442). Does this mean that Nixon, alive right now at their time of writing (Nixon went on to die on April 22, 1994), was alive and deceased simultaneously? This in itself would appear to constitute an absurdity, because an object cannot be both one thing and its polar opposite. According to our commonsense view, a person is either alive or dead. Either Nixon is alive or he is dead. How can the ex-president be both at once? As we have seen, Boethius claims that in the eternal life, all processes are given simultaneously, but this most definitely should not be taken to imply that they consist in a coincidence of opposites. We may resolve this apparent paradox by duplicating temporality. The fullness of the eternal life embraces both the flowing *duratio* and the stillness of the *duratio stans*. As Boethius maintains, "He beholdeth all things in His eternal present" (Boethius 1968: 405). Simplicity contains all moments within immediacy, knowing nothing of vacancy. For us, involved and finite beings, temporal processes are transient. We shall become past, but from the perspective of the atemporal, we have passed already. And yet, miraculously, the former and the antecedent are both preserved within the non-sequentially given current of the mobile Absolute. Every single being persists, shattered into a medley of perspectives. In Stump and Kretzmann's words, "What the concept of eternity implies (. . .) is that there is one objective reality that contains two modes of real existence in which two different sorts of duration are measured by two irreducibly different sorts of measure: time and eternity" (Stump and Kretzmann 1981: 443). It is not then a matter of choosing between the temporal and the atemporal. For Stump and Kretzmann, writing in 1981, Nixon is very much alive, while from the eternalized perspective, Nixon is both alive in 1981 and deceased in 1994, the futurity of his death depending on which perspective we happen to inhabit. Past and future alike are relative to a limited, situated consciousness. Certainly, one could object that such a transposition of our experience into the mode of the Absolute is impossible. Through intuition, however, atemporality can be accessed. Instead of viewing the present as a durationless instant, the idea of atemporal duration modifies, in a real way, how we ourselves can think of the actual, real present. The idea of the present as a durationless instant is an illusion, which may be corrected by expanding our perception into the divine mode. Bracketing the status of the concept of God, we refer these considerations back to our own time experience, enriching the perception of time, inundating the sense of the present with a new depth. We realize that "genuine duration is fully realized duration—not only extended existence (. . .) but also existence none of which is already gone and none of which is yet to come—and such fully realized duration must be atemporal duration" (Stump and Kretzmann 1981: 445). A most intriguing prospect: is our present too an eternity? Is the immanent flow of time itself already a commitment to the core enlivening all times from within? The fully realized duration is immediate coincidence with enlightenment, felt as the presence of eternity. Time redeemed is duration regained through the sacrifice of both spatial extension and linear succession. Deform the straight lines, introduce curvature, until all directions are united in the circumference. Nothing is lost, all is well, everything is present. The unconventionality of the phrase "atemporal duration" should not bother us in the least, for, as Stump and Kretzmann comment revealingly, "an attempt to convey a new philosophical or scientific concept by adapting familiar expressions is not to be rejected on the basis of its violation of ordinary usage" (Stump and Kretzmann 1981: 445–446). The authors affirm that their use is no more unconventional that the adaptation of the common word *aion* (αἰών) by Hellenic philosophers, the latter phrase denoting as it does a view of time as unbounded.

Another possible objection, somewhat more theological in nature, is also countenanced by the authors. This relates to the omnipotence of God. We could omit this from our altogether more ontologically minded discussion altogether, but we fear another important philosophical insight would then be lost. If there all is given for an eternal, unbounded consciousness, then God "cannot deliberate, anticipate, remember, or plan ahead," because all these activities relate to contingent processes which are not yet or, conversely, have already elapsed, being forms of unknowledge, the lack of knowledge (Stump and Kretzmann 1981: 446). Because it contains nothing of deficiency, knowledge, as a realized actuality, can have nothing to do with temporality. To know means, in a sense, to free oneself of time, at least to a certain degree. Therefore, an eternal duration which has completely sloughed off any attachment to temporality would have a perfect knowledge of all events, thereby lacking time-bound mental activities such as those enumerated above. Does this not mean that we have introduced a deficiency into divinity? What must be realized, Stump and Kretzmann argue, is that not all negativities are truly deficient. The absence of a limitation, for example, cannot be viewed as a real negativity, because such a use of language masks an underlying absolute positivity. That which appears from a finite perspective as a limitation is revealed to constitute a limitless affirmation, for absolute affinity with a completed duration is also the completion of power, the realization of movement through a stillness that changes perpetually. Eternity, as activity without end, is capable of acting in the world of temporality because it *is* every activity. The footprints of the dead are already in the beginning, as if every happening were already ceasing through its accomplishment. Through a power of will, that which is omniscient is revealed through even the most mangled of beginnings. The making which is, the "-ing" without prefix, is making itself simultaneously absent, emergent, alive, passing away, and erased. Past, present, and future are united in the durational core. By willing it, the limitless gives occasion for everything, even the impossible: "It is not impossible for an omnipotent, eternal entity to act in eternity (by atemporarily willing) in such a way that" a "temporal entity continues to exist during a particular temporal interval" (Stump and Kretzmann 1981: 450). Within the intersection, presence arrives. Clarity is the consolation, grounding the truth of simultaneity. From this meditation upon the infinite nature of God, it follows that we ourselves can, choosing this reflection of the divine as our signpost, return to duration through an act of will. Animated by love, the enflamed, enlivened will knows no bounds. Through the Incarnation, an intersection has been created, in the context of which "an eternal entity itself entered time" (Stump and Kretzmann 1981: 451). Could anything more absurd be imagined than a coincidence of opposites? The multilevel simultaneity of the perpetual present is knowledge, no longer intended or aimed for, but achieved through an experience of duration. Eternal duration, change without time, motion without a subject. *Amongst them, you were not: present, sweeping the day, in the summer midnight, thou art. The voice of them, naught.* Thy *voice of them, are they.* Uniting the continuity of things with the multiplicity of their durations, we distort our own attention, mangling our words, until our life has been mended, reattached, without injury, to the enduring present. As the divine invasion attests, an eternal entity can "become temporal without having ceased (*per impossibile*) to exist eternally" (ibid.). Ing has no beginning, because the ing—emanating from the outside which is also internal to every time—is beyond failure, its efficacy guaranteed by the atemporal crystalline core gifting it with nourishments. If the infinity permeating the temporal flow, penetrating us, is authentically omniscient, then "an eternal entity cannot foreknow anything," because "an omniscient eternal entity is aware" of all events "as they are present" (Stump and Kretzmann 1981: 454). Even in their absence, something is known of the beginnings, endurances, and endings, for all these persist as constituents elements of each duration. The power of the liberated will dive into the pure perception of material life. Our authors also maintain that "an eternal entity cannot alter a past event," because "the past is solely a feature of the experience of temporal entities" (ibid.). To fix ourselves to the past would be to concentrate our attentions on the defeated, the instants which have already made themselves absent. Similarly, the future is also a product of the deficiency of the time-bound mind. A consciousness directing itself toward awareness of the scenery would flatten significance, until ease of mind is joined with ease of movement. The past event is made present once more, in the eternal consciousness which has rendered itself aware of the occurrent scene. Renewed, transfigured by exposure to a prayer not in time, our own profane experience is brightened by the intuition of atemporal duration. As Moravec comments on Stump and Kretzmann's insights, "God's *durée* is analogical to ours since our *durée* shares with His complete indivisibility" (Moravec 2019: 208). The result of an hour hence does not evaporate into nothingness. Rather, each sense is preserved, recorded, by the omnipresence of the middle inherent to all moments. The whole is immanent to each interval, "from the eternal viewpoint every temporal event is actually happening" (Stump and Kretzmann 1981: 457). *Thou shalt satisfy my desire to make heaven bestowed.* The whole of life is in all things, the part contains the entirety, the integral being indigenous to the partial. Divided senses are melted down into a supreme, eternal referentiality of unchanging strength. *Suddenly, the Master teacheth. Reverse thy silence, until the reordering unseats the illusion.* No part of the whole of creatures, of temporal events, of matters, is foreign to the core. Shimmering, the stillness dances.

NOTES

1. We are indebted to Matyáš Moravec's invaluable article on the topic of Stump and Kretzmann's view and the broader controversy surrounding their position, as well as the relation of atemporal duration to Bergson's philosophy. As Moravec has addressed the latter topic more than adequately, we focus here in our own Conclusion upon the broader implications of the atemporal mode of duration. (Moravec 2019: 197–224).
2. https://www.accesstoinsight.org/tipitaka/kn/ud/ud.8.01.than.html.
3. For example, following Boethius's lead, the thirteenth-century theologian Eustace of Arras distinguishes between a merely temporal duration and an atemporal eternal duration, calling the latter *duratio stans* (Dales 1989: 63).

Bibliography

WORKS BY HENRI BERGSON

(1935 [1932]) *Two Sources of Morality and Religion.* Trans. R. Ashley Audra, Cloudesley Brereton, and W. Horsfall Carter. London: Macmillan. [abbreviated as TS]

(1944 [1907]) *Creative Evolution.* Trans. Arthur Mitchell. New York: Random House. [abbreviated as CE]

(1946 [1934]) *The Creative Mind.* Trans. Mabelle L. Andison. New York: The Philosophical Library Inc. [abbreviated as CM]

(1950 [1889]) *Time and Free Will: An Essay on the Immediate Data of Consciousness.* Trans. F. L. Pogson. London: George Allen & Unwin. [abbreviated as TFW]

(1965 [1922]) *Duration and Simultaneity: With Reference to Einstein's Theory.* Trans. Leon Jacobson. Indianapolis, New York and Kansas City: The Bobbs-Merrill Company. [abbreviated as DS]

(2005 [1896]) *Matter and Memory.* Trans. Nancy Margaret Paul and W. Scott Palmer. New York: Zone Books. [abbreviated as MM]

SECONDARY LITERATURE AND OTHER WORKS

Al-Saji, Alia (2004) "The memory of another past: Bergson, Deleuze and a new theory of time." *Continental Philosophy Review* 37.2. pp. 203–239.

Al-Saji, Alia (2009) "A phenomenology of critical-ethical vision: Merleau-Ponty, Bergson, and the question of seeing differently." *Chiasmi International* 11. pp. 375–398.

Al-Saji, Alia (2010) "Life as vision: Bergson and the future of seeing differently." In Michael R. Kelly (ed.) *Bergson and Phenomenology.* Basingstoke and New York: Palgrave Macmillan. pp. 148–174.

Alberic, Marie, Cayla A. Stifler, Zhaoyong Zou, Chang-Yu Sun, Christopher E. Killian, Sergio Valencia, Mohamad-Assaad Mawass, Luca Bertinetti, Pupa UPA Gilbert, and Yael Politi (2019) "Growth and regrowth of adult sea urchin spines involve hydrated and anhydrous amorphous calcium carbonate precursors." *Journal of Structural Biology* X.1. pp. 1–9.

Alexander, Samuel (1966 [1920]) *Space, Time, and Deity: The Gifford Lectures at Glasgow, 1916–1918.* Volume I. London: Palgrave Macmillan.

Alfaro, Adriana Altamirano (2017) *The Belief in Intuition: A Politico-Philosophical Reading of Henri Bergson and Max Scheler.* Doctoral dissertation, Harvard University, Graduate School of Arts & Sciences.

Anderson, Elizabeth (2012) "Burnt and blossoming: Material mysticism in *Trilogy* and *Four Quartets.*" *Christianity & Literature* 62.1. pp. 121–142.

Ansell-Pearson, Keith (2002) *Philosophy and the Adventure of the Virtual: Bergson and the Time of Life.* London and New York: Routledge.

Ansell-Pearson, Keith (2018) *Bergson: Thinking Beyond the Human Condition.* London and New York: Bloomsbury.

Ansell-Pearson, Keith, and John Mullarkey (2002) "Introduction." In Keith Ansell-Pearson and John Mullarkey (eds) *Henri Bergson: Key Writings.* London: Athlone. pp. 1–49.

Antoniaou, Ioannis, and T. Christidis (2010) "Bergson's Time and the Time Operator." *Mind and Matter* 8.2. pp. 185–202.

Artuso, M. C., J. S. Roldán, L. A. Scolaro, and M. J. Carlucci (2016) "Viruses: As mediators in 'Élan vital' of the 'creative' evolution." *Infection, Genetics and Evolution* 46. pp. 78–84.

Bachelard, Gaston (2000 [1950]) *The Dialectic of Duration.* Trans. Mary McAllaster Jones. Manchester: Clinamen Press.

Balsillie, David (1911) "Prof. Bergson on time and free will." *Mind* 20.79. pp. 357–378.

Barnard, G. William (2011) *Living Consciousness: The Metaphysical Vision of Henri Bergson.* Albany: State University of New York Press.

Barr, Nann Clark (1913) "The dualism of Bergson." *The Philosophical Review* 22.6. pp. 639–652.

Barth, Karl (1968 [1922]) *Epistle to the Romans.* Trans. Edwyn C. Hoskyns. Oxford: Oxford University Press.

Barthélemy-Madaule, Madeleine (1963) *Bergson et Teilhard de Chardin.* Paris: Éditions du Seuil.

Barthélemy-Madaule, Madeleine (1966) *Bergson: Adversaire de Kant.* Paris: Presses Universitaires de France.

Barthélemy-Madaule, Madeleine (1970) *L'idéologie du hasard et de la nécessité.* Paris: Éditions du Seuil.

Bay-Petersen, Ole (1985) "T. S. Eliot and Einstein: The fourth dimension in the *Four Quartets.*" *English Studies* 66.2. pp. 143–155.

Bazzano, Manu (2014) "Togetherness: Intersubjectivity revisited." *Person-Centered & Experiential Psychotherapies* 13.3. pp. 203–216.

Becker, Robert, and Gary Selden (1998) *The Body Electric: Electromagnetism and the Foundation of Life.* New York: William Morrow & Company.

Belot, Gordon (2013) "Time in classical and relativistic physics." In Heather Dyke and Adrian Bardon (eds) *A Companion to the Philosophy of Time*. Chichester: John Wiley & Sons. pp. 185–201.

Ben-Menahem, Yemima (2009) *Conventionalism*. Cambridge and New York: Cambridge University Press.

Bennett, Jane (2010) *Vibrant Matter: A Political Ecology of Things*. Durham and London: Duke University Press.

Benson, Bruce Ellis (2003) *The Improvisation of Musical Dialogue: A Phenomenology of Music*. Cambridge and New York: Cambridge University Press.

Bergson, Henri, Albert Einstein, and Henri Piéron (1969 [1922]) "Remarks concerning relativity theory." In Pete A. Y. Gunter (ed.) *Bergson and the Evolution of Physics*. Knoxville: University of Tennessee Press. pp. 123–135.

Berlin, Isaiah (1958) "Two concepts of liberty." In Isaiah Berlin (ed.) *Liberty*. Oxford and New York: Oxford University Press. pp. 166–218.

Bernet, Rudolf (2005) "A present folded back on the past (Bergson)." Trans. Basil Vassilicos. *Research in Phenomenology* 35.1. pp. 55–76.

Bernet, Rudolf (2010) "Bergson on the driven force of consciousness and life." Trans. Trevor Perri. In Michael R. Kelly (ed.) *Bergson and Phenomenology*. Basingstoke and New York: Palgrave Macmillan. pp. 42–63.

Berteval, W. (1969 [1943]) "Bergson and Einstein." In Pete A. Y. Gunter (ed.) *Bergson and the Evolution of Physics*. Knoxville: University of Tennessee Press. pp. 214–227.

Bhattacharya, Abhoy Chandra (1972) *Sri Aurobindo and Bergson: A Synthetic Study*. Gyanpur: Jagabandhu Prakashan.

Bizzarri, Mariano, Noemi Monti, Mirko Minini, and Andrea Pensotti (2019) "Field-dependent effects in biological systems." *Organisms: Journal of Biological Sciences* 3.1. pp. 35–42.

Blumenthal, George A., S. M. Faber, Joel R. Primack, and Martin J. Rees (1984) "Formation of galaxies and large-scale structure with cold dark matter." *Nature* 311.11. pp. 517–525.

Boethius (1968) *The Theological Tractates*. Trans. H. D. Stewart and E. K. Rand. Cambridge: Harvard University Press.

Bognon-Küss, Cécilia, Bohang Chen, and Charles T. Wolfe (2018) "Metaphysics, function and the engineering of life: The problem of vitalism." *Kairos: Journal of Philosophy & Science* 20.1. pp. 113–140.

Bogost, Ian (2012) *Alien Phenomenology, or What It's Like to be a Thing*. Minneapolis and London: University of Minnesota Press.

Bönke, H. (1915) *Bergson Plagiator*. Charlottenburg: Huth.

Booker, M. Keith (2010) *Encyclopedia of Comic Books and Graphic Novels*. Westport: Greenwood.

Borges, José Luis (1964) *Labyrinths: Selected Stories & Other Writings*. Eds Donald A. Yates and James E. Irby. New York: New Directions.

Brockdorff, Baron Cay von (1916) *Die Wahrheit über Bergson*. Berlin: Curtius.

Brouwer, René (2015) "Stoic sympathy." In Eric Schliesser (ed.) *Sympathy: A History*. Oxford and New York: Oxford University Press. pp. 15–36.

Bryant, Levi R. (2008) *Difference and Givenness: Deleuze's Transcendental Empiricism and the Ontology of Immanence*. Evanston: Northwestern University Press.
Bryant, Levi R. (2011) *The Democracy of Objects*. Ann Arbor: Open Humanities Press.
Bryant, Levi R., Graham Harman, and Nick Srnicek (2011) *The Speculative Turn: Continental Materialism and Realism*. Melbourne: re.press.
Buddhaghosa, Bhadantācariya (2010) *The Path of Purification (Visuddhimagga)*. Trans. Bhikku Nāṇamoli. Kandy: Buddhist Publication Society.
Burton, James (2015) *The Philosophy of Science Fiction: Henri Bergson and the Fabulations of Philip K. Dick*. London and New York: Bloomsbury.
Canales, Jimena (2015) *The Physicist and the Philosopher: Einstein, Bergson, and the Debate That Changed Our Understanding of Time*. Princeton and Oxford: Princeton University Press.
Canguilhem, Georges (1992 [1952]) "Machine and organism." Trans. Mark Cohen and Randall Cherry. In Jonathan Crary and Sanford Kwinter (eds) *Incorporations*. New York: Zone Books. pp. 44–69.
Ĉapek, Miliĉ (1971) *Bergson and Modern Physics: A Reinterpretation and Re-evaluation*. Dordrecht: D. Reidel Publishing Company.
Ĉapek, Miliĉ (1990) *The New Aspects of Time: Its Continuity and Novelties*. Dordrecht: Springer.
Cariou, Marie (1995) *Bergson et Bachelard*. Paris: Presses Universitaires de France.
Carr, H. Wildon (1911) *Henri Bergson: The Philosophy of Change*. London: T. C. and E. C. Jack.
Carr, H. Wildon, R. A. Sampson, and Alfred North Whitehead (1923) "Symposium: The problem of simultaneity: Is there a paradox in the principle of relativity in regard to the relation of time measured to time lived?" *Proceedings of the Aristotelian Society, Supplementary Volumes, Vol. 3, Relativity, Logic, and Mysticism*. pp. 15–41.
Cather, Willa (2006 [1922]) *One of Ours*. Lincoln: University of Nebraska Press.
Chevalier, Jacques (1928 [1926]) *Henri Bergson*. Trans. Lilian A. Clare. London: Rider & Co.
Childs, Donald J. (1991) "TS Eliot's Rhapsody of matter and memory." *American Literature* 63.3. pp. 474–488.
Clarke, Melissa (2002) "The space-time Image: The case of Bergson, Deleuze, and *Memento*." *The Journal of Speculative Philosophy* 16.3. pp. 167–181.
Cleeremans, Axel (2011) "The radical plasticity thesis: How the brain learns to be conscious." *Frontiers in Psychology* 86. pp. 1–12.
Clynes, Manfred E., and Nathan S. Kline (1960) "Cyborgs and space." *Astronautics* 5.9. pp. 26–27, 74–76.
Cochran, Gregory, and Henry Harpending (2009) *The 10,000 Year Explosion: How Civilization Accelerated Human Evolution*. New York: Basic Books.
Coole, Diana, and Samantha Frost (2010) *New Materialisms: Ontology, Agency, and Politics*. Durham and London: Duke University Press.
Costa de Beauregard, Raphaëlle (2010) "Green apples and red prawns: The colour of time in Peter Greenaway's *A Zed & Two Noughts (Z.0.0.)*." *Journal of British Cinema and Television* 7.1. pp. 82–94.

Crelinsten, Jeffrey (2006) *Einstein's Jury: The Race to Test Relativity*. Princeton and Oxford: Princeton University Press.

Croissant, Jennifer L. (2014) "Agnotology: Ignorance and absence or towards a sociology of things that aren't there." *Social Epistemology* 28.1. pp. 4–25.

Dales, Richard C. (1989) *Medieval Discussions of the Eternity of the World*. Leiden: Brill.

Dawkins, Richard (2006 [1976]) *The Selfish Gene*. Oxford and New York: Oxford University Press.

DeLanda, Manuel, and Graham Harman (2017) *The Rise of Realism*. Cambridge and Malden: Polity.

Deleuze, Gilles (1994 [1968]) *Difference and Repetition*. Trans. Paul Patton. New York: Columbia University Press.

Deleuze, Gilles (1997 [1966]) *Bergsonism*. Trans. Hugh Tomlinson and Barbara Habberjam. New York: Zone Books.

Deleuze, Gilles (1997 [1983]) *Cinema 1: The Movement-Image*. Trans. Hugh Tomlinson and Barbara Habberjam. London: Athlone Press.

Deleuze, Gilles (1997 [1989]) *Cinema 2: The Time-Image*. Trans. Hugh Tomlinson and Robert Galeta. Minneapolis and London: University of Minnesota Press.

Deleuze, Gilles (2000 [1964]) *Proust and Signs*. Trans. Richard Howard. London: Athlone Press.

Deleuze, Gilles, and Félix Guattari (1987 [1980]) *A Thousand Plateaus: Capitalism and Schizophrenia*. Volume 2. Trans. Brian Massumi. Minneapolis and London: University of Minnesota Press.

DePalma, Robert A., Jan Smit, David A. Burnham, Klaudia Kuiper, Philip L. Manning, Anton Oleinik, Peter Larson, Florentin J. Maurrasse, Johan Vellekoop, Mark A. Richards, Loren Gurche, and Walter Alvarez (2019) "A seismically induced onshore surge deposit at the KPg boundary, North Dakota." *PNAS* 116.17. pp. 8190–8199.

Deppe, Sonja (2016) "The mind-dependence of the relational structure of time (or: What Henri Bergson would say to B-theorists)." *Kriterion* 30.2. pp. 107–124.

DiFrisco, James (2015) "*Élan Vital* revisited: Bergson and the thermodynamic paradigm." *The Southern Journal of Philosophy* 53.1. pp. 54–73.

Dolbeault, Joël (2002) "L'idée que la matière a une mémoire, dans L'Évolution créatrice." *Annales Bergsoniennes IV*. pp. 529–537.

Dolbeault, Joël (2012) "From mind to matter: How Bergson anticipated quantum ideas." *Mind and Matter* 10.1. pp. 25–45.

Dolbeault, Joël (2018) "Bergson's panpsychism." *Continental Philosophy Review* 51.4. pp. 549–564.

Dolphijn, Rick, and Iris Van der Tuin (2011) "Pushing dualism to an extreme: On the philosophical impetus of a new materialism." *Continental Philosophy Review* 44.4. pp. 383–400.

Douglass, Paul (1986) *Bergson, Eliot, and American Literature*. Lexington: The University Press of Kentucky.

Douglass, Paul (1998) "Deleuze, Cinema, Bergson." *Social Semiotics* 8.1. pp. 25–35.

Dow, Arthur Wesley (1913 [1899]) *Composition: A Series of Exercises in Art Structure for the Use of Students and Teachers.* New York: Doubleday, Page & Company.

Dupré, Louis (1998) "Kant's theory of history and progress." *The Review of Metaphysics* 51.4. pp. 813–828.

Durand, Guillhaume, and Michel Weber (2007) *Le principes de la connaissance naturelle d'Alfred North Whitehead.* Nantes: Départment de philosophie de l'Université de Nantes.

Durie, Robin (2000) "Splitting time. Bergson's philosophical legacy." *Philosophy Today* 44(2). pp. 152–168.

Durie, Robin (2002) "Creativity and life." *The Review of Metaphysics* 56.2. pp. 357–383.

During, Elie (2007) "Philosophical twins? Bergson and Whitehead." In Guillhaume Durand and Michel Weber (eds) *Le principes de la connaissance naturelle d'Alfred North Whitehead.* Nantes: Départment de philosophie de l'Université de Nantes. pp. 79–105.

Dyke, Heather (2013) "Time and tense." In Heather Dyke and Adrian Bardon (eds) *A Companion to the Philosophy of Time.* Malden: John Wiley & Sons. pp. 328–345.

Dyson, Frank Watson, Arthur Stanley Eddington, and C. Davidson (1920) "A determination of the deflection of light by the Sun's gravitational field, from observations made at the total eclipse of May 29, 1919." *Philosophical Transactions of the Royal Society A: Mathematical, Physical and Engineering Sciences* 220(571–581). pp. 291–333.

Eden, Amnon H., James H. Moor, Johnny H. Soraker, and Eric Steinhart (2013) *Singularity Hypotheses: A Scientific and Philosophical Assessment.* Cham: Springer.

Einstein, Albert (1921 [1917]) *Relativity: The Special and General Theory.* Trans. Robert W. Lawson. New York: Henry Holt & Company.

Eliot, Thomas Stearns (1963) *Collected Poems, 1909–1962.* New York: Harcourt, Brace, & World.

Essen, Louis (1971) *The Special Theory of Relativity: A Critical Analysis.* Oxford: Clarendon Press.

European Southern Observatory (2020) "Exoplanet where it rains iron discovered." 11 March. https://www.sciencedaily.com/releases/2020/03/200311121832.htm.

Fell, Elena (2012) *Duration, Temporality, Self: Prospects for the Future of Bergsonism.* Frankfurt am Main: Peter Lang.

Feneuil, Anthony (2012) "De l'imméediatement donné au 'detour de l'expérience mystique' remarques sur l'unité de la méthode intuitive chez Bergson." *PhilóSophos: Revista De Filosofia.* 17.1. pp. 31–54.

Feneuil, Anthony (2015) *Le Serpent D'Aaron: Sur L'Expérience Religieuse chez Karl Barth et Henri Bergson.* Paris: L'Age D'Homme.

Ferraris, Maurizio (2014 [2012]) *Manifesto of New Realism.* Trans. Sarah De Sanctis. Albany: State University Press of New York.

Feuillet, Lionel, Henry Dufour, and Jean Pelletier (2007) "Brain of a white-collar worker." *Lancet* 370. p. 262.

Fiant, Océane (2018) "Canguilhem and the machine metaphor in the life sciences: History of science and philosophy of biology at the service of sciences." *Transversal* 4. pp. 149–162.

Fog, Agner (1999) *Cultural Selection.* Dordrecht: Kluwer Academic Publishers.

Foot, Philippa (1967) "The problem of abortion and the doctrine of the double effect." In Philippa Foot (ed.) *Virtues and Vices and Other Essays in Moral Philosophy.* Oxford and New York: Oxford University Press. pp. 19–33.

François, Arnaud (2005) "Bergson plagiaire de Schopenhauer? Analyse d'une polémique." *Etudes germaniques* 60.3. pp. 469–490.

Frank, Jacob (2004) *The Collection of the Words of the Lord [Jacob Frank] from the Polish manuscripts.* Trans. Harris Lenowitz. Independently Published.

Frost, Robert (1964) *Complete Poems of Robert Frost.* New York, Chicago and San Francisco: Holt, Rinehart and Winston.

Fujita, Hisashi (2007) "Bergson's hand: Toward a history of (non)-organic vitalism." Trans. Roxanne Lapidus. *Substance* 36.3. pp. 115–130.

Gabriel, Markus (2015) *Fields of Sense: A New Realist Ontology.* Edinburgh: Edinburgh University Press.

Gal, Ram, and Frederic Libersat (2010) "A wasp manipulates neuronal activity in the sub-esophageal ganglion to decrease the drive for walking in its cockroach prey." *PLoS One* 5.4. pp. 1–10.

Gallagher, Idella J. (1970) *Morality in Evolution: The Moral Philosophy of Henri Bergson.* Cham: Springer.

Garre, Arran (2020) "Henri Bergson and the mind body problem: Overcoming cartesian dualism." *Cosmos & History* 16.2. pp. 165–181.

Gayon, Jean (2005) "Bergson's spiritualist metaphysics and the sciences." In Gary Gutting (ed.) *Continental Philosophy of Science.* Malden: Blackwell. pp. 43–59.

Gayon, Jean (2010) "Defining life: Synthesis and conclusions." *Origins of Life and Evolution of Biospheres* 40.2. pp. 231–244.

Georges, Michel, Michel (2011 [1914]) "Henri Bergson talks to us about cinema." Trans. Louis-Georges Schwartz. *Cinema Journal* 50.3. pp. 79–82.

Gibney, Elizabeth (2017). "A matter of time: The quest to crystallize time." *Nature* 543.7644. pp. 164–166.

Gilbert, Margaret (1992) *On Social Facts.* Princeton and Oxford: Princeton University Press.

Gilbert, Scott F (2016) "Developmental plasticity and developmental symbiosis: The return of Eco-Devo." *Current Topics in Developmental Biology* 116. pp. 415–433.

Gildenhard, Ingo, and Andrew Zissos (2017a) "General introduction: Metamorphosis. A phenomenology." In Ingo Gildenhard and Andrew Zissos (eds) *Transformative Change in Western Thought: A History of Metamorphosis from Homer to Hollywood.* London and New York: Routledge. pp. 1–36.

Gildenhard, Ingo and Andrew Zissos (2017b) *Transformative Change in Western Thought: A History of Metamorphosis from Homer to Hollywood.* London and New York: Routledge.

Gillies, Mary Anne (1996) *Henri Bergson and British Modernism.* Montreal: McGill-Queen's University Press.

Gilmour, Darren, Martina Rembold, and Maria Leptin (2017) "From morphogen to morphogenesis and back." *Nature* 541.7637. pp. 311–320.
Goodwin, Brian (2001) *How the Leopard Changed Its Spots: The Evolution of Complexity*. Princeton: Princeton University Press.
Gould, Stephen Jay (1996) *Full House: The Spread of Excellence from Plato to Darwin*. New York: Harmony Books.
Gratton, Peter (2014) *Speculative Realism: Problems and Prospects*. London: Bloomsbury.
Gribov, I. A. (2019) "Periodic waveguided multiverse as a source of dark matter and dark energy." *Journal of Physics: Conference Series* 1147.1. pp. 1–25.
Gribov, I. A., and S. A. Trigger (2016) "Gravitationally neutral dark matter–dark antimatter universe crystal with epochs of decelerated and accelerated expansion." *Journal of Physics Conference Series* 774. pp. 1–10.
Grogin, Paul (1988) *The Bergsonian Controversy in France 1900–1914*. Calgary: University of Calgary Press.
Grosz, Elizabeth (2004) *The Nick of Time: Politics, Evolution and the Untimely*. Crows Nest: Allen & Unwin.
Grosz, Elizabeth (2005) "Bergson, Deleuze and the becoming of unbecoming." *Parallax* 11.2. pp. 4–13.
Grosz, Elizabeth (2013) "Habit today: Ravaisson, Bergson, Deleuze and us." *Body & Society* 19.2–3. pp. 217–239.
Guerlac, Suzanne (2006) *Thinking in Time: An Introduction to Henri Bergson*. Ithaca and London: Cornell University Press.
Gunning, Tom (2014) "Animation and alienation: Bergson's critique of the cinématographe and the paradox of mechanical motion." *The Moving Image* 14.1. pp. 1–9.
Gunter, Pete A. Y. (1969a) "Introduction: Bergsonian method and the evolution of science." In Pete A. Y. Gunter (ed.) *Bergson and the Evolution of Physics*. Knoxville: University of Tennessee Press. pp. 3–43.
Gunter, Pete A. Y. (1969b) *Bergson and the Evolution of Physics*. Knoxville: University of Tennessee Press.
Gunter, Pete A. Y. (2005) "Temporal hierarchy in Bergson and Whitehead." *Interchange* 36.1–2. pp. 139–157.
Gunter, Pete A. Y. (2009) "Gilles Deleuze, Deleuze's Bergson and Bergson himself." In Keith Robinson (ed.) *Deleuze, Whitehead, Bergson: Rhizomatic Connections*. Basingstoke and New York: Palgrave Macmillan. pp. 167–181.
Gunter, Pete A. Y. (n.d.) "The actual indefinite in Bergson and Whitehead." http://www.fieldbeing.org/doc/Gunter_Actual%20Indefinite%20in%20Bergson%20and%20Whitehead.pdf.
Gutting, Gary (2005) *Continental Philosophy of Science*. Malden: Blackwell.
Hacking, Ian (1998) "Canguilhem amid the cyborgs." *Economy and Society* 27.2–3. pp. 202–216.
Hafele, Joseph Carl, and Richard E. Keating (1972a) "Around-the-world atomic clocks: Predicted relativistic time gains." *Science* 177.4044. pp. 166–168.
Hafele, Joseph Carl, and Richard E. Keating (1972b) "Around-the-world atomic clocks: Observed relativistic time gains." *Science* 177.4044. pp. 168–170.

Harman, Graham (2002) *Tool-Being: Heidegger and the Metaphysics of Objects*. La Salle and Chicago: Open Court.

Harman, Graham (2005) *Guerrilla Metaphysics: Phenomenology and the Carpentry of Things*. La Salle and Chicago: Open Court.

Harman, Graham (2016) *Immaterialism: Objects and Social Theory*. Cambridge and Malden: Polity.

Harman, Graham (2017) *Object-Oriented Ontology: A New Theory of Everything*. New York: Pelican Books.

Headley, Clevis (2019) "Bergson, Senghor, and the philosophical foundations of Négritude. intellect, intuition, and knowledge." In Andrea J. Pitts and Mark William Westmoreland (eds) *Beyond Bergson: Examining Race and Colonialism Through the Writings of Henri Bergson*. Albany: State University Press of New York. pp. 79–121.

Heidegger, Martin (1992 [1924]) *The Concept of Time*. Trans. William McNeil. Cambridge: Blackwell.

Heidegger, Martin (2001 [1927]) *Being and Time*. Trans. John Macquarrie and Edward Robinson. Oxford and Cambridge: Blackwell.

Hinchliff, Mark (1996) "The puzzle of change." *Philosophical Perspectives* 10. pp. 119–136.

Hollander, Jacob (1914) *Economic Liberalism*. New York and Cincinnati: The Abingdon Press.

Hoquet, Thierry (2018) "Cyborgs, between organology and phenomenology: Two perspectives on artifacts and life." In Sacha Loeve, Yavier Guchet, and Bernadette Bensaude Vincent (eds) *French Philosophy of Technology: Classical Readings and Contemporary Approaches*. Cham: Springer. pp. 257–279.

Horvath, Mark (2016) *Darkening Places: Essays on Excess*. Charleston: VoidFront Press.

Horvath, Mark, and Adam Lovasz (2016) "Absentology and epistemology." https://socfss.blog.respekt.cz/absentology-and-epistemology/.

Horvath, Mark, and Adam Lovasz (2019) "A brief introduction to absentology." *Sūdō Journal* 1.1. pp. 46–47.

Hulleman, Johan, and Christian N. L. Olivers (2015) "The impending demise of the item in visual search." *Behavioral and Brain Sciences* 40. https://doi.org/10.1017/S0140525X15002794.

Husserl, Edmund (1983 [1913]) *Ideas Pertaining to a Pure Phenomenology and to a Phenomenological Philosophy: First Book: General Introduction to a Pure Phenomenology*. Trans. F. Kersten. The Hague, Boston and Lancaster: Martinus Nijhoff Publishers.

Huxen, Keith (2019) "Why we remember D-Day and the fight for freedom." 6 June. https://thehill.com/opinion/national-security/446990-why-we-remember-d-day-and-the-fight-for-freedom.

Huxley, Julian (1912) *The Individual in the Animal Kingdom*. Cambridge: Cambridge University Press.

Ingold, Tim (2011) *Being Alive: Essays on Movement, Knowledge, and Description*. London and New York: Routledge.

Ivakhiv, Adrian J. (2013) *Ecologies of the Moving Image: Cinema, Affect, Nature.* Waterloo: Wilfrid Laurier University Press.

Jaffro, Laurent (2010) "Infinity, intuition, and the relativity of knowledge: Bergson, Carrau, and the Hamiltonians." *British Journal for the History of Philosophy* 18.1. pp. 91–112.

James, William (1950 [1890]) *The Principles of Psychology.* Volume I. Cambridge: Harvard University Press.

Janicaud, Dominique (1969) *Une Généalogie du Spiritualisme Français. Aux sources du bergsonisme: Ravaisson et la métaphysique.* La Haye: Martinus Nijhoff.

Jankélévich, Vladimir (2015 [1959]) *Henri Bergson.* Trans. Nils F. Schott. Durham and London: Duke University Press.

Jones, Allen Thomas (2016) "The force of the present: A Bergsonian challenge to psychophysics." *International Journal of Philosophy and Theology* 77.4–5. pp. 252–272.

Kaliff, Anders, and Terje Oestigaard (2017) *Cremation, Corpses and Cannibalism: Comparative Cosmologies and Centuries of Cosmic Consumption.* Newcastle-upon-Tyne: Cambridge Scholars Publishing.

Kant, Immanuel (1959 [1785]) *Foundations of the Metaphysics of Morals & What is Enlightenment?* Trans. A. W. Beck. Indianapolis: Bobs-Merrill Company.

Kant, Immanuel (1998 [1781]) *Critique of Pure Reason.* Trans. P. Guyer and A. W. Wood. Cambridge: Cambridge University Press.

Kauffman, Stuart A. (1993) *The Origins of Order: Self-Organization and Selection in Evolution.* Oxford and New York: Oxford University Press.

Kebede, Messay (2019) *Bergson's Philosophy of Self-Overcoming: Thinking Without Negativity or Time as Striving.* Basingstoke and New York: Palgrave Macmillan.

Keiler, Jonas, Christian S. Wirkner, and Stefan Richter. (2017). "One hundred years of carcinization–the evolution of the crab-like habitus in Anomura (Arthropoda: Crustacea)." *Biological Journal of the Linnean Society* 121.1. pp. 200–222.

Kelly, Michael R. (2010a) "A phenomenological (Husserlian) defense of Bergson's 'idealistic concession.'" *Epoché: A Journal for the History of Philosophy* 14.2. pp. 399–415.

Kelly, Michael R. (2010b) *Bergson and Phenomenology.* Basingstoke and New York: Palgrave Macmillan.

Khandker, Wahida (2013) "The idea of will and organic evolution in Bergson's philosophy of life." *Continental Philosophy Review* 46.1. pp. 57–74.

Klein, Colin, and Andrew B. Barron (2016) "Insects have the capacity for subjective experience." *Animal Sentience* 1.9. pp. 1–19.

Kleinherenbrink, Arjen (2014) "Time, duration and freedom–Bergson's critical move against Kant." *Diametros* 39. pp. 203–230.

Kleinherenbrink, Arjen (2019) "The two times of objects: A solution to the problem of time in object-oriented ontology." *Open Philosophy* 2. pp. 539–551.

Kreps, David (2015) *Bergson, Complexity and Creative Emergence.* Basingstoke and New York: Palgrave Macmillan.

Kumar, Shiv K. (1962) *Bergson and the Stream of Consciousness Novel.* London and Glasgow: Blackie & Son Limited.

Kuznetsov, B. G. (1987) *Reason and Being*. Cham: Springer.
Lacey, A. R. (1989) *Bergson: The Arguments of the Philosophers*. London and New York: Routledge.
Lalande, André (1898) *L'idée directrice de la dissolution opposée à celle de l'évolution dans la méthode des sciences physiques et morales*. Alcan: Paris.
Lambert, Ben, Georgies Kontonatsios, Matthias Mauch, Theodore Kokkoris, Matthew Jockers, Sophia Ananiadou, and Armand M. Leroi (2020) "The pace of modern culture." *Nature Human Behaviour* 4. pp. 352–360.
Langevin, Paul (1911) "L'Évolution de l'espace et du temps." *Scientia* 10. pp. 31–54.
Lapoujade, David (2005) "The normal and the pathological in Bergson." *MLN* 120.5. pp. 1146–1155.
Lapoujade, David (2018 [2010]) *Powers of Time: Versions of Bergsonism*. Trans. Andrew Goffey. Minneapolis and London: University of Minnesota Press/ Univocal.
Latour, Bruno (2002) "Gabriel Tarde and the end of the social." In Patrick Joyce (ed.) *The Social in Question: New Bearings in History and the Sciences*. London and New York: Routledge. pp. 117–133.
Lawlor, Leonard (2003) *The Challenge of Bergsonism: Phenomenology, Ontology, Ethics*. London and New York: Continuum.
Lawlor, Leonard (2004) "What immanence? What transcendence? The prioritization of intuition over language in Bergson." *Journal of the British Society for Phenomenology* 35.1. pp. 24–41.
Lawlor, Leonard (2010) "Intuition and duration: An introduction to Bergson's 'introduction to metaphysics.'" In Michael R. Kelly (ed.) *Bergson and Phenomenology*. Basingstoke and New York: Palgrave Macmillan. pp. 25–42.
Lee, A. M., L. H. Tai, A. Zador, and L. Wilbrecht (2015) "Between the primate and 'reptilian' brain: Rodent models demonstrate the role of corticostriatal circuits in decision making." *Neuroscience* 296. pp. 66–74.
Lewis, Michael, and Linda Michalson (1983) *Children's Emotions and Moods: Developmental Theory and Measurement*. Cham: Springer.
Liu, D., W. Li, C. Ma, W. Zheng, Y. Yao, C. F. Tso, P. Zhong, X. Chen, J. H. Song, W. Choi, S. B. Paik, H. Han, and Y. Dan (2020) "A common hub for sleep and motor control in the substantia Nigra." *Science* 367.6476. pp. 440–445.
Loeve, Sacha, Yavier Guchet, and Bernadette Bensaude Vincent (2018) *French Philosophy of Technology: Classical Readings and Contemporary Approaches*. Cham: Springer.
Lombrozo, Tania, Joshua Knobe, and Shaun Nichols (2014) *Oxford Studies in Experimental Philosophy*. Volume I. Oxford and New York: Oxford University Press.
Lorand, Ruth (2000) *Aesthetic Order: A Philosophy of Order, Beauty and Art*. London and New York: Routledge.
Lotze, Hermann (1884 [1879]) *Metaphysic*. Trans. Bernard Bosanquet. Oxford: Clarendon Press.
Lovasz, Adam (2016) *The System of Absentology in Ontological Philosophy*. Newcastle-upon-Tyne: Cambridge Scholars Publishing.

Lovasz, Adam (2020) "Enlivening society: Life as elasticity in Henri Bergson's Le Rire." *Cosmos & History* 16.2. pp. 198–244.

Lovejoy, Arthur Oncken (1913) "The problem of time in recent French philosophy II." *The Philosophical Review* 21.3. pp. 322–343.

Luo, Xinyue, Ruije Song, David F. Moreno, Hong-Yeoul Ryu, Mark Hochstrasser, and Murat Acar (2020) "Epigenetic mechanisms contribute to evolutionary adaptation of gene network activity under environmental selection." *Cell Reports* 33. pp. 1–13.

McCarthy, Harold E. (1952) "TS cliot and buddhism." *Philosophy East and West* 2.1. pp. 31–55.

McEwan, Ian (1977) "Dead as they come." *The Iowa Review* 8.4. pp. 71–82.

McKenna, Michael, and Derk Pereboom (2016) *Free Will: A Contemporary Introduction.* London and New York: Routledge.

McNamara, Patrick (1996) "Bergson's matter and memory and modern selectionist theories of memory." *Brain and Cognition* 30.2. pp. 215–231.

McShea, Daniel W., and Robert N. Brandon (2010) *Biology's First Law: The Tendency for Diversity and Complexity to Increase in Evolutionary Systems.* Chicago: University of Chicago Press.

MacWilliam, John (1928) *Criticism of the Philosophy of Bergson.* Edinburgh: T & T Clark.

Madrigal, Alexis (2008) "The best cure for Hiccups: Remind your brain you're not a fish." 25 February. https://www.wired.com/2008/02/evolution-expla/

Maiorino, Giancarlo (1990) *The Cornucopian Mind and the Baroque Unity of the Arts.* University Park and London: University of Pennsylvania Press.

Maitra, S. K. (1988) *The Meeting of East and West in Sri Aurobindo's Philosophy.* Puducherry: Sri Aurobindo Ashram Publications Department.

Manzotti, Riccardo (2011) "The spread mind. Seven steps to situated consciousness." *Journal of Cosmology* 14. pp. 4526–4541.

Marchesini, Paula (2018) "The end of time or time reborn? Henri Bergson and the metaphysics of time in contemporary cosmology." *Philosophy and Cosmology* 21. pp. 140–152.

Marder, Michael (2013) *Plant-Thinking: A Philosophy of Vegetal Life.* New York: Columbia University Press.

Maritain, Jacques (1955 [1914]) *Bergsonian Philosophy and Thomism.* Trans. Mabelle L. Andison and J. Gordon Andison. New York: The Philosophical Library.

Marrati, Paola (2005) "Time, life, concepts: The newness of Bergson." *MLN* 120.5. pp. 1099–1111.

Martz, Louis L. (1948) "The wheel and the point: Aspects of imagery and theme in Eliot's later poetry." In Leonard Unger (ed.) *T. S. Eliot: A Selected Critique.* New York: Rinehart. pp. 444–462.

Massumi, Brian (2002) *Parables for the Virtual: Movement, Affect, Sensation.* Durham and London: Duke University Press.

Megay, Joyce N. (1976) *Bergson et Proust. Essai de Mise au Point de la Question de l'influence de Bergson Sur Proust.* Paris: J. Vrin.

Meillassoux, Quentin (2007) "Subtraction and contraction: Deleuze, immanence, and matter and memory." *Collapse* III. pp. 63–107.

Meillassoux, Quentin (2008 [2006]) *After Finitude: An Essay on the Necessity of Contingency.* Trans. Ray Brassier. London and New York: Continuum.

Menke, Richard (2005) "Media in America, 1881: Garfield, Guiteau, Bell, Whitman." *Critical Inquiry* 31.3. pp. 638–664.

Merleau-Ponty, Maurice (1959) "Bergson in the making." Trans. Richard C. McCleary. In Maurice Merleau-Ponty (1977 [1960]) *Signs.* Trans. Richard C. McCleary. Evanston: Northwestern University Press. pp. 182–192.

Merleau-Ponty, Maurice (2003 [1968]) *Nature: Course Notes from the Collège de France.* Trans. Robert Vallier. Evanston: Northwestern University Press.

Metz, Andre, and Henri Bergson (1969 [1924]) "Exchanges concerning Bergson's new edition of duration and simultaneity." In Pete A. Y. Gunter (ed.) *Bergson and the Evolution of Physics.* Knoxville: University of Tennessee Press. pp. 135–190.

Meyerson, Emile (1962 [1908]) *Identity and Reality.* Trans. Katie Loewenberg. New York: Gordon and Breach.

Minkowski, Hermann (1920 [1908]) "Principle of relativity." In Hermann Minkowski and Albert Einstein (eds) *The Principle of Relativity: Original Papers by Albert Einstein and Hermann Minkowski.* Trans. M. N. Saha and S. N. Bose. Kolkata: University of Calcutta. pp. 1–88.

Minkowski, Hermann (2012 [1907]) "The principle of relativity." In Hermann Minkowski (ed.) *Space and Time: Minkowski's Papers on Relativity.* Quebec: Minkowski Institute Press. pp. 39–51.

Montebello, Pierre (2007) "Matter and light in Bergson's creative evolution." Trans. Roxanne Lapidus. *Substance* 36.3. pp. 91–99.

Moore, F. C. T. (1996) *Bergson: Thinking Backwards.* Cambridge: Cambridge University Press.

Moravec, Matyaš (2019) "Perpetual present: Henri Bergson and atemporal duration." *European Journal for Philosophy of Religion* 11.3. pp. 197–224.

Morgan. C. Lloyd (1927) "New books: Holism and evolution. By general the right Honourable Jan Smuts (London: Macmillan & Co. 1926. Pp. 361+ix. Price: 18s)." *Philosophy* 2.5. pp. 93–97.

Moulard-Leonard, Valentine (2008) *Bergson-Deleuze Encounters: Transcendental Experience and the Thought of the Virtual.* Albany: State University of New York Press.

Mounk, Yascha (2018) "What an Audacious Hoax reveals about academia." 5 October. https://www.theatlantic.com/ideas/archive/2018/10/new-sokal-hoax/572212/.

Mourélos, Georges (1964) *Bergson et le Niveaux de Réalité.* Paris: Presses Universitaires de France.

Mullarkey, John (1995) "Bergson's method of multiplicity." *Metaphilosophy* 26.3. pp. 230–259.

Mullarkey, John (1996) "Bergson: The philosophy of Durée-Différence." *Philosophy Today* 40.3. pp. 367–380.

Mullarkey, John (2000) *Bergson and Philosophy: An Introduction.* New York: University of Notre Dame Press.

Mullarkey, John (2004) "Forget the virtual: Bergson, actualism, and the refraction of reality." *Continental Philosophy Review* 37.4. pp. 469–493.

Murray, Paul (1991) *T.S. Eliot and Mysticism: The Secret History of Four Quartets.* Basingstoke: Macmillan.

Nolan, Christopher (2000) *Memento.* Newmarket Capital Group, Team Todd, I Remember Productions.

Oancea, Mihaela-Cristina (2018) "The galaxy within: Space, quantum physics and the fantasy of bodily unity in Jeanette Winterson's *Gut Symmetries.*" *Analele Universităţii Ovidius din Constanţa. Seria Filologie* 29.1. pp. 30–39.

Ó Maoilearca, John (2014) "Spirit in the materialist world: On the structure of regard." *Angelaki* 19.1. pp. 13–29.

Palmer, Daryl W. (2009) "Ripening Claude: Willa Cather's *one of ours* and the philosophy of Henri Bergson." *American Literary Realism* 41.2. pp. 112–132.

Papanicolaou, Andrew C., and Pete A. Y. Gunter (eds) (1989) *Bergson and Modern Thought: Towards a Unified Science.* Reading: Harwood Publishers.

Patrides, C. A. (1973) "The renascence of the renaissance: T. S. Eliot and the pattern of time." *Michigan Quarterly Review* XII. pp. 172–196.

Penrose, Roger (2004) *The Road to Reality: A Complete Guide to the Laws of the Universe.* London: Jonathan Cape.

Perri, Trevor (2014) "Bergson's philosophy of memory." *Philosophy Compass* 9.12. pp. 837–847.

Perry, Ralph Barton (1911a) "Notes on the philosophy of Henri Bergson. I. Anti-intellectualism and immediatism." *The Journal of Philosophy, Psychology and Scientific Methods* 8.25. pp. 673–682.

Perry, Ralph Barton (1911b) "Notes on the philosophy of Henri Bergson. II. Indeterminism and dynamism." *The Journal of Philosophy, Psychology and Scientific Methods* 8.26. pp. 713–721.

Pickering, Andrew (1995) "Cyborg history and the world war II regime." *Perspectives on Science* 3.1. pp. 1–48.

Pigot, Alex L., Catherine Sheard, Eliot T. Miller, Tom P. Bregman, Benjamin G. Freeman, Uri Roll, Nathalie Seddon, Christopher H. Trisos, Brian C. Weeks, and Joseph A. Tobias (2020) "Macroevolutionary convergence connects morphological form to ecological function in birds." *Nature Ecology & Evolution* 4. pp. 230–239.

Poincaré, Henri (1905) *Science and Hypothesis.* Trans. W. J. G. London and Newcastle-upon-Tyne: Walter Scott Publishing Co.

Poulet, Georges (2011) "Bergson: The theme of the panoramic vision of the dying and juxtaposition." Trans. Mark Cirino and William J. Hemminger. *PMLA* 126.2. pp. 483–499.

Power, Nina (2006) "Bachelard contra Bergson: Ancient atomism and the debate over continuity." *Angelaki* 11.3. pp. 117–123.

Prigogine, Ilya (1999) *The Arrow of Time: Inaugural Lecture.* Pescara, Italy: International Center for Relativistic Astrophysics.

Proctor, William (2018) "Reboots and retroactive continuity." In Mark J. P. Wolf (ed.) *The Routledge Companion to Imaginary Worlds.* London and New York: Routledge. pp. 224–236.

Quick, Oliver (1913) "Bergson's 'creative evolution' and the individual." *Mind* 22.86. pp. 217–230.

Radhakrishnan, Sarvepalli (1917) "Is Bergson's philosophy monistic?" *Mind* 26.103. pp. 329–339.

Raghumaraju, A. (2019) *Calibrating Western Philosophy for India: Rousseau, Derrida, Deleuze, Guattari, Bergson and Vaddera Chandidas.* London and New York: Routledge.

Ravaisson, Félix (2008 [1838]) *Of Habit.* Trans. Claire Carlisle and Mark Sinclair. London and New York: Bloomsbury.

Robbins, Stephen E. (2000) "Bergson, perception and Gibson." *Journal of Consciousness Studies* 7.5. pp. 23–45.

Robson, Mark Ian Thomas (2017) "Is ultimate moral responsibility metaphysically impossible? A Bergsonian critique of Galen Strawson's argument." *Philosophy* 92.4. pp. 519–538.

Rolle, Richard (1918) *Richard Rolle of Hampole's Mending of Life.* Cleveland: Western Reserve University Press.

Rovelli, Carlo (2018) *The Order of Time.* New York: Riverhead Books.

Rudrum, Alan (1974) "An aspect of Vaughan's Hermeticism: The doctrine of cosmic sympathy." *Studies in English Literature, 1500–1900* 14.1. pp. 129–138.

Ruse, Michael (1989) "Teleology in biology. Is it a cause for concern?" *Trends in Ecology & Evolution* 4.2. pp. 51–54.

Ruse, Michael (2002) "The critique of intellect: Henri Bergson's prologue to an organic epistemology." *Continental Philosophy Review* 35.3. pp. 281–302.

Ruse, Michael (2009) *Monad to Man: The Concept of Progress in Evolutionary Biology.* Cambridge: Harvard University Press.

Ruse, Michael (2016) "Evolutionary biology and the question of teleology." *Studies in History and Philosophy of Science Part C: Studies in History and Philosophy of Biological and Biomedical Sciences* 58. pp. 100–106.

Russell, Bertrand (1912) "The philosophy of Bergson." *The Monist* 22.3. pp. 321–347.

Saint Augstine of Hippo (1900) *The Confessions of St. Augustine.* Trans. J. G. Pilkington. New York: International Collector's Library.

Saint Robert Southwell (1872) *The Complete Poems of Robert Southwell S. J.* The Fuller Worthies' Library.

Santamaria, Ulysses, and Anne M. Bailey (1984) "A note on Braudel's structure as duration." *History and Theory* 23.1. pp. 78–83.

Scarcelli, N., S. Tostain, Y. Vigouroux, C. Agbangla, O. Daïnou, and J. L. Pham (2006) "Farmers' use of wild relative and sexual reproduction in a vegetatively propagated crop: The case of yam in Benin." *Molecular Ecology* 15.9. pp. 2421–2431.

Scheler, Max (2017 [1928]) *The Human Place in the Cosmos.* Trans. Karin S. Frings. Evanston: Northwestern University Press.

Schiller, F. C. S. (1910) *Riddles of the Sphinx: Studies in Humanism.* New York: Macmillan.

Schmid, Peter F. (1998) "'Face to face.' The art of encounter." In Brian Thorne and Elke Lambers (eds) *Person-Centred Therapy: A European Perspective.* London, Thousand Oaks and New Delhi: Sage Publications. pp. 74–91.

Schmidtke, Charles (1989) "Bergson and a pulsational-wave model of temporality: A way to disentangle theories in gerontology." In Andrew C. Papanicolaou and Pete A. Y. Gunter (eds) *Bergson and Modern Thought: Towards a Unified Science.* Reading: Harwood Publishers. pp. 223–250.

Scott, David (2006) "The 'concept of time' and the 'being of the clock': Bergson, Einstein, Heidegger, and the interrogation of the temporality of modernism." *Continental Philosophy Review* 39.2. pp. 183–213.

Scott, J. W. (1913) "The pessimism of creative evolution." *Mind* 22.87. pp. 344–360.

Scott, J. W. (1914) "Ethical pessimism in Bergson." *The International Journal of Ethics* 24.2. pp. 147–167.

Shapere, Alfred and Frank Wilczek (2012). "Classical time crystals." *Physical Review Letters* 109(16). pp. 160402. *arXiv*:1202.2537v2.

Shores, Corry (2014) "In the still of the moment: Deleuze's phenomena of motionless time." *Deleuze Studies* 8.2. pp. 199–229.

Simon, Matt (2015) "Absurd creature of the week: The insect that devours its mother from the inside out." 23 January. https://www.wired.com/2015/01/absurd-creature-of-the-week-strepsiptera/.

Sinclair, Mark (2011) "Is habit 'the fossilised residue of a spiritual activity'? Ravaisson, Bergson, Merleau-Ponty." *Journal of the British Society for Phenomenology* 42.1. pp. 33–52

Sinclair, Mark (2014) "Inheritance, originality and the will: Bergson and Heidegger on creation." *International Journal of Philosophical Studies* 22.5. pp. 655–675.

Sinclair, Mark (2020) *Bergson.* London and New York: Routledge.

Sipfle, David A. (1969) "Henri Bergson and the epochal theory of time." In Pete A. Y. Gunter (ed.) *Bergson and the Evolution of Physics.* Knoxville: University of Tennessee Press. pp. 275–295.

Smuts, Jan (1926) *Holism and Evolution.* London: Macmillan.

Sokal, Alan, and Jean Bricmont (1997) *Impostures Intellectuelles.* Paris: Odile Jacob.

Sokol, B. J. (1990) "Bergson, instinct, and frost's 'The White-Tailed Hornet.'" *American Literature* 62.1. pp. 44–55.

Soud, David (2014) "The greedy dialectic of time and eternity. Karl Barth, T. S. Eliot, and *Four Quartets.*" *ELH* 81.4. pp. 1363–1391.

Sparrow, Tom (2014) *The End of Phenomenology: Metaphysics and the New Realism.* Edinburgh: Edinburgh University Press.

Speck, Oliver C. (2011) "Overcoming metaphysics in three easy steps: Tom Tykwer's *Lola rennt.*" In Jill E. Twark (ed.) *Strategies of Humor in Post-Unification German Literature, Film, and Other Media.* Newcastle-upon-Tyne: Cambridge Scholars Publishing. pp. 182–202.

Spyksma, Arie P. J., Richard B. Taylor, and Nick T. Shears (2017) "Predation cues rather than resource availability promote cryptic behaviour in a habitat-forming sea urchin." *Oecologia* 183. pp. 821–829.

Starr, Michelle (2020) "This exoplanet disappeared from Hubble observations. Now, we finally know why." 21 April. https://www.sciencealert.com/astronomers-have-solved-the-mystery-of-a-disappearing-exoplanet.

Steinmetz, Julia (2019) "In recognition of their desperation: Sonic relationality and the work of deep listening." *Studies in Gender and Sexuality* 20.2. pp. 119–132.

Strawson, Galen (2012) "The bounds of freedom." In Robert Kane (ed.) *The Oxford Handbook of Free Will*. Oxford and New York: Oxford University Press. pp. 441–461.

Stump, Eleonore, and Norman Kretzmann (1981) "Eternity." *The Journal of Philosophy* 78.8. pp. 429–458.

Stump, Eleonore, and Norman Kretzmann (1987) "Atemporal duration: A reply to Fitzgerald." *The Journal of Philosophy* 84.4. pp. 214–219.

Teixeira, Maria Teresa Martins Vieira (2009) "Epochal time and the creativity of thinking: Henri Bergson and Alfred North Whitehead." *Concrescence* 10. pp. 77–85.

Tollefsen, Deborah, Roger Kreuz, and Rick Dale (2014) "Flavors of 'togetherness': Experimental philosophy and theories of joint action." In Tania Lombrozo, Joshua Knobe and Shaun Nichols (eds) *Oxford Studies in Experimental Philosophy*. Volume I. Oxford and New York: Oxford University Press. pp. 232–253.

Trifonova, Temenuga (2007) *The Image in French Philosophy*. Amsterdam and New York: Rodopi.

Twark, Jill E. (2011) *Strategies of Humor in Post-Unification German Literature, Film, and Other Media*. Newcastle-upon-Tyne: Cambridge Scholars Publishing.

Underhill, Evelyn (1912) *Mysticism: A Study—The Nature and Development of Man's Spiritual Consciousness*. New York: E. P. Dutton and Company.

Van der Tuin, Iris. (2011) ""A different starting point, a different metaphysics": Reading Bergson and Barad diffractively." *Hypatia* 26.1. pp. 22–42.

Van Flandern, Tom (1998) "What the global positioning system tells us about relativity." In Franco Selleri (ed.) *Open Questions in Relativistic Physics*. Quebec: Apeiron. pp. 81–93.

Van Inwagen, Peter (1989) "When is the will free?" *Philosophical Perspectives* 3. pp. 399–422.

Van Sant, Gus (2002) *Gerry*. Epsilon Motion Pictures, My Cactus, Tango Films.

Vaughan, Henry (1678) *Thalia rediviva the pass-times and diversions of a countrey-muse, in choice poems on several occasions: With some learned remains of the eminent Eugenius Philalethes, never made publick till now*. London: Robert Pawlet.

Vieillard-Baron, Jean-Louis (1993) "Bergson, Professeur de Khâgne." *Les Études philosophiques* 1. pp. 87–92.

Villarreal, Luis P. (2009) *Origin of Group Identity: Viruses, Addiction and Cooperation*. Cham: Springer.

Wallace, David Foster (2018 [1996]) *Infinite Jest*. New York: Little, Brown Book Group.

Weitz, Morris (1952) "T. S. Eliot: Time as a mode of salvation." *The Sewanee Review* 60.1. pp. 48–64.

Westerman, David, and Ron Tamborini (2010) "Scriptedness and televised sports: Violent consumption and viewer enjoyment." *Journal of Language and Social Psychology* 29.3. pp. 321–337.

Whitehead, Alfred North (1933) *Adventures of Ideas.* New York and London: Macmillan.

Whitehead, Alfred North (1978 [1929]) *Process and Reality: An Essay in Cosmology.* Gifford Lectures Delivered in the University of Edinburgh During the Session 1927–8. New York: The Free Press.

Wight, Doris T. (1990) "Metaphysics through Paradox in Eliot's *Four Quartets.*" *Philosophy & Rhetoric* 23.1. pp. 63–69.

Williams, Clifford (1998) "A Bergsonian approach to A-and B-time." *Philosophy* 73.3. pp. 379–393.

Wilm, Emil Carl (1914) *Henri Bergson: A Study in Radical Evolution.* New York: Sturgis & Walton Company.

Winterson, Jeanette (1997) *Gut Symmetries.* London: Granta.

Wisden, William, and Nicholas P. Franks (2020) "The stillness of sleep." *Science* 367.6476. pp. 366–367.

Wittgeinstein, Ludwig (1986 [1953]) *Philosophical Investigations.* Trans. G. E. M. Anscombe. Oxford: Basil Blackwell.

Witzany, Guenther (2012) *Viruses: Essential Agents of Life.* Cham: Springer.

Wolosky, Shira (1995) *Language Mysticism: The Negative Way of Language in Eliot, Beckett and Celan.* Stanford: Stanford University Press.

Woodcock, George (1944) *The Tyranny of the Clock.* Philadelphia: Wooden Shoe Books. [reprint edition]

Woolf, Virginia (1992) *The Collected Novels of Virginia Woolf. Mrs. Dalloway: To the Lighthouse. The Waves.* Basingstoke: Macmillan.

Worms, Frédéric (1997) *Introduction à Matière et mémoire de Bergson.* Paris: Presses Universitaires de France.

Worms, Frédéric (2010) "Consciousness or life? Bergson between phenomenology and metaphysics." Trans. Mark Sentesy. In Michael R. Kelly (ed.) *Bergson and Phenomenology.* Basingstoke and New York: Palgrave Macmillan. pp. 245–258.

Wu, Zhiqiang, Gus Waneka, Amanda K. Broz, Connor R. King, and Daniel B. Sloan (2020) "MSH1 is required for maintenance of the low mutation rates in plant mitochondrial and plastid genomes." *PNAS.* https://www.pnas.org/content/early/2020/06/26/2001998117, 1–8.

Yirka, Bob (2017) "Virgin female spiders found willing to give themselves up to being eaten alive by spiderlings." 19 September. https://phys.org/news/2017-09-virgin-female-spiders-eaten-alive.html.

Zahavi, Dan (2010) "Life, thinking and phenomenology in the early Bergson." In Michael R. Kelly (ed.) *Bergson and Phenomenology.* Basingstoke and New York: Palgrave Macmillan. pp. 118–134.

Zilczer, Judith (1999) "'Light coming on the plains.' Georgia O'Keeffe's sunrise series." *Artibus et Historiae* 20.40. pp. 191–208.

Index

About the Author

Adam Lovasz is an Australian-born philosopher based in Hungary. Presently Lovasz is a researcher affiliated with Eötvös Loránd University, Budapest. His interests include Continental philosophy, embodiment, phenomenology, posthumanism, and speculative realism. Lovasz is author and coauthor of numerous scholarly works, such as the first Hungarian-language textbook about Posthumanism, and is also cofounder (with Mark Horvath) of Absentology, a center for international collaboration and interdisciplinary philosophical inquiry. Lovasz's work has been published in a number of journals, including *Chiasma*, *Cosmos & History*, *Helikon*, *Horizon: Studies in Phenomenology*, *Horror Studies*, *Iowa Journal of Cultural Studies*, *Rhizomes*, *Textual Practice*, and more.

www.ingramcontent.com/pod-product-compliance
Ingram Content Group UK Ltd.
Pitfield, Milton Keynes, MK11 3LW, UK
UKHW041841031026
15587UKWH00002B/926

9 781793 640819